HIGHLIGHTS OF SPANISH ASTROPHYSICS

HIGHLIGHTS OF SPANISH ASTROPHYSICS

Proceedings of the third Scientific Meeting of the Spanish Astronomical Society (SEA) held in La Laguna, Tenerife, Spain, September 21–24, 1998

Edited by

JAVIER GORGAS and JAIME ZAMORANO
Department of Astrophysics, The Complutense University, Madrid, Spain

Reprinted from *Astrophysics and Space Science*
Volume 263, Nos. 1–4, 1998

KLUWER ACADEMIC PUBLISHERS
DORDRECHT / BOSTON / LONDON

Library of Congress Cataloging-in-Publication Data

Highlights of Spanish astrophysics / edited by Javier Gorgas and Jaime Zamorano.
p. cm.
Includes index.

1. Astrophysics Congresses. 2. Astronomy Congresses. I. Gorgas, Javier. II. Zamorano, Jaime.
QB460.H55 1999
523.01--dc21 99-38091

ISBN 978-90-481-5288-9

Published by Kluwer Academic Publishers,
P.O. Box 17, 3300 AA Dordrecht, The Netherlands.

Kluwer Academic Publishers incorporates
the publishing programmes of
D. Reidel, Martinus Nijhoff, Dr W. Junk and MTP Press.

Sold and distributed in the U.S.A. and Canada
by Kluwer Academic Publishers,
101 Philip Drive, Norwell, MA 02061, U.S.A.

In all other countries, sold and distributed
by Kluwer Academic Publishers,
P.O. Box 322, 3300 AH Dordrecht, The Netherlands.

Printed on acid-free paper

TABLE OF CONTENTS

Preface xi

Organizing Committees xiii

List of Participants xv–xvii

J. GALLEGO / The Evolution of Star-Forming Galaxies: The SFR Density of the Universe (Invited Review) 1

A.J. CASTRO-TIRADO / Cosmic Gamma-Ray Bursts: The most energetic phenomenon in the Universe (Invited Review) 15

A. RAIG, E. SALVADOR-SOLÉ and G. GONZÁLEZ-CASADO / Modelling the Growth of Dark-Matter Halos 27

G. YEPES, D. ELIZONDO and Y. ASCASIBAR / Star Formation and Cosmological Simulations 31

R. DOMÍNGUEZ-TENREIRO, P.B. TISSERA and A. SÁIZ / Galaxy Disk Formation in Hierarchical Hydrodynamical Simulations 35

A. MANRIQUE and E. SALVADOR-SOLÉ / Analytical Modelling of Galaxy Formation 39

A. SÁIZ, P.B. TISSERA and R. DOMÍNGUEZ-TENREIRO / The Stabilizing Role of Stellar Bulges in Galaxy Disk Formation 43

R. GIL-MERINO, L.J. GOICOECHEA, M. SERRA, A. OSCOZ, E. MEDIAVILLA and J. BUITRAGO / Analysis of the Difference Light Curve of the Gravitational Mirage QSO 0957+561 47

J.A. MUÑOZ, E.E. FALCO, C.S. KOCHANEK, J. LEHÁR, B.A. MCLEOD, C.D. IMPEY, H.-W. RIX and C.Y. PENG / The CASTLES Project 51

R. PELLÓ, J.P. KNEIB and G. BRUZUAL / Identification and Study of Distant Galaxies Through Cluster Lenses 55

E. ROS, J.A. ZENSUS and A.P. LOBANOV / VLBA Monitoring of the QSO 3C 345 59

R. CARBALLO, J.I. GONZÁLEZ-SERRANO, S. SÁNCHEZ, C.R. BENN and M. VIGOTTI / The Shape of the Blue/UV Continuum of B3-VLA Radio Quasars 63

M. VILLAR-MARTÍN, C. TADHUNTER, R. MORGANTI, D. AXON and A. KOEKEMOER / PKS 2250–41 and the Role of Jet Cloud Interactions in Powerful Radio Galaxies 67

A.I. DÍAZ, M. MOLLÁ and F. FERRINI / Chemical Evolution in the Early Universe: Can Damp Lyα Systems Be Spiral Galaxies? 71

X. BARCONS, K.M. LANZETTA, H.-W. CHEN, N. YAHATA, J.K. WEBB, A. FERNÁNDEZ-SOTO and A. ORTIZ-GIL / QSO Lyman-Alpha Absorbers and Galaxy Halos 75

M. CENTURIÓN, P. BONIFACIO, P. MOLARO and G. VLADILO / On the Nature of Damped Lyα Systems: Clues from Determinations of Elemental Abundance Ratios 79

N. CARDIEL, J. GORGAS and A. ARAGÓN-SALAMANCA / The Role of Cooling Flows in the Star Formation History of Central Cluster Galaxies 83

F.J. SÁNCHEZ-SALCEDO, A. BRANDENBURG and A. SHUKUROV / Turbulence and Magnetic Fields in Clusters of Galaxies 87

F.J. CASTANDER (FOR THE SDSS COLLABORATION) / The Sloan Digital Sky Survey 91

C.E. GARCÍA DABÓ, J. GALLEGO and J. ZAMORANO / Optimizing Results from Automatic Objective-Prism Surveys 95

C.E. GARCÍA DABÓ and J. GALLEGO / OPERA: Objective-Prism Reduction Package 99

A.M. PÉREZ GARCÍA and J.M. RODRÍGUEZ ESPINOSA / Seyfert Galaxies: A Perspective with the Infrared Space Observatory (ISO) 103

M. SÁNCHEZ PORTAL, A.I. DÍAZ, M. ÁLVAREZ ÁLVAREZ, R. TERLEVICH, E. TERLEVICH and I. ARETXAGA / CCD Surface Photometry of Nearby Active Galaxies 107

L. JIMÉNEZ-BENITO, A.I. DÍAZ, R. TERLEVICH and E. TERLEVICH / Kinematics and Stellar Populations of Seyfert Galaxies 111

X. BARCONS, R. CARBALLO, F.J. CARRERA, M.T. CEBALLOS, J.I. GONZÁLEZ-SERRANO, A.C. FABIAN and R.S. WARWICK / X-ray Observations of Active Galactic Nuclei in the Cosmological Context 115

M. SÁNCHEZ PORTAL and A.I. DÍAZ / A Method for Obtaining the Nuclear Component in Active Galaxies 119

L. TOMAS-CALDERON, J.A. ACOSTA-PULIDO, B. VILA-VILARO and I. PEREZ-FOURNON / Spectroscopy of the Extended Emission Line Region in NGC 4388 123

R. GUZMÁN / Blue Compact Galaxies at Redshifts $z = 0.2$–1.3 127

A. ALONSO-HERRERO, M.J. RIEKE and G.H. RIEKE / The Starburst Properties of the Interacting System Arp 299 (IC 694 + NGC 3690) 131

M. CASTELLANOS, A.I. DÍAZ, E. TERLEVICH and M.L. GARCÍA-VARGAS / Spectrophotometric Analysis of HII Regions in NGC 4258 135

A.I. DÍAZ, M. ÁLVAREZ ÁLVAREZ, M. SÁNCHEZ PORTAL, E. TERLEVICH, I. ARETXAGA and R. TERLEVICH / Star Formation in Circumnuclear Rings 139

A.I. DÍAZ / Functional and Physical Parameters of HII Galaxies 143

A. GIL DE PAZ, J. GALLEGO, J. ZAMORANO, A. ALONSO-HERRERO and A. ARAGÓN-SALAMANCA / NIR Properties of the UCM Star-Forming Galaxies 147

A. GIL DE PAZ, J. ZAMORANO and J. GALLEGO / Mrk 86: A 'Not-So-Dark' Galaxy 151

M.A. GÓMEZ-FLECHOSO / Dwarf Spheroidal Galaxies: Dynamical Evolution 155

S. PEDRAZ, J. GORGAS, N. CARDIEL and R. GUZMÁN / Stellar Populations in Dwarf Elliptical Galaxies 159

J.C. VEGA BELTRÁN, E. PIGNATELLI, W.W. ZEILINGER, A. PIZZELLA, E.M. CORSINI, F. BERTOLA and J.E. BECKMAN / Kinematics of Gas and Stars in Spiral Galaxies 163

J. GORGAS, N. CARDIEL and S. PEDRAZ / Towards an Understanding of the λ4000Å Break Behaviour in Old Stellar Populations 167

M.I. MONTEVERDE and A. HERRERO / Stellar α Elements Abundance Gradient in M33 171

J. CERNICHARO / Infrared Astrophysics in the Space with ISO: From ISO to FIRST (Invited Review) 175

C. ESTEBAN / Temperature Fluctuations in Galactic HII Regions 193

J. TRAPERO, E.J. ALFARO and D. DE MIGUEL / Spectroscopic Study of the Cygnus Superbubble 197

J. PALACIOS and C. EIROA / Observations Towards Star Formation Regions: H_2 Near-Infrared Emission Lines 201

M.J. SEMPERE, J. CERNICHARO, E. GONZÁLEZ-ALFONSO, B. LEFLOCH, S. PÉREZ-MARTÍNEZ and S.J. LEEKS / Far-Infrared CO Rotational Lines in the Orion Molecular Cloud 205

M.W. FEAST / Distance Scales After Hipparcos (Invited Review) 209

X. LURI, J. TORRA, F. FIGUERAS, A.E. GÓMEZ, M.J. GOUPIL and J.P. BEAULIEU / Calibration of the Classical Cepheid Period-Luminosity Relation from Hipparcos Data 215

E. SOLANO / The Absolute Magnitude of RR Lyraes: From Hipparcos Parallaxes and Proper Motions 219

P. FABIANI BENDICHO / Radiative Transfer Under Inhomogeneous Configurations 223

E. ORTIZ and C. EIROA / Wind and Disk Diagnostics in PMS Stars. A Critical Analysis 227

D. MONTES / Chromospheric Activity of Weak-Lined T Tauri Stars 231

D. BARRADO Y NAVASCUÉS / The Age of Beta Pic Type Stars: Vega, Fomalhaut, β Pic and HR 4796 235

D. BARRADO Y NAVASCUÉS, J.R. STAUFFER and J. BOUVIER / Brown Dwarfs and Very Low Mass Stars: Towards a New Age Scale for Young Open Clusters 239

J.M. TORREJÓN, G. BERNABEU and J. FABREGAT / On the Ionization Structure of Circumstellar Disks Around Be Type Stars 243

M.R. VILLAMARIZ and A. HERRERO / The Most Luminous Stars in Our Galaxy 247

F. NAJARRO and D.F. FIGER / The Pistol Star 251

L.J. CORRAL, A. HERRERO, R. VILLAMARIZ and E. MARTÍN / Spectroscopic Analysis of Stars in Cyg OB2 255

J.J. RODES, G. BERNABEU and J.FABREGAT / Radial Velocity Measurements with FXCOR 259

E. VERDUGO, A. TALAVERA and A.I. GÓMEZ DE CASTRO / Stellar Winds and Mass Loss in A-Type Supergiants 263

E. SOLANO and O.I. PINTADO / Spectroscopic Survey of δ Sct Stars 267

E. SOLANO and E. PAUNZEN / Identification of λ Bootis Stars Using IUE Spectra 271

D. MONTES / Libraries of High and Mid-Resolution Spectra of F, G, K, and M Field Stars 275

D. MONTES, J. SANZ-FORCADA, M.J. FERNÁNDEZ-FIGUEROA, E. DE CASTRO, M. CORNIDE and L.W. RAMSEY / Multiwavelength Optical Observations of Chromospherically Active Binary Systems 279

O. SUÁREZ, M. MANTEIGA, A. ULLA, P. GARCÍA-LARIO and A. MANCHADO TORRES / Optical Classification of IRAS Post-AGB Candidates 283

N. SERICHOL, D. GARCÍA-SENZ and E. BRAVO / Consequences of the Collison Between the Gas Ejected in the Collapse of a White Dwarf and a Low-Mass Star 287

J.A. MIRALLES, J.A. PONS and S. REDDY / Thermal Evolution of Neutron Stars 291

G. GÓMEZ and R. LÓPEZ / Standard Spectral Evolutionary Sequence for Type Ia Supernovae 295

A. MARCO, G. BERNABEU, J.M. TORREJÓN and J. FABREGAT / CCD Photometry in Very Young Open Clusters 299

D. BARRADO Y NAVASCUÉS, J.R. STAUFFER, J. BOUVIER and E.L. MARTÍN / The Mass Function at the End of the Main Sequence: the M35 Open Cluster 303

D. GALADÍ-ENRÍQUEZ, C. JORDI and E. TRULLOLS / Astrometry and Photometry of Open Clusters: NGC 1746, NGC 1750 and NGC 1758 307

M.A. PÉREZ-TORRES, J.M. MARCAIDE, J.C. GUIRADO and E. ROS / Submilliarcsecond Astrometry Over Large Sky Separations 311

J. TORRA, X. LURI, F. FIGUERAS and C. JORDI / GAIA: Fundamental Parameters from the Space 315

E. ROS, J.M. MARCAIDE, J.C. GUIRADO and M.A. PÉREZ-TORRES / High Precision Astrometry with VLBI: From the Triangle 1803+784/1928+ 738/2007+777 to the Complete S5 Polar Cap Sample 319

B. MONTESINOS and J.H. THOMAS / The Evershed Effect in Sunspots: A Theoretical Explanation 323

J. PATRÓN, I. GONZÁLEZ HERNÁNDEZ, A. JIMÉNEZ, S.J. JIMÉNEZ REYES, I. MARTÍN, P.L. PALLÉ, F. PÉREZ HERNÁNDEZ, C. RÉGULO and T. ROCA-CORTÉS / 3-D Helioseismology: Rings and Horizontal Flows 327

B. RUIZ COBO / SIR: an Inversion Technique of Spectral Lines 331

I. GONZÁLEZ HERNÁNDEZ, J. PATRÓN, T. ROCA-CORTÉS, F. PÉREZ HERNÁNDEZ, A. JIMÉNEZ, S.J. JIMÉNEZ REYES, I. MARTÍN, P.L. PALLÉ and C. RÉGULO / Probing the Solar Subphotospheric Layers Using Ring Analysis 335

I. MARTÍN MATEOS, I. GONZÁLEZ HERNÁNDEZ, A. JIMÉNEZ, S.J. JIMÉNEZ REYES, J. PATRÓN, P.L. PALLÉ, F. PÉREZ HERNÁNDEZ, C. RÉGULO and T. ROCA-CORTÉS / A New Observational Strategy to Search for Solar *g*-Modes 339

M. SÁNCHEZ CUBERES, J.A. BONET, M. VÁZQUEZ A. and D. WITTMANN / Centre-To-Limb Variation of the Solar Granulation 343

A. EFF-DARWICH, S.G. KORZENNIK, T. ROCA-CORTÉS, F. PÉREZ HERNÁNDEZ and P. PALLÉ / The Solar Rotation Rate $\Omega\ (r,\theta)$ 347

A. SÁNCHEZ-LAVEGA, J.R. ACARRETTA, R. HUESO, J.F. ROJAS, J. LECACHEUX, F. COLAS and J.M. GÓMEZ / An Overview of Saturn's Equatorial Storms: 1990–1997 351

J.M. RODRÍGUEZ ESPINOSA, P. ALVAREZ and F. SÁNCHEZ / The GTC: An Advanced 10 m Telescope for the ORM 355

M. BALCELLS / The EMIR Near-Infrared Multi-Object Spectrograph for GTC 361

R.J. GARCÍA LÓPEZ / An Optical Intermediate Dispersion Spectrograph for the GTC (ODIN: Espectrógrafo Óptico de Dispersión INtermedia) 365

J. CEPA / OSIRIS. Imaging and Spectroscopy for the GTC 369

A.M. VARELA, C. MUÑOZ-TUÑÓN and T. MAHONEY / Site-Testing Results at La Palma Observatory 373

V. FONSECA (HEGRA Collaboration) / Status and Results from the HEGRA Air Shower Experiment 377

J.M. TORRELLES, A. ALBERDI, A. RIUS, J.M. PAREDES, X. BARCONS, F. BRIONES, J. CEPA, J. VILCHEZ, C. EIROA, J.M. MARCAIDE and J.M. MAS / Spanish Participation in the Millimeter Array 381

J. GÓMEZ-GOMAR, J. ISERN, E. BRAVO, F. GIOVANNELLI and C. DE LA PADULA / SIXE: A Payload for MINISAT-02 389

C. MORALES, J.F. GÓMEZ, J. TRAPERO, S. BOWYER, J. EDELSTEIN and E. KORPELA / EURD: First Year Results 393

J.A. BERNÁ, M. PÉREZ and G. BERNABEU / Dynamics and Control for a Small Telescope 397

M. ÁLVAREZ ÁLVAREZ and A.I. DÍAZ / Women in the History of Astronomy 401

E. STENGLER and C.S. HANSEN RUIZ / Considerations on the Teaching and the Popularization of Cosmology 405

Author Index 409

PREFACE

The *Spanish Astronomical Society* (*Sociedad Española de Astronomía*, SEA) was founded on November 25, 1991. It celebrated its first society meeting (*asamblea constituyente*) in October 1992. Presently, its members include over 70% of all Spanish astronomers. The main aim of the SEA is to encourage the development of Spanish astronomy by providing a forum for discussion of issues of interest, scientific and technical, present and future. For this purpose, among other things, the SEA organizes a scientific meeting every other year.

The Third Scientific Meeting of the SEA was held at the *Museo de la Ciencia y el Cosmos* in La Laguna, Tenerife, from September 21 to 24, 1998, under the auspices of the SEA and the Instituto de Astrofísica de Canarias (IAC). We thank the *Museo* for its hospitality. During these four days, 109 astronomers from 19 Spanish and 10 foreign institutions gathered to present their latest results and to discuss ongoing research on many different topics, ranging from the solar system to observational cosmology, with special emphasis on the development of the next generation of instruments to propel Spanish research into the new century. The contents of these Proceedings thus reflect the broad interests of the Spanish astronomical community.

Following general guidelines of the Executive Committee of the SEA, the Scientific Organizing Committee (SOC), made sure that many of the speakers were young astronomers, albeit not inexperienced ones, and yet holding further promise. Undoubtedly, this generation of young astronomers constitutes the driving force of present Spanish astronomical research.

This meeting could not have taken place without the efforts of the Local Organizing Committee and the logistical support of the IAC. We thank them. We also thank the members of the SOC for their decisive contribution to the scientific success of the meeting and gratefully acknowledge the financial support received from several universities, research centers, and governmental institutions. We thank A. Gil de Paz and N. Cardiel for their efficient help in preparing these proceedings at the Departamento de Astrofísica of the Universidad Complutense de Madrid. Last, but not least, we would like to acknowledge the generous, and usually timely, help of the 33 anonymous referees who insured the high standard of quality of the contributions in this volume.

The symbolic significance of these proceedings for the SEA can hardly be exaggerated, since here, for the first time, the society publishes the proceedings of its own scientific meeting. Let us wish for the SEA a long life and may it publish many more proceedings of its future scientific meetings.

Juan-María Marcaide
SEA President

Javier Gorgas
Jaime Zamorano
Guest editors

Astrophysics and Space Science is the original source of publication of his article. It is recommended that this article is cited as: *Astrophysics and Space Science* **263**: xi, 1999.

SCIENTIFIC ORGANIZING COMMITTEE

Rosa Domínguez Tenreiro	Universidad Autónoma de Madrid (UAM)
Xavier Barcons	Instituto de Física de Cantabria (CSIC–UC)
Javier Gorgas	Universidad Complutense de Madrid (UCM)
José María Ibáñez	Universidad de Valencia (UVEG)
José Miguel Rodríguez Espinosa	Instituto de Astrofísica de Canarias (IAC)
Antonio Talavera	Lab. de Astrofísica Espacial y Física Fundamental (LAEFF)

LOCAL ORGANIZING COMMITTEE

José Miguel Rodríguez Espinosa (IAC)
Jordi Cepa Nogué (IAC)
Manuel Collados Vera (IAC)
Gabriel Gómez Velarde (IAC)
Artemio Herrero Davó (IAC)
Beatriz Mederos (IAC)
Mónica Murphy (IAC)
Ana María Pérez García (IAC)

SPONSORED BY:

Sociedad Española de Astronomía (SEA)
Instituto de Astrofísica de Canarias (IAC)
Consejo Superior de Investigaciones Científicas (CSIC)
Universidad Autónoma de Madrid (UAM)
Universidad Complutense de Madrid (UCM)
Dirección General de Universidades e Investigación (DGUI)
(Gobierno Autónomo de Canarias)
Dirección General de Enseñanza Superior (DGES)
(Ministerio de Educación y Cultura)

Astrophysics and Space Science is the original source of publication of his article. It is recommended that this article is cited as: *Astrophysics and Space Science* **263**: xiii, 1999.

LIST OF PARTICIPANTS

Alonso Herrero, Almudena. Steward Observatory (aalonso@as.arizona.edu)
Alvarez, Pedro. Instituto de Astrofísica de Canarias (pam@iac.es)
Alvarez Alvarez, Mar. Universidad Autónoma de Madrid (malvar4@roble.pntic.mec.es)
Aparicio, Antonio. Instituto de Astrofísica de Canarias (aaj@iac.es)
Ascasibar, Yago. Universidad Autónoma de Madrid (yago.ascasibar@uam.es)
Balcells, Marc. Instituto de Astrofísica de Canarias (balcells@iac.es)
Barcons, Xavier. Instituto de Física de Cantabria (barcons@ifca.unican.es)
Barrado y Navascues, David. MPIA, Heidelberg (barrado@mars.harvard.edu)
Bernabeu, Guillermo. Universidad de Alicante (bernabeu@disc.ua.es)
Cabrera, Fernando. Instituto de Astrofísica de Canarias (fcabrera@iac.es)
Carballo, Ruth. Instituto de Física de Cantabria (carballo@ifca.unican.es)
Cardiel, Nicolás. Universidad Complutense de Madrid (ncl@astrax.fis.ucm.es)
Castander, F. Javier. University of Chicago (fcj@oodjob.uchicago.edu)
Castellanos, Marcelo. Universidad Autónoma de Madrid (marcelo@astro4.ft.uam.es)
Castro Tirado, Alberto. LAEFF, Madrid (ajct@laeff.esa.es)
Centurión, Miriam. Instituto de Astrofísica de Canarias (mcm@iac.es)
Cepa, Jordi. Instituto de Astrofísica de Canarias (jcn@iac.es)
Cernicharo, José. IEM-CSIC, Madrid (cerni@oan.es)
Chueca, Sergio. Instituto de Astrofísica de Canarias (chueca@iac.es)
Corral, Luis. Instituto de Astrofísica de Canarias (lcorral@iac.es)
Díaz, Angeles. Universidad Autónoma de Madrid (angeles@astro4.ft.uam.es)
Domínguez Tenreiro, Rosa. Universidad Autónoma de Madrid (rosa@astroph.ft.uam.es)
Eiroa, Carlos. Universidad Autónoma de Madrid (carlos@xiada.ft.uam.es)
Esteban, César. Instituto de Astrofísica de Canarias (cel@iac.es)
Fabiani Bendicho, Peña. Instituto de Astrofísica de Canarias (pfb@iac.es)
Feast, Michael. University of Cape Town (mwf@uctvms.uct.ac.za)
Fonseca, Victoria. Universidad Complutense de Madrid (fonseca@eucmdx.gae.ucm.es)
Galadí Enríquez, David. Universidad de Barcelona (dgaladi@am.ub.es)
Gallego, Jesús. Universidad Complutense de Madrid (jgm@astrax.fis.ucm.es)
García Dabó, C. Enrique. Universidad Complutense de Madrid (ceg@astrax.fis.ucm.es)
García López, Ramón J.. Instituto de Astrofísica de Canarias (rgl@iac.es)
García Vargas, María Luisa. GRANTECAN (mgarcia@iac.es)
Génova, Ricardo. Instituto de Astrofísica de Canarias (rgg@iac.es)

Astrophysics and Space Science is the original source of publication of his article. It is recommended that this article is cited as: *Astrophysics and Space Science* **263**: xv–xvii, 1999.

Gil de Paz, Armando. Universidad Complutense de Madrid (gil@astrax.fis.ucm.es)

Goicoechea, Luis Julián. Universidad de Cantabria (goicol@besaya.unican.es)

Gómez Flechoso, Mª Angeles. Observatoire de Gèneve (maria.gomez@obs.unige.ch)

Gómez Gomar, Jordi. Institut d'Etudis Espacials de Catalunya (jgomez@ieec.fcr.es)

Gómez Velarde, Gabriel. Instituto de Astrofísica de Canarias (ggv@iac.es)

González Solares, Eduardo. Instituto de Astrofísica de Canarias (eglez@iac.es)

Gorgas, Javier. Universidad Complutense de Madrid (fjg@astrax.fis.ucm.es)

Guzmán, Rafael. University of Yale (rguzman@astro.yale.edu)

Hansen Ruiz, Cristina Silvia. Tenerife (cshrc@correo.rcanaria.es)

Herrero, Artemio. Instituto de Astrofísica de Canarias (ahd@iac.es)

Hueso, Ricardo. Universidad del País Vasco (wuahualr@bi.ehu.es)

Jiménez Benito, Luis. Universidad Autónoma de Madrid (luis@delta.ftuam.es)

Kidger, Mark. Instituto de Astrofísica de Canarias (mrk@iac.es)

Koo, David. UCO/Lick Observatory, University of California, (koo@ucolick.org)

López, Rosario. Universidad de Barcelona (rosario@mizar.am.ub.es)

Luri, Xavier. Universidad de Barcelona (xluri@mizar.am.ub.es)

Manrique, Alberto. Universidad de Barcelona (alberto@pcess2.am.ub.es)

Marcaide, Jon. Universidad de Valencia (jmm@vlbi.uv.es)

Marco, Amparo. Universidad de Alicante (tobarra@disc.ua.es)

Martín Fleitas, Juan Manuel. GRANTECAN (jmartin@iac.es)

Martín Mateos, Isabel. Instituto de Astrofísica de Canarias (imm@iac.es)

Martínez Delgado, David. Instituto de Astrofísica de Canarias (ddelgado@iac.es)

Mayor, Michel. Observatoire de Géneve (michel.mayor@obs.unige.es)

Mediavilla, Evencio. Instituto de Astrofísica de Canarias (emg@iac.es)

Millares, Juan Antonio. Universidad de Valencia (juan.a.miralles@uv.es)

Monreal, Ana. Universidad de La Laguna

Montes, David. Universidad Complutense de Madrid (dmg@astrax.fis.ucm.es)

Montesinos, Benjamín. LAEFF/INTA, Madrid (bmm@laeff.esa.es)

Monteverde, Ilusión. Instituto de Astrofísica de Canarias (imonteve@iac.es)

Morales, Carmen. LAEFF, Madrid, (morales@laeff.esa.es)

Moreno Insertis, Fernando. Instituto de Astrofísica de Canarias (fmi@iac.es)

Muñoz, Jose A.. Harvard-Smithsonian Center for Astrophysics (jmunoz@cfa.harvard.edu)

Najarro, Francisco. IEM-CSIC, Madrid (najarro@astro.iem.csic.es)

Ortiz, Elena. Universidad Autónoma de Madrid (elena@astro1.ft.uam.es)

Palacios, Javier. Universidad Autónoma de Madrid (javier@xiada.ft.uam.es)

Pallé, Pere Lluis. Instituto de Astrofísica de Canarias (plp@iac.es)

Patrón, Jesús. Instituto de Astrofísica de Canarias (jpr@iac.es)

Pedraz, Santos. Universidad Complutense de Madrid (spm@astrax.fis.ucm.es)

Pello, Roser. Observatoire Midi-Pyrénées (roser@obs-mip.fr)

Pérez Acosta, Nieves. Tenerife

Pérez Fournon, Ismael. Instituto de Astrofísica de Canarias (ipf@iac.es)

Pérez García, Ana María. Instituto de Astrofísica de Canarias (apg@iac.es)

Pérez Moreno, Javier. Universidad de Alicante (jperez@astronomia.disc.ua.es)

Pérez Ortiz, Juan Vicente. Círculo Astronómico del Mediterraneo (tobarra@disc.ua.es)

Pérez Torres, Miguel Angel. Universidad de Valencia (miguel@vlbi.uv.es)

Raig, Andrés. Universidad de Barcelona (araig@mizar.am.ub.es)

Roca, Teodoro. Instituto de Astrofísica de Canarias (trc@iac.es)

Rodes, José Joaquín. Universidad de Alicante (rodes@astronomia.disc.ua.es)

Rodríguez Espinosa, José Miguel. Instituto de Astrofísica de Canarias (jre@iac.es)

Rodríguez Pascual, Pedro. Univ. Europea de Madrid (p_miguel.rodriguez@fis.cie.uem.es)

Ros, Eduardo. MPI, Bonn (ros@mpifr-bonn.mpg.de)

Ruiz Cobo, Basilio. Instituto de Astrofísica de Canarias (brc@iac.es)

Ruiz Morales, Jorge. Tribuna de Astronomía (astronomia@mad.servicom.es)

Saiz, Alejandro. Universidad Autónoma de Madrid (asaiz@delta.ft.uam.es)

Salvador Solé, Eduard. Universidad de Barcelona (eduard@faess0.am.ub.es)

Sánchez Lavega, Agustín. Universidad del País Vasco (wupsalaa@bi.ehu.es)

Sánchez Salcedo, F. Javier. University of Newcastle (f.j.sanchez-salcedo@newcastle.ac.uk)

Sempere, María Jesús. IEM–CSIC, Madrid (sempere@astro.iem.csic.es)

Serichol, Nuria. Universidad Politécnica de Cataluña (nuria@montcau.upc.es)

Solano, Enrique. Universidad Europea de Madrid (esm@vilspa.esa.es)

Stengler, Erik. Universidad de La Laguna (erik@adwspain.com)

Suárez, Olga. Universidad de Vigo (olga@uvigo.es)

Tomás Calderón, Laura. ISO Science Operation Centre (ltomas@iso.vilspa.esa.es)

Torra, Jordi. Universidad de Barcelona (jordi@mizar.am.ub.es)

Torrejón, José Miguel. Universidad de Alicante (jmt@castor.daa.uv.es)

Torrelles, José María. Instituto de Astrofísica de Andalucía (torrelles@iaa.es)

Trapero, Joaquín. LAEFF, Madrid (trapero@laeff.esa.es)

Urbaneja, Miguel Alejandro. Instituto de Astrofísica de Canarias

Varela, Antonia María. Instituto de Astrofísica de Canarias (avp@iac.es)

Vázquez, Manuel. Instituto de Astrofísica de Canarias (mva@iac.es)

Vega Beltrán, Juan Carlos. Osservatorio Astronomico di Padova (jvega@pd.astro.it)

Verdugo, Eva. ISO Science Operation Centre (ev@iso.vilspa.esa.es)

Villamariz, Charo. Instituto de Astrofísica de Canarias (ccid@iac.es)

Villar Martín, Montse. University of Sheffield (m.villar-martin@sheffield.ac.uk)

Yepes, Gustavo. Universidad Autónoma de Madrid (gustavo@astrosg.ft.uam.es)

Zamorano, Jaime. Universidad Complutense de Madrid (jaz@astrax.fis.ucm.es)

THE EVOLUTION OF STAR-FORMING GALAXIES: THE SFR DENSITY OF THE UNIVERSE

JESÚS GALLEGO
Dpto Astrofísica
Universidad Complutense de Madrid, E-28040, Spain

Abstract. The evolution of the Star Formation Rate (SFR) density of the Universe as a function of look-back time is a fundamental parameter in order to understand the formation and evolution of galaxies. The current picture, only outlined in the last years, is that the global SFR density has dropped by about an order of magnitude from a redshift of z~1.5 to the current value at z=0. Because these SFR density studies are now extended to the whole range in redshift, it becomes mandatory to combine data from different SFR tracers. At low redshifts, optical emission lines are the most widely used. Using Hα as current-SFR tracer, the Universidad Complutense de Madrid (UCM) Survey provided the first estimation of the global SFR density in the Local Universe. The Hα flux in emission is directly related to the number of ionizing photons and, modulo IMF, to the total mass of stars formed. Metallic lines like [OII]λ3727 and [OIII]λ5007 are affected by metallicity and excitation. Beyond redshifts z~0.4, Hα is not observable in the optical and [OII]λ3727 or UV luminosities have to be used. The UCM galaxy sample has been used to obtain a calibration between [OII]λ3727 luminosity and SFR specially suitable for the different types of star-forming galaxies found by deep spectroscopic surveys in redshifts up to z~1.5. These calibrations, when applied to recent deep redshift surveys confirm the drop of the SFR density of the Universe since z~1 previously infered in the UV. However, the fundamental parameter that determines galactic evolution is mass, not luminosity. The mass function for local star-forming galaxies is critical for any future comparison with other galaxy populations of different evolutionary status. Hα velocity-widths for UCM galaxies indicate that besides a small fraction of 10^{10}–10^{11} $M_\odot$ starburst nuclei spirals, the majority have dynamical masses in the $\sim 10^9$ $M_\odot$ range. A comparison with published data for faint blue galaxies suggests that star-forming galaxies at z~1 would have SFR per unit mass and burst strengths similar to those at z=0, but being intrinsically more massive.

Astrophysics and Space Science is the original source of publication of his article. It is recommended that this article is cited as: *Astrophysics and Space Science* **263**: 1–14, 1999.

1. Introduction

The evolution of the global Star Formation Rate (SFR) of the Universe is a fundamental parameter in order to understand the formation and evolution of galaxies. The current picture is that the global SFR density observed has dropped by about an order of magnitude from a peak at redshift of z=1-2 to the current value at z=0 (Madau et al. 1996; 1997b, see also § 3). Up to date, values at z<0.4 rely on [OII]λ3727 and Hα luminosities of selected samples (Gallego et al. 1995, Tresse & Maddox 1998, Hogg et al. 1998). For intermediate redshifts z$\sim$0.5-1 the determinations are based on broad-band continuum luminosities at 2800 Å and 4400 Å (Lilly et al., 1996) and [OII]λ3727 luminosities (Guzmán et al. 1997; Hogg et al. 1998; Hammer et al. 1997) scaled with the SFR. Beyond z=1, UV broad-band measurements from photometric redshifts are the only SFR indicators available (Madau et al. 1996; Madau 1997; Connolly et al. 1997), given that spectroscopic redshifts are still highly incomplete (Lowenthal et al., 1997).

A few pioneering works have estimated average SFR densities measuring in the nIR Hα luminosities for small samples at z=1 (Glazebrook et al., 1998) and Hβ luminosities for a sample of five objects at z=3 (Pettini et al., 1998). These preliminary results indicate that SFRs from Balmer lines are found to be 2-3$\times$ higher than those inferred from the UV at all redshifts, pointing to the idea that the behavior of the SFR density at medium-high redshift is still not clear. Detailed theoretical works are starting to seed light into the problem, with some groups reproducing the observed data (Baugh et al., 1998) and other authors predicting a much higher SFR density at high redshift (Somerville & Primack, 1998b).

2. The Local Reference of Star-forming Galaxies

One of the most representative samples of current star-forming galaxies for the Local Universe is the Universidad Complutense de Madrid (UCM) Survey. The objects were selected in photographic IIIa-F objective-prism plates obtained with Schmidt Telescope via the presence of Hα+[NII]λ6584 emission in very low resolution spectra. The observational procedure and the sample of galaxies are more detailed in the three discovery lists (Zamorano et al. 1994; 1996; Alonso et al. 1998). Follow-up spectroscopy (Gallego et al. 1996; 1997), photometry (Vitores et al. 1996a; 1996b) and IRAS data (Rego et al., 1993) provide with a complete database of physical parameters for these objects.

The UCM survey constitutes an ideal tool for studying the properties of star-forming galaxies at low redshift, since they were selected exclusively according to their Hα emission. The sample consists of 264 sources in 500 deg^2 (z<0.045) with EW(Hα+[NII])>10Å. It is dominated by a

TABLE 1. Average properties of the local star-forming galaxies

Type	M_r (H=50)	$L_{H\alpha}$ 10^8 $L_\odot$	Equivalent width (Å) Hα	[OII]	[OIII]	Re kpc	σ km/s	M 10^{10} $M_\odot$
Sy 1	-21.3	5.22	260	13	38	–	–	–
Sy 2	-22.0	3.78	86	26	90	–	–	–
SBN	-21.1	2.24	77	27	11	4.6	86	3.4
DANS	-20.3	0.58	54	38	13	1.5	44	0.7
HIIH	-20.4	2.32	150	63	74	2.3	44	0.7
DHIIH	-19.1	0.38	107	64	78	2.9	25	0.2
BCD	-18.1	0.35	294	89	458	2.8	26	0.1

heterogeneous population of galaxies, with 66% being intermediate to low luminosity late-type galaxies. The line emission comes largely from nuclear regions. The median luminosity is about 1 mag fainter than galaxies found in blue surveys. Up to 59% of the sample have low ionization or high extinction properties, being these galaxies difficult to identify in blue surveys for the lack of strong [OII]λ3727, Hβ or [OIII]λ5007 in emission (Gallego et al., 1997). In table 1 have been summarized the average properties of the different spectroscopic types. They include r Gunn-Thuan absolute magnitudes, Hα luminosities, equivalent widths of the most important emission lines, effective radius, Hα velocity widths and estimated dynamical masses (see Section 5). The subcategories are the spectroscopic types as defined by Gallego et al. (1996) and correspond to Seyfert 1 and Seyfert 2, low-ionization StarBurst Nuclei (SBN) galaxies, Dwarf Amorphous Nuclear Starbursts (DANS), high-ionization HII-like Blue Compact Dwarfs (BCD), HII galaxies (HIIH) and Dwarf HII galaxies (DHIIH).

From a complete sample of emission line galaxies (ELGs) from the UCM survey, Gallego et al. (1995) computed the Hα luminosity function for the current star-forming galaxies in the Local Universe. An aperture correction was applied to all the spectroscopic data transforming the observed Hα equivalent widths obtained from spectroscopy using very wide slits to total Hα luminosities scaling with the broad-band Gunn r images. Finally, observed luminosities were corrected for extinction using Balmer decrements. The average extinction resulted to be 1.5 mag. A Schechter function with the following parameters: $\alpha = -1.3 \pm 0.2$; $L^*(\mathrm{H}\alpha) = 10^{42.15 \pm 0.08}\,\mathrm{erg\,s^{-1}}$ and $\phi^* = 10^{-3.2 \pm 0.2}\,\mathrm{Mpc^{-3}}$ for $H_0 = 50\,\mathrm{km\,s^{-1}\,Mpc^{-1}}$ provides a good fit to the present-epoch SFR function, which describes the number of star-forming galaxies as a function of their ongoing SFR. Integrating over the full range of luminosities, an Hα luminosity density of $10^{39.1 \pm 0.1}\,\mathrm{erg\,s^{-1}\,Mpc^{-3}}$ is obtained. Using the Hα emission as SFR estimator, this translates into a SFR

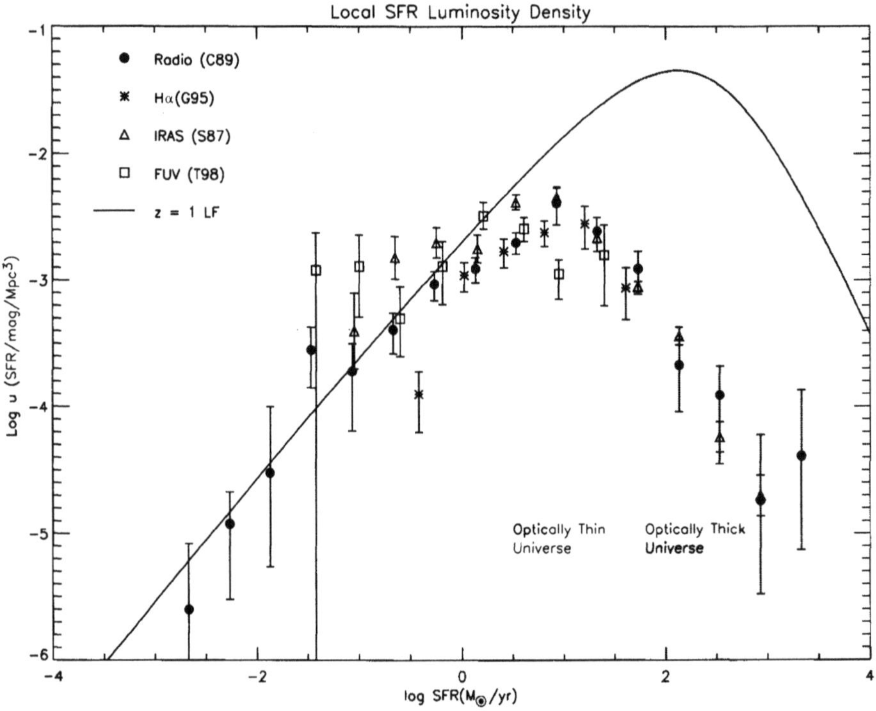

Figure 1. Local SFR luminosity functions in the radio, IRAS, Hα and far-UV as taken from Richards (astro-ph/9811120). The shaded region represents what may be a dust screen beyond which optical surveys are blind to star formation. The solid line represents the SFR luminosity function at z=1.

density for the Local Universe of $10^{-2.4\pm0.1}\,M_\odot\,\mathrm{yr}^{-1}\,\mathrm{Mpc}^{-3}$ in star-forming galaxies with $EW(\mathrm{H}\alpha+[\mathrm{NII}]) > 10$Å and $z < 0.045$, for a Salpeter IMF. It is worth noting that if all the Hα luminosity detected from AGNs in the UCM survey was also interpreted as originated in star formation processes, they would contribute with an extra 15% to the total amount.

A similar result to the one obtained by the UCM survey can be derived applying a luminosity-weighted average color to the observed luminosity function in the B band, and then converting the UV flux into an instantaneous SFR. Tinsley & Danley (1980) also inferred a similar density from the distribution of colors of spiral galaxies. Gronwall (1998) has obtained a higher value from preliminary results of the deeper KISS survey, but given so many corrections needed it will be necessary to wait for the follow-up spectroscopy before deriving any further conclusion.

Despite all these results, it should be kept in mind that if some fraction of newly formed stars is completely hidden by dust, they would not contribute to the observed Hα luminosity, implying that the SFR derived from this quantity should be considered a lower limit (Rowan-Robinson et al., 1997). In this way, Serjeant et al. (1998) have computed the 1.4 GHz radio luminosity function of local spiral galaxies. The best-fit Schechter function reproduces the faint-end slope obtained from the UCM survey but gives a 2-3× larger density normalization (see Table 2 and Figure 1), with data suggesting that optically selected surveys could be missing the most extreme and dust obscured starbursts (Richards, 1998). However, There is a significative difference in the results depending on the selection of the sample (Cram 1998; Dunlop 1997). Despite the far infrared and 1.4 GHz luminosities are very promising tools to estimate SFRs, some work has still to be done in order to improve the calibration and completeness of the samples.

3. The Evolution of the SFR Density of the Universe

The Hα+[NII]λ6584 fluxes of the I-selected Canada-France Redshift Survey (CFRS) galaxies provided the Hα luminosity function at redshift z<0.3 (Tresse & Maddox, 1998). The CFRS survey is a magnitude limited (M_B>-21 mag for z<0.3) sample fairly representative of field galaxies. Because of the I-band selection, the galaxies have been selected by light from the old stellar population, rather than on their young stars. Consequently, the CFRS sample is less sensitive to recent star formation than Hα selected samples. An aperture correction based in the V-band flux and an average extinction correction (except for a fraction of objects with Hβ detected) were applied. Due to the low dispersion of the spectroscopic data an average correction for the [NII]λ6584 was used. A Schechter function provides a good fit with the following parameters: $\alpha = -1.35 \pm 0.06$; $L^*(\mathrm{H}\alpha) = 10^{42.13\pm0.13}\,\mathrm{erg\,s^{-1}}$ and $\phi^* = 10^{-2.8\pm0.1}\,\mathrm{Mpc^{-3}}$ for $H_0 = 50\,\mathrm{km\,s^{-1}\,Mpc^{-1}}$. They obtained a total Hα luminosity density of at least $10^{39.44\pm0.04}$ erg s^{-1} Mpc^{-3} at a mean z=0.2 for galaxies with rest-frame EW(Hα+[NII]) > 10Å. This is twice the value found at z=0 by Gallego et al. (1995).

They have also obtained that at least ~40% of the ELGs present in the sample are relatively red galaxies that contribute to the overall luminosity as well as the blue galaxies. Most of these objects are spiral galaxies and correspond to the StarBurst Nuclei (SBN) and Dwarf Amorphous Nuclear Starbursts (DANS) that were already found within the UCM survey. However, meanwhile AGNs where only 15% of the UCM sample, Hammer et al. (1997) conclude that a large fraction of the z<0.3 CFRS galaxies are low-luminosity objects with AGN activity. The discrepancy observed in Figure 2 between UV and Hα data can be solved assuming a dust extinction correction of ~ 1 mag for the UV data. New information has been added by a spectroscopic survey of galaxies at z~0.2 selected in the rest-frame

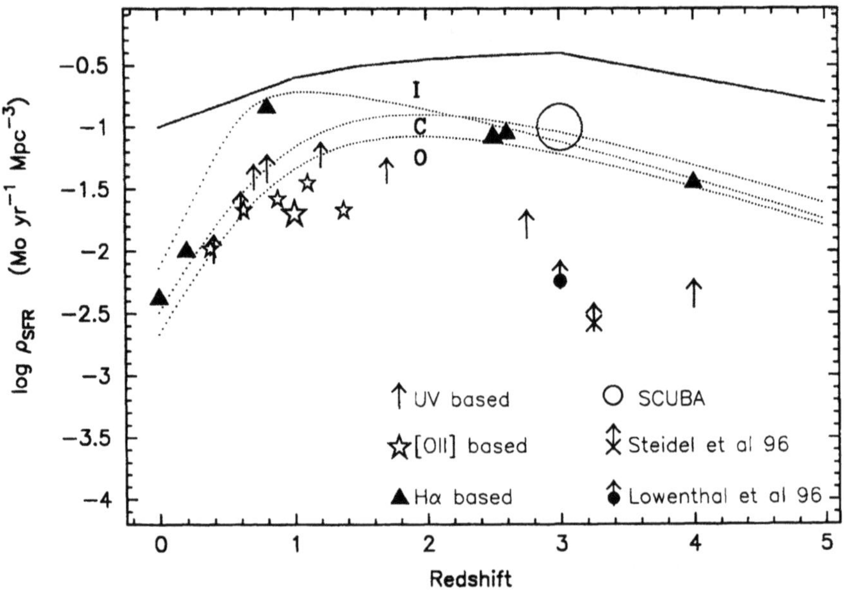

Figure 2. Volume averaged SFR density as a function of redshift (Salpeter IMF, q_o=0.5 and $H_0 = 50\,\mathrm{km\,s^{-1}\,Mpc^{-1}}$ cosmology). The dotted lines represent Pei & Fall (1995) models. C=Closed box model, I=Inflow of metal-free gas and O=Outflow of metal-enriched gas model. The line that runs on top of all the data is the one predicted by Somerville & Primack (1998). Errors bars have been omitted for clarity. Arrows are from UV CFRS (Lilly et al. 1996) and HDF data (Madau et al. 1996; Connolly et al. 1997). The filled triangles are from local Hα (Gallego et al. 1995) Hα CFRS (Tresse & Maddox 1998) and Hα in the nIR (Bechtold 1998; Glazebrook et al. 1998; Pettini et al. 1997). Stars are from [OII] data from Guzmán et al. (1997); Hogg et al. (1998). Note that Hα data are reddening corrected, while UV data are not. See text for a more detailed description of the various data sets.

ultraviolet (Treyer et al., 1998), with a similar SFR density.

Several authors have applied an empirical calibration between SFR and [OII]λ3727 luminosity to the spectroscopic data obtained from deep redshift surveys. Hogg et al. (1998) have applied the Kennicutt (1992) calibration to the galaxies of the Caltech Faint Galaxy Redshift Survey. It corresponds to a sample of 375 galaxies (~75% spectroscopically complete) down to R=23.3 eight arc-minutes around the HDF. They used the [OII]λ3727 equivalent widths, R fluxes and spectral energy distributions (from R-K color) to estimate the [OII] luminosity. Overall, the line luminosity is higher by an order of magnitude in the higher-redshift sample thn for the low redshift objects.

Figure 2 represents all the most important observations and data sets to date. UV luminosities (arrows) for the CFRS sample (Lilly et al., 1996) and for the HDF (Madau et al., 1996) have been also plotted for reference. Guzmán et al. (1997) applied their own calibration to the Hawaii data set (Cowie et al., 1995) (open stars in Figure 2), which corresponds to a Keck spectroscopic survey ~90% complete down to I_{AB}<22.5. Up to z~1, where incompleteness affects more seriously to the data, the general increase observed for the CFRS sample using UV luminosities (Lilly et al., 1996) is confirmed. Moreover, they follow very tightly a general relation

$$SFR \sim (1+z)^3 \quad (1)$$

Cowie et al. (1997) analyzed an extended to B<24.5 redshift sample to study the evolution of the SFR luminosity function. Correcting for the different SFR-[OII]λ3727 luminosity their results are consistent with a SFR density for the 0.1-0.4 redshift range as twice the Gallego et al. (1995) value for the Local Universe. this 2× increase is in agreement with previous works (Guzmán et al. 1997; Tresse & Maddox 1998).

Hammer et al. (1997) applied the Kennicutt (1992) calibration to study the SFR density evolution from the [OII]λ3727 luminosity density of the CFRS sample. They found an increase for field galaxies at z=0.375 of a factor 1.6 respect to the local value (Gallego et al., 1995), and a factor of 8.4 between z=0.375 and z=0.85. The latter increase is even larger than the one obtained at 2800 Å(Lilly et al., 1996), although the overall increase over the whole redshift range from z=0 to z=1 is comparable. The UV luminosities for the CFRS sample are observed values. It is necessary to apply one mag correction for extinction to recover the total luminosity densities (Hammer et al., 1997). Photometric redshifts applied to the HDF (Connolly et al., 1997) are in agreement with the scenario exposed above (arrows in the z~1.5 of Figure 2).

A few pioneering works have estimated average SFR densities measuring in the nIR Hα luminosities for a sample of 13 objects at z=1 (Glazebrook et al., 1998) and Hβ luminosities for a sample of five objects at z=3 (Pettini et al., 1998). These results indicate that SFRs from Balmer lines are found to be 2-3× higher than those inferred from the UV at all redshifts. Finally, Bechtold (1998) have carried out a spectroscopic survey in the near infrared looking for Hα from high redshift galaxies. Using the IRTF and Palomar 5m nIR cameras they have detected a population of Hα emitters at z~2.5. Despite the spectroscopic confirmation is still in the way, preliminary results based on the Hα flux observed in tuned nIR narrow band images are very promising. It means a SFR density at z~2.5 (filled triangle in Figure 2) roughly 20 times larger than the local value. This result is consistent with the results derived from UV luminosities by Madau et al. (1996).

Hu et al. (1998) have estimated lower limits for the SFR density of a sample of strong emission-line galaxies in the redshift range z=3-6. Within

TABLE 2. Characteristics of the several SFR density estimations

Survey	Tracer	SFR LF α	SFR LF ϕ^*	SFR LF SFR*	Sample	Spec	AGN Corr	Dust Corr
UCM	Hα	-1.3	$6.3\ 10^{-4}$	9.4	270	Yes	Yes	Yes
KISS	Hα	-1.2	$2.5\ 10^{-3}$	4.7	800	No	No	Av.
CFRS	Hα	-1.3	$1.5\ 10^{-3}$	9.0	110	Yes	No	Av.
CFRS	UV	...	...	...	1000	Yes	No	No
Treyer	2000Å	-1.6	$9.1\ 10^{-3}$	...	100	Yes	Yes	Yes
Connolly	2800Å	...	...	...	219	No	No	No
CFRS	[OII]λ3727	...	...	...	400	Yes	Yes	Yes
CFGRS	[OII]λ3727	-1.3	$6.3\ 10^{-3}$	9.5	375	Yes	Yes	Av.
Sergeant	1.4 GHz	-1.3	$1.5\ 10^{-3}$	5.3	231	Yes	Yes	...

the substantial uncertainties of small numbers and extinction corrections, the results suggest that the SFRs for emission-line objects are constant or may be possibly increasing with redshift, in contrast with the color-based samples where the rate is declining at higher redshifts (Madau et al. 1996; Madau et al. 1998; Arnauts et al. 1998). There are some hints that the discrepancy could be due to the effect of cosmological $(1+z)^3$ surface brightness dimming (Pascarelle et al., 1998).

However, there is mounting evidence that the true SFR history may be considerably different from the original estimates. The SFRs obtained from far-infrared luminosities for 13 galaxies detected by ISO in the Hubble Deep Field (Rowan-Robinson et al., 1997) are higher than the results based in optical luminosities by a factor of 10× or so. Finally, a sub-mm survey using SCUBA and centered on the HDF (Hughes et al., 1998) implies a SFR density over the 2-4 redshift range at least five times higher than the inferred from the UV-selected objects (Open circle at z=3 in Figure 2). It is clear that it can not yet be clearly established whether the SFR falls off or remains constant at high redshift.

In a parallel effort, semi-analytic models developed in the last years can now predict global galaxy properties such as the Tully-Fisher relation, the B-band luminosity function, gas contents and the global evolution of the volume-averaged SFR density of the Universe at different redshifts (Pei & Fall 1995; Kauffmann et al. 1997; Baugh et al. 1998; Somerville & Primack 1998b). Most of the models reproduce the observed increase from z=0 to z$\sim$1.5, where the SFR density would decrease down to the local value at z$\sim$5 or would stay at a high value up to very high redshift. On the contrary, there is still controversy respect to the galaxy populations involved in this SFR density evolution. Some authors invoke new kind of galaxies, meanwhile others prefer mild evolution of the already known star populations (see Ellis 1997 for a nice review on this subject). It is worth to mention that when all this information is contrasted with the local star formation record encoded in the present-day stellar mass function, some incongruities still do appear (Renzini 1998; Fields 1998).

Given so many SFR tracers and indicators used for different galaxy populations, all of them have to be properly scaled and corrected for dust extinction, incompleteness, etc to provide a globally coherent picture. Any comprehensive study of global galaxy properties at high redshift have to be compared with well established local reference populations. In this way, current efforts are being oriented to obtain star-forming galaxy populations for which several different SFR estimators can be used simultaneously.

4. Star Formation Rate Tracers up to z∼1.5

The Hα luminosity, related to the Lyman-continuum flux of the embedded stars, is also a direct measurement of the current SFR modulo the IMF, since it is directly related to the number of massive stars (Kennicutt 1983; Kennicutt 1992; Glazebrook et al. 1998; Charlot 1998). It is better than other optical Balmer lines such as Hβ or Hγ, which are more affected by stellar absorption and reddening. Metallic nebular lines like [OII]λ3727, [OIII]λ5007 (affected by excitation and metallicity) are SFR indicators rather than quantitative measurements (Charlot, 1998). Despite the Hα line is one of the most reliable indicators of SFR, it is observable in the optical only out to redshifts $z<0.4$. In order to compare in a direct way the results for local samples and the results for $z>0.4$ star-forming galaxies, several SFR indicators have been considered.

Because the UV continuum emission from a galaxy with significant ongoing star formation is dominated by hot stars, the unreddened UV luminosity scales (modulo IMF) very well with the current SFR of the burst, independently of the past star formation history. This UV continuum, usually measured at 1500Å , 2000Å or 2800Å, lies in the optical for intermediate to high redshift objects but is not easily measurable for low redshift objects. The Lyα emission has been proved not to be a good star formation estimator because resonant scattering of Lyα photons by neutral atomic hydrogen dramatically affects their relation to the SFR in a galaxy (Charlot & Fall, 1993).

Far-IR fluxes (affected by the dust abundance and properties) could be very useful indicators but still need a better theoretical knowledge (Rowan-Robinson et al., 1997). However, they are heavily affected by dust abundance and properties and, moreover, they are not easily observable for high-z sources. The decimetric radio continuum has been recently proposed as an useful SFR tracer (Cram et al., 1998). It appears to be directly proportional to the rate of formation of supernovae in a galaxy. Overall, It shows a good concordance with SFR estimated from U-band, Hα and infrared luminosities (see Figure 1), but serious discrepancies have been found (Cram et al. 1998; Serjeant et al. 1998).

Despite metallic nebular lines such as [OII]λ3727 and [OIII]λ5007 are affected by excitation and metallicity, they could be used as SFR indicators when the population of galaxies used for the calibration is similar to the one we want to study (Gallagher et al. 1989; Kennicutt 1992). Figure 3 shows the relation between the equivalent widths of the [OII]λ3727 line versus Hα+[NII]λ6584 for the whole UCM sample of galaxies. All the star-forming galaxies follow a mean relation EW([OII]) = 0.2 EW(Hα+[NII]). Still a fraction of mainly SBN-like galaxies show no [OII]λ3727 emission at all but their contribution is in average ~26% of the total amount.

The most promising property of the [OII]λ3727 is that it is observable in the optical up to z~1.4. In this way, the [OII] luminosity provides with a useful SFR calibration for distant star-forming galaxies found within deep spectroscopic surveys in the optical. If we assume that Hα luminosity scales with [OII] luminosity, the calibration between SFR and Hα can be transformed to [OII] luminosity. Using the UCM galaxies database as a whole, we obtain the following general relation

$$\mathrm{SFR\ M_{\odot}\,yr^{-1}} = 8.1 \times 10^{-41}\ \mathrm{L_{[OII]}\ erg\,s^{-1}} \tag{2}$$

A tighter correlation can be obtained if we split the total sample of star forming galaxies into two main classes for red objects (SBN-like) and blue galaxies (HII-like), following the classifications used for Table 1. In general, all spiral galaxies that host a nucleus with an star-forming process and low ionization and dim or null emission lines in the blue would classify as SBN-like objects. Bluer objects, with strong [OIII]λ5007 in emission and compact or idly defined morphologies classify as HII-like objects. The corresponding calibrations result

$$\mathrm{SFR_{SBN}\ M_{\odot}\,yr^{-1}} = 7.9 \times 10^{-41}\ \mathrm{L_{[OII]}\ erg\,s^{-1}} \tag{3}$$

$$\mathrm{SFR_{HII}\ M_{\odot}\,yr^{-1}} = 9.9 \times 10^{-41}\ \mathrm{L_{[OII]}\ erg\,s^{-1}} \tag{4}$$

Gallagher et al. (1989) and Kennicutt (1992) also derived SFR calibrations for L([OII]) based on spectrophotometry of blue irregular and normal spiral galaxies respectively. The first one is close to the obtained for the UCM HII-like galaxies. The second one is close to the one obtained for SBN-like galaxies. In order to avoid flux calibrations and provide a calibration suitable for redshift surveys of distant galaxies it is desirable to introduce a

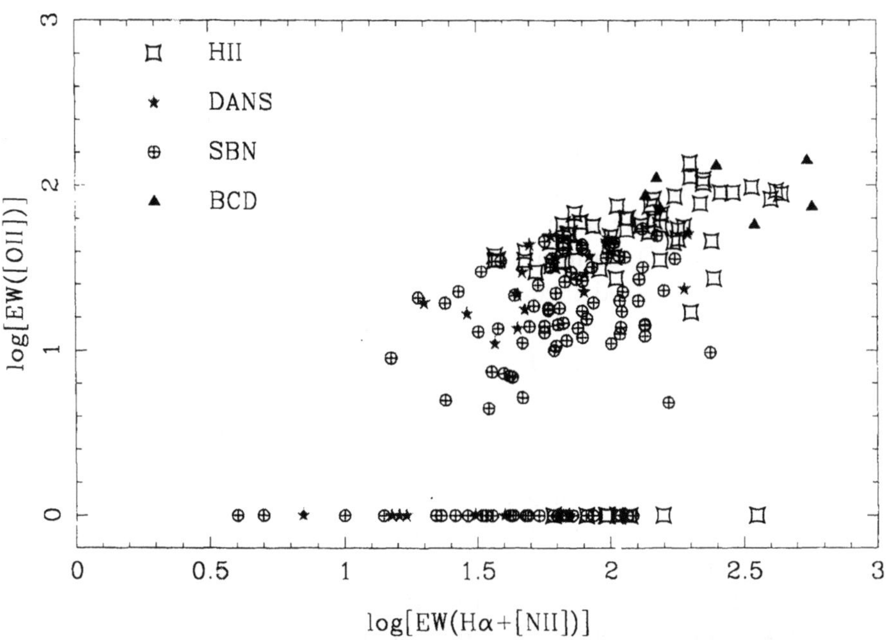

Figure 3. Relation between the equivalent widths of the [OII]λ3727 line versus Hα+[NII] for UCM galaxies. Symbols follow the Gallego et al. (1996) classification used for Table 1. It is worth noting the significative fraction of star-forming objects detected in Hα with no emission in the blue.

relation between blue absolute magnitude M_B, [OII]λ3727 luminosity and [OII]λ3727 equivalent width.

$$\mathrm{SFR_{SBN}}\ \mathrm{M_\odot\ yr^{-1}} \sim 8.2 \times 10^{-12} \times 10^{-0.4\,(M_B - M_{B\odot})}\ EW_{[\mathrm{OII}]} \qquad (5)$$

$$\mathrm{SFR_{HII}}\ \mathrm{M_\odot\ yr^{-1}} \sim 10.2 \times 10^{-12} \times 10^{-0.4\,(M_B - M_{B\odot})}\ EW_{[\mathrm{OII}]} \qquad (6)$$

These expressions are ~1.1 and ~1.25 times larger than the one derived for normal spirals (Kennicutt, 1992), ~6 times larger than that obtained for local irregulars (Gallagher et al., 1989) and ~4 times larger than that used for compact galaxies in the HDF (Guzmán et al., 1997). The dispersion in the main correlations introduces a significant error that is amplified by the errors inherent to the [OII]λ3727 flux itself as SFR tracer. Hammer et al. (1997) already pointed out that the Kennicutt (1992) calibration, when applied to CFRS sample, would predict more mass in the form of stars that the quantity observed in the Local Universe. This problem no longer exists when the calibration we have obtained is applied.

5. Dynamical Masses of UCM Galaxies

One of the major problems that arise when analyzing galactic evolution is that the fundamental parameter that determines evolution over long periods is mass, not luminosity. Pushing the idea of providing the spectroscopic surveys of distant galaxies with a reference in the local universe, it is worth to carry out the estimation of masses for the star-forming galaxies of the UCM Survey. Dynamical masses of strong star-forming galaxies can be obtained via the virial theorem using the effective radius as measured in broad-band images and the velocity dispersions obtained from optical emission lines. A significative fraction of the UCM galaxies (82 object so far) have been observed at high spectroscopic resolution. In Figure 1 the $R_e - \sigma$ diagram for the UCM sample is shown. The zones where Ellipticals (E), spirals (S), Irregulars (Irr) and Dwarf Spheroidals (Sph) lie have been delimited. Average values of effective radius, velocity width and infered dynamical masses are shown in Table 1 for the different spectroscopical types. Despite the emission-line velocity widths could depend on the space distribution of the ionized gas and turbulence due to stellar winds, various evidences favor gravity as the mechanism that dominates the gas kinematics. Terlevich & Melnick (1981) discussed this subject and concluded that for HII galaxies the similarity of their $L - \sigma$ and $R_e - \sigma$ correlations with those of self-gravitating systems pointed to the gravitational origin. More recent models of star-forming regions (Tenorio-Tagle et al., 1993) show that stellar winds provide an average turbulent motion similar to the velocity dispersion of the stars. Comparison between 21cm and Hα line widths further strengthens this conclusion. For a sample of 14 HII galaxies Gallego et al. (in prep) derive $\langle\sigma([OIII])/\sigma(21\text{cm})\rangle = 0.7 \pm 0.1$. The 30% difference may be the consequence of the larger distribution of neutral gas than for the ionized region.

As expected, the majority of local star-forming galaxies have dynamical masses in the $\sim 10^9$ M$_\odot$ range. Looking to Figure 4, local star-forming galaxies lie on top of the distribution of spirals and Irregulars, with nearly all of the HII-like objects them in the low mass ($<10^{10}$ M$_\odot$) region. Except a fraction of SBN-like objects with $\sim 10^{11}$ M$_\odot$), the majority of SBN-like galaxies also lie below the 10^{10} M$_\odot$ line. It is important to mention here that those galaxies with obvious rotation curves and supposedly the most massive, have been not included in this analysis. Higher stellar masses have been obtained for a sample of bright UCM objects from nIR broad-band observations (Gil de Paz et al., 1998).

When comparing the velocity widths for the local sample with similar data for compact star-forming systems at intermediate (0.4<z<1.4) redshift (Guzmán et al., 1997), star-forming galaxies at z=0 show SFR per unit

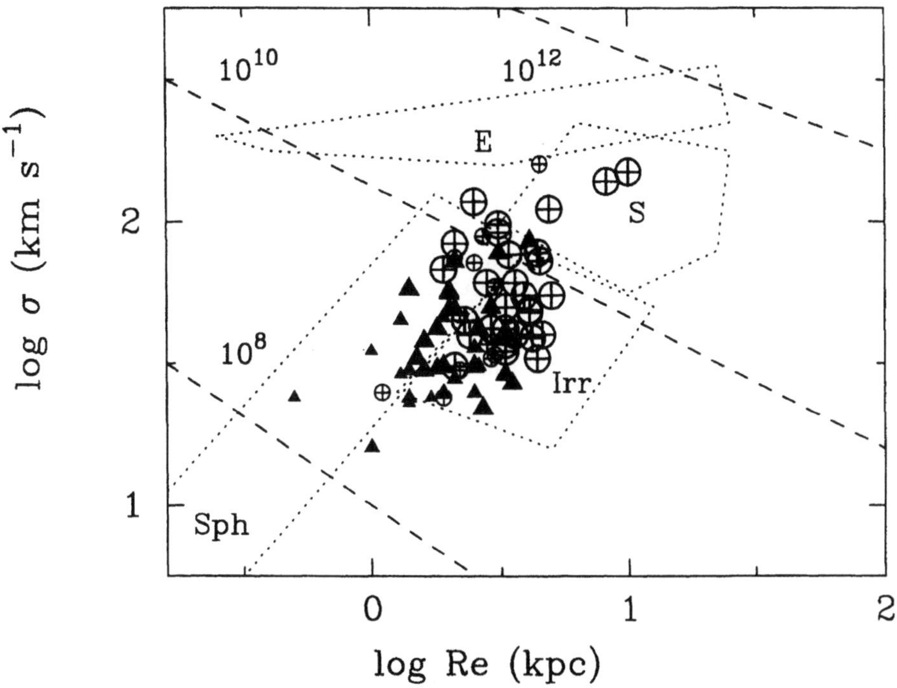

Figure 4. Half-light radius (R_e) versus velocity width (σ) diagram for some UCM star-forming galaxies. The dashed lines correspond to constant-mass lines (in solar masses). Boundaries of typical regions occupied by different morphological types have been delimited. Sph: Spheroidals, Irr: Irregulars, S: Spirals and E: Ellipticals. Filled triangles correspond to BCD, DHIIH and HIIH. Open circles are for SBN and DANS.

mass and burst strengths similar to those at z$\sim$1, but are intrinsically less massive. These results agree qualitatively with a "downsizing" scenario, in which more massive galaxies form at higher redshift. Instead of indicating a new population of galaxies, the results suggest that these high-z star-forming objects may be related to local luminous starbursts. This scenario points out to an increase in the number of massive star-forming systems with redshift. To finally assess whether significant evolution has actually occurred will require more information on the relative numbers of emission-line galaxies as a function of mass in distant and nearby galaxy samples.

Acknowledgements

My collaborators on the papers relevant to this review were A. Aragón-Salamanca, C. García-Dabó, R. Guzmán, D. Koo, M. Rego, A. Vitores and J. Zamorano. This work was supported in part by the Spanish "Programa Sectorial de Promoción del Conocimiento" under grant PB96-0645.

References

Alonso, O., García-Dabó, C.E., Zamorano J., Gallego J., Rego M., 1998, ApJS in press
Arnouts, S., et al., 1998, A&A, astro-ph/9810135
Baugh, C.M., Cole, S., Frenk, C.S., Lacey, C.G., 1998, ApJ 498, 504
Bechtold, J., 1998, in The Young Universe, ASP Conference Series, 146
Connolly, A.J., Szalay, A.S., Dickinson, M.E. et al., 1997, ApJ 486, L11
Cowie, L.L., Hu, E.M., Songaila, A., 1995, Nature 377, 603
Cowie, L.L., Hu, E.M., Songaila, A., 1997, ApJ 481, L9
Cram, L., Hopkins, A., Mobasher, B., Rowan-Robinson, M., 1998, astro-ph/9805327
Cram, L., 1998, ApJL 506, 85, astro-ph/9808228
Charlot, S., 1998, 34th Liege Astroph Colloq, eds. Benvenuti, P. et al, ESA-SP 429, 135
Charlot, S. & Fall, S.M., 1993, ApJ 415, 580
Dunlop, J.S., 1997, Observational Cosmology with Radio Surveys, astro-ph/9704294
Ellis, R.S., 1997, ARA&A 35, 389
Fields, B.D., 1998, APJL astro-ph/9806186
Gallagher, J.S., Bushouse, H., Hunter, D.A., 1989, AJ 97, 700
Gallego J., Zamorano J., Aragón-Salamanca A., Rego M. 1995, ApJ 455, L1
Gallego J., Zamorano J., Rego M., Alonso, O., Vitores, A.G., 1996, A&A 120, 323
Gallego J., Zamorano J., Rego M., Vitores A.G. 1997, ApJ 475, 502
Gil de Paz, A., et al., 1998, this volume
Glazebrook, K., et al., 1999, MNRAS, astro-ph/9808276
Gronwall, C., 1998, *Dwarf Galaxies and Cosmology* eds. Thuan et al., Editions Frontieres
Guzmán, R., Gallego, J., Koo, D.C. et al., 1997, ApJ 489, 559
Hammer, F., et al., 1997, ApJ 481, 49
Hogg, D.W., Cohen, J.G., Blandford, R., Pahre, M.A., 1998, ApJ 504, 622
Hu, E.M., Cowie, L., McMahon, R.G., 1998, ApJL 502, 99
Hughes, D.H., et al., 1998, Nature 394, 241
Kauffmann, G., Nusser, Steinmetz, 1997, MNRAS 286, 795
Kennicutt, R.C., 1983, ApJ 272, 54
Kennicutt, R.C., 1992, ApJ 388, 310
Lilly, S.J., Le Fevre, O., Hammer, F., & Crampton, D., 1996, ApJ 46, L1
Lowenthal, J.D., Koo, D.C., Guzmán, R., Gallego, J., et al., 1997, ApJ 481, 673
Madau, P., Ferguson, H.C., et al., 1996, MNRAS 283, 1388
Madau, P., 1997a, in SF Near and Far, ed. Holt & Mundy, astro-ph/9612157
Madau, P., 1997b, The HDF, ed. M. Livio, et al., STScI Symp Series, astro-ph/9709147
Madau, P., Pozzetti, L., Dickinson, M.E., 1998, ApJ 498, 106
Pascarelle, S., et al., 1998, Xth Rencontres de Blois, astro-ph/9809295
Pei, Y.C., Fall, S.M., 1995, ApJ 454, 69
Pettini, M., et al., 1998, ApJ, astro-ph/9806219
Rego, M., Cordero-Gracia, M., Zamorano, J., Gallego, J., 1993, AJ 105, 427
Renzini, A., in The Young Universe, ASP Conference Series, 146 astro-ph/9801209
Richards, E.A., 1998, Maryland October Astrophysics Conf astro-ph/9811120
Rowan-Robinson, M., et al., 1997, MNRAS 289, 490
Serjeant, S., Gruppioni, C., Oliver, S., 1998, MNRAS in press, astro-ph/9808259
Somerville, R.S., Primack, J.R., 1998, Xth Recontre de Blois astro-ph/9811001
Tenorio-Tagle, G., Muñoz-Tuñón, C., & Cox, D.P., 1993, ApJ 418, 767
Terlevich, R., Melnick, J., 1981, MNRAS 195, 839
Tinsley, B.M., Danly, L., 1980, ApJ 242, 435
Tresse, L., Maddox, S.J., 1998, ApJ 495, 691
Treyer, M.A., Ellis, R.S., Milliard, B., Donas, J., Bridges, T.J., 1998, MNRAS 300, 303
Vitores A.G., Zamorano J., Rego, et al., 1996a, A&AS 118, 7
Vitores A.G., Zamorano J., Rego, et al., 1996b, A&AS 120, 385
Zamorano J., Rego M., Gallego J., et al., 1994, ApJ 95, 387
Zamorano J., Gallego J., Rego M., et al., 1996, ApJS 105, 343

COSMIC GAMMA-RAY BURSTS

The most energetic phenomenon in the Universe

A.J. CASTRO-TIRADO
Laboratorio de Astrofísica Espacial
y Física Fundamental (LAEFF-INTA)
P.O. Box 50727, E-28080, Madrid, Spain

AND

Instituto de Astrofísica de Andalucía (IAA-CSIC)
P.O. Box 03004, E-18080, Granada, Spain

Abstract. Since their discovery in 1973, Gamma-Ray Bursts (GRBs) have remained for many years one of the most elusive mysteries in High Energy-Astrophysics. Despite the fact that ~800 GRBs are detected every year in the full sky, only a few of them can be localized accurately to less than half a degree. Thus the main problem regarding the nature of GRBs has usually been the lack of knowledge of their distance scale. For many years, follow-up observations by other satellites and ground-based telescopes were conducted, but no counterparts at other wavelengths were found. The breakthrough took place in 1997, thanks to the observation by BSAX and RXTE of the fading X-ray emission that follows the more energetic gamma-ray photons once the GRB event has ended. This emission (the afterglow) extends at longer wavelengths, and the good accuracy in the position determination has led to the discovery of the first counterparts at other wavelengths, greatly improving our understanding of these puzzling sources. Now it is widely accepted that GRBs originate at cosmological distances but the central engines that power these extraordinary events remain still unknown.

1. Introduction. General characteristics.

In 1967-73, the four VELA spacecraft (named after the spanish verb *velar*, to keep watch), that where originally designed for verifying whether the

Astrophysics and Space Science is the original source of publication of his article. It is recommended that this article is cited as: *Astrophysics and Space Science* **263**: 15–26, 1999.

former Soviet Union abided by the Limited Nuclear Test Ban Treaty of 1963, observed 16 peculiarly strong events (Klebesadel, Olson and Strong 1973, Bonnell and Klebesadel 1996). On the basis of arrival time differences, it was determined that they were related neither to the Earth nor to the Sun, but they were of cosmic origin. Therefore they were named cosmic Gamma-Ray Bursts (GRBs hereafter).

GRBs appear as brief flashes of cosmic high energy photons, emitting the bulk of their energy above ≈ 0.1 MeV. They are detected by instruments somewhat similar to those used by the particle physicists at their laboratories. The difference is that GRB detectors have to be placed on-board balloons, rockets or satellites.

The KONUS instrument on Veneras 11 and 12 gave the first indication that GRB sources were isotropically distributed in the sky (Mazets et al. 1981, Atteia et al. 1987). Based on a much larger sample, this result was nicely confirmed by BATSE on board the CGRO satellite, launched in the spring of 1991, an instrument that has revolutioned the GRB field (Meegan et al. 1992). About 800 GRBs are detected on a yearly basis, but only few of them are localized accurately. The apparent isotropy of the bursts in the sky ruled out the models dealing with neutron stars in the Galactic Plane, and it was rather interpreted in terms of GRBs arising at cosmological distances, although the possibility of a small fraction of the sources lying nearby, within a galactic disc scale of few hundred pc, or in the halo of the Galaxy, could not be discarded by that time. Another result was that the time profiles of the bursts are very different, with some GRBs lasting a few ms and others lasting for several minutes. In general, there was no evidence of periodicity in the time histories of GRBs. However there was indication of a bimodal distribution of burst durations, with $\sim$25% of bursts having durations around 0.2 s and $\sim$75% with durations around 30 s. A deficiency of weak events was noticed in the log N–log S diagram, as the GRB distribution deviates from the $-3/2$ slope of the straight line expected for an homogeneous distribution of sources assuming an Euclidean geometry. All these observational data led many researchers to believe that GRBs are indeed at cosmological distances. However, the origin of GRB remains unknown for more than 30 years. A review of the observational characteristics can be found in Fishman and Meegan (1995).

2. The search and detection of counterparts at other wavelengths

It is well known that an important clue for solving the GRB puzzle is the detection of transient emission –at longer wavelengths– associated with the bursts. A recent review on the unsuccessful search for counterparts prior to 1997 can be seen in Castro-Tirado (1998) and references therein. Here I will summarize some results concerning a few selected bursts detected by the BeppoSAX (BSAX) and RossiXTE (RXTE) satellites in 1996-98. See also Piran (1999).

2.1. GRB 960720

This is the first GRB for which a precise position (5′ error radius) was obtained by BSAX (in′t Zand et al. 1997). The bright radio-loud quasar 4C 49.29 at $z = 1.038$ lies within the tiny GRB error box. The probability to find such object within the error box is 2×10^{-4} (Greiner and Heise 1997, Piro et al. 1997a), but no firm relationship can be established.

2.2. GRB 970111

The position for this GRB was rapidly diseminated by the BSAX team, allowing to perform deep optical imaging only 19 hr after the high energy event. No optical variability was found within the error box down to B = 23, R = 22.5 (Castro-Tirado et al. 1997a, Gorosabel et al. 1998a). A careful analysis of the X-ray data have revealed a weak fading X-ray source within the GRB error box (Feroci et al. 1998a), but its association to GRB 970111 cannot be definetively proven.

2.3. GRB 970228

Thanks to BSAX, it was possible on 28 Feb 1997 to detect the first *clear* evidence of a long X-ray tail –the X-ray afterglow– following GRB 970228. A previously unknown X-ray source was seen to vary by a factor of 20 on a 3 days timescale. The X-ray fluence was ~ 40 % of the gamma-ray fluence, as reported by Costa et al. (1997b), implying that the X-ray afterglow was not only the low-energy tail of the GRB, but also a significant channel of energy dissipation of the event on a completely different timescale. Another important result was the non-thermal origin of the burst radiation and of the X-ray afterglow (Frontera et al. 1998a). This led to the discovery of the first optical transient (OT) associated to a GRB, identified on 28 Feb 1997, 20 hr after the event, by Groot et al. (1997). The OT was afterwards found on earlier images taken by Pedichini et al. (1997) and Guarnieri et al. (1997), in the rising phase of the light curve. The maximum was reached ~ 20 hr after the event (V ~ 21.3), and followed by a power-law decay $F \propto t^{-1.2}$ (Galama et al. 1997, Bartolini et al. 1998). An extended source was seen at the OT position since the very beginning by both ground-based and HST observations (van Paradijs et 1997, Sahu et al. 1997). New HST observations taken 6 months after the event were reported by Fruchter et al. (1997). Both the OT (at V = 28) and the extended source (V = 25.6) were seen. The extended source surrounding the point-source was interpreted as a galaxy, according to the similarities (apparent size, magnitude) with objects in the HST Deep Field. Despite exhaustive efforts in order to get spectroscopy of the extended object, no emission lines were found, implying that the redshift of this galaxy should lie in the range $1.3 \leq z \leq 2.5$ (Fruchter et al. 1999).

2.4. GRB 970402

Following the detection by BSAX of GRB 970402, no OT was detected within the GRB error, with the exception of the variable star BL Cir (Hudec et al. 1999). The ESA Infrared Space Observatory (ISO) observed the field 55-hours after the event, in the 8-15 μm and 174 μm bands. A further observation was carried out 8 days later. Only BL Cir was found to vary in the 8-15 μm band, and upper limits for any new object were $F_{(12\mu m)} \leq 0.14$ mJy and $F_{(174\mu m)} \leq 350$ mJy (Castro-Tirado et al. 1998a).

2.5. GRB 970508

The second OT associated to a GRB was discovered by Bond (1997). It was related to GRB 970508, and observed only 3 hr after the burst in unfiltered images (Pedersen et al. 1998) and 4-hr after in both the U and R-bands. The optical light curve reached a peak two days after (R = 19.7, Castro-Tirado et al. 1998b, Djorgovski et al. 1997, Galama et al. 1998a) and was followed by a power-law decay $F \propto t^{-1.2}$. The long-lasting OT was observed at many ground-based observatories (Chevalier et al. 1997, Garcia et al. 1997, Mignoli et al. 1997, Schaefer et al. 1997). Optical spectroscopy was performed at Calar Alto and La Palma (Castro-Tirado et al. 1997b) and Hawaii (Metzger et al. 1997a). The latter spectrum allowed a direct determination of a lower limit for the redshift of GRB 970805 ($z \geq 0.835$) and was the first proof that at least a fraction of the GRB sources lie at cosmological distances (Metzger et al. 1997b). The flattening of the decay in late August 1997 (Pedersen et al. 1998, Sokolov et al. 1998a) revealed the contribution of a constant brightness source –the host galaxy–. The galaxy was revealed in late-time imaging obtained in 1998 (Bloom et al. 1998a, Castro-Tirado et al. 1998c, Zharikov et al. 1998). The luminosity of the galaxy is well below the knee of the galaxy luminosity function, $L \approx 0.12\ L^{\star}$, and the detection of deep Mg I absorption (during the bursting episode) and strong [O II] 3727 Å emission (the latter mainly arising in H II regions within the host galaxy) confirmed $z = 0.835$ and suggested that the host could be a normal dwarf galaxy (Pian et al. 1998a), with a star formation rate (SFR) of $\sim 1.0\ M_{\odot}$ year^{-1} (Bloom et al. 1998a). Prompt VLA observations of the GRB 970508 error box allowed detection of a variable radio source at 1.4, 4.8 and 8.4 GHz (Frail et al. 1997). VLBI observations did not resolve the object (Taylor et al. 1997). Frail et al. proposed that the fluctuations could be the result of strong scattering by the irregularities in the ionized Galactic interstellar gas, with the damping of the fluctuations with time indicating that the source expanded to a significantly larger size. The transient was also detected at 15 GHz (Pooley and Green 1997) and as a continuum point source at 86 GHz with the IRAM PdBI. A $\sim$ 2 mJy flux was observed on 19-21 May 1997, but GRB 970508 was not detected any more after 28 May. No significant intra-day variation was seen, and it was excluded that the flux was modulated significantly by interstellar scintillation (Bremer et al. 1998). A flux of $\sim$ 65 μJy was measured by ISO in the 8-15 μm band (Hanlon et al. 1999). A recent result has been the detection of a Fe Kα line redshifted at $z = 0.835$ in the X-ray afterglow spectrum (Piro et al. 1999), which could be attributed to a thick torus surrouding the central engine (Meszaros and Rees 1998).

2.6. GRB 970828

This burst was detected by RXTE (Remillard et al. 1997) and was followed up by ASCA and ROSAT (Murakami et al. 1997, Greiner et al. 1997). The X-ray spectrum as seen by ASCA is strongly absorbed (Yoshida et al. 1999), suggesting that the event occurred in a dense medium. The fact that no optical counterpart down to R = 23.8 was detected between 4 hr and 8 days after the event, could support the idea that the non-detection was due to photoelectric absorption (Groot et al. 1998). An excess at 6.7 keV has been foud by ASCA in the X-ray afterglow spectrum. If this is due to highly ionized Fe, then $z \sim 0.33$ (Yoshida et al. 1999).

2.7. GRB 971214

The third optical transient was related to GRB 971214 and identified by Halpern et al. (1997) as an I = 21.2 object that faded 1.4 mag in 1 day. Independently, this source was also noticed as suspicious by Itoh et al. (1997). Further observations proved that the decay followed a power-law decline (Diercks et al. 1997, Castander et al. 1997) with $t^{-1.4}$, similar to, but steeper than, the previous GRBs with optical counterparts. GRB 971214 was detected in the K-band 4 hr after the burst (Gorosabel et al. 1998b) and decreasing from J = 20.3 on Dec 15 to J = 21.5 on Dec 16 (Tanvir et al. 1997). In the case of GRB 971214, no source was detected at 850 μm at the James Clerk Maxwell Telescope. The upper limit was 1 mJy on Dec 17-22 (Smith et al. 1999). Spectroscopy of the host galaxy (R = 25.6) one month later reveled a strong emission-line attributted to Lyα redshifted at $z = 3.42$ (Kulkarni et al. 1998a), implying a SFR of 1.0 ± 0.5 $M_\odot$ year^{-1}, similarly to other galaxies at comparable redshift. The total energy budget, assuming isotropic emission, was 3×10^{53} erg.

2.8. GRB 980329

Together with GRB 970111, this burst is among the top 2% burst with larger gamma-ray fluences as detected by BATSE. The detection of a variable radiosource (Taylor et al. 1998) within the GRB error box led to the identification of an optical transient (Klose et al. 1998, Palazzi et al. 1998). The evidence for a dusty host (R − K = 4.7) was confirmed by a SCUBA detection at 850 μm (Smith et al. 1999). Fruchter (1999) has proposed that the host could lie at $z \sim 5$. In that case, if the gamma-rays were radiated isotropically, the implied energy would be 5×10^{54} erg.

2.9. GRB 980425

A peculiar supernova (SN 1998bw) has been found in the WFC error box for this soft GRB (Galama et al. 1998b). The SN lies in the galaxy ESO 184-82 (at $z = 0.0085$). The fact that the SN event occurred within ± 1 day of the GRB event, together with the relativistic expansion speed derived from the radio observation (Kulkarni et al. 1998b) strengths such a relationship. In that case, the total energy budget would be 8×10^{47} erg. However, the fact that a fading X-ray source –as in *all* the previous cases– unrelated to the SN was detected by BSAX in the GRB error box (Pian et al. 1998b,c, Piro et al. 1998) cast some doubts on the SN/GRB association (Graziani et al. 1999).

2.10. GRB 980703

A variable radiosource was found by Frail et al. (1998), and the optical counterpart was independently discovered by Frail et al. (1998) and Zapatero-Osorio et al. (1998). The host galaxy is the brightest one so far detected (R = 22.5, H = 20.5) according to Bloom et al. (1998b) and Castro-Tirado et al. (1999). Optical spectroscopy revealed the [O II] emission line at $z = 0.966$, as well as some Fe II y Mg II absorption ones. The derived SFR is $\sim 63\ M_{\odot}\ \mathrm{year}^{-1}$ (Djorgovski et al. 1998). The released energy during the GRB event amounts to $\sim 3 \times 10^{52}$ erg. Table 1 summarizes the properties of the host galaxies found so far.

Further X-ray afterglows were observed by BSAX and RXTE in 1998 (GRB 980706, GRB 981220 and GRB 981226). Exponents for the power-law decay in the X-rays and in the optical are in the range $\alpha = 1.10$-2.25 for a dozen of bursts. These results are summarized on Table 2.

TABLE 1. GRB host galaxies

GRB	R (host)	z	SFR ($M_\odot$ year^{-1})	References
970228	25.2	$1.3 \leq z \leq 2.5$		Fruchter et al. (1998)
980828	$\geq$24	0.33 ?		Yoshida et al. (1999)
970508	25.7	0.835	1	Bloom et al. (1998a)
971214	25.6	3.418	5	Kulkarni et al. (1998a)
980326	$\geq$27.3			Bloom and Kulkarni (1998)
980329	$\geq$25.5	$\sim$5 ?		Fruchter (1999)
980519	$\sim$26			Hjorth et al. (1999)
980613	23.8	1.096	3	Djorgovski et al. (1999)
980703	22.5	0.966	63	Djorgovski et al. (1998)

TABLE 2. GRBs detected by BeppoSAX and RXTE in 1996-98

GRB	X-rays	optical-IR	radio	References[1]
960720				Piro et al. (1996)
970111	yes?	no	no	Costa et al. (1997a)
970228	yes	yes	no	Costa et al. (1997b)
970402	yes	no		Piro et al. (1997b)
970508	yes	yes	yes	Piro et al. (1997c)
970616	yes?	no	no	Marshall et al. (1997)
970815	yes?	no	no	Smith et al. (1997)
970828	yes	no		Remillard et al. (1997a)
971214	yes	yes	no	Antonelli et al. (1997)
971227	yes?			Sofitta et al. (1997)
980109		yes?		in't Zand et al. (1998)
980326	yes	yes		Celidonio et al. (1998)
980329	yes	yes	yes	Frontera et al. (1998b)
980425	yes?	yes?	yes?	Sofitta et al. (1998)
980515	yes?			Feroci et al. (1998b)
980519	yes	yes		Muller et al. (1998)
980613	yes	yes		Smith et al. (1998a)
980703	yes	yes	yes	Levine et al. (1998)
980706	yes?			Hurley et al. (1998)
981220	yes			Smith et al. (1998b)
981226	yes	yes?	yes?	Di Ciolo et al. (1998)

[1] only given for the first X-ray detection

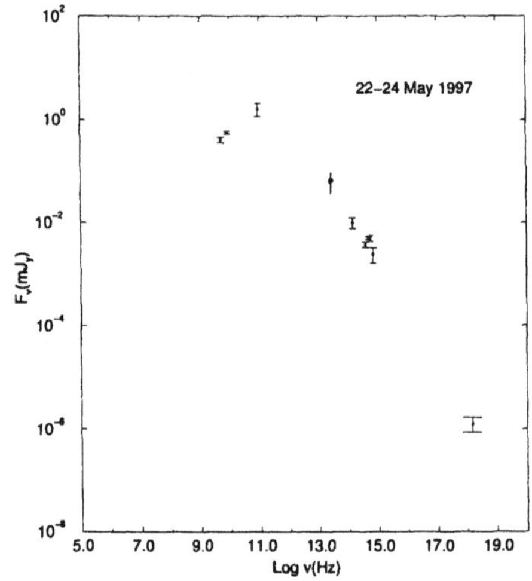

Figure 1. The multiwavelength spectrum of GRB 970508, on May 22-24, 1997

3. Theoretical models

All the previous observational efforts have provided the complete picture of the GRB multiwavelength spectra (see Figure 1).

The observational characteristics of the GRB counterparts can be accommodated in the framework of the relativistic fireball models, first proposed by Goodman (1986) and Paczyński (1986), in which a compact source releases 10^{53} ergs of energy within dozens of seconds. See also Mészáros et al. (1994), Mészáros and Rees (1997), Tavani (1997), Vietri (1997), Waxman (1997) and Wijers et al. (1997). The opaque radiation-plasma accelerates to relativistic velocities (the fireball). The blast wave is moving ahead of the fireball, and sweeps up the interstellar matter, producing an afterglow at frequencies gradually declining from X-rays to visible and radio wavelenghts. See Piran (1999) for an extensive discussion.

The properties of the blast wave can be derived from the classical synchrotron spectrum (Ginzburg and Syrovatskii 1965) produced by a population of electrons with the addition of self absorption at low frequencies and a cooling break (Sari, Piran and Narayan 1998). Thus, the four main quantities involved are:

i) the synchrotron frequency: the population of electrons has a power-law distribution of Lorentz factors γ_e following $\mathrm{d}N/\mathrm{d}\gamma_e \propto \gamma_e^{-p}$ above a minimum Lorentz factor $\gamma_e \geq \gamma_m$, corresponding to the synchrotron frequency ν_m,

ii) the break frequency: if the electrons are energetic they will cool rapidly for a break frequency $\nu_c < \nu_m$, and low energy electrons will have always slow cooling for $\nu_c > \nu_m$,

iii) the self-absorption frequency: synchrotron self absorption becomes important below a critical frequency ν_a, and

iv) the maximum flux $F_{\nu,\max}$.

After some time, the evolution of the blast wave is adiabatic, and $\nu_c > \nu_m$ implying that $F_\nu \propto \nu^{-(p-1)/2}$ for $\nu_m \leq \nu < \nu_c$; $F_\nu \propto \nu^{-p/2}$ for $\nu > \nu_c$ and $F_\nu \propto \nu^{1/3}$ for $\nu < \nu_m$. For $\nu_a \ll \nu_m$, $F_\nu \propto \nu^2$.

In any case, the time evolution of the flux as a function of a given frequency is a function of ν_c ($\propto t^{-1/2}$) and ν_m ($\propto t^{-3/2}$). For $\nu \leq \nu_c$ the flux decays following a power-law with exponent α, $F \propto t^\alpha$ with $\alpha = (2-3p)/4$, thus allowing to determine p.

The determination of the six observables ν_m, ν_c, ν_a, $F_{\nu,\max}$, p and z for every burst, allow to obtain the total energy per solid angle (E_{52}), the fraction of the shock energy in electrons (ϵ_e), the fraction of the shock energy in post-grb magnetic fields (ϵ_B) and the density of the ambient medium (n). See for example, Wijers and Galama (1998).

But what is causing the GRB? The fireball model is based on the existence of a "central engine". However, current observations have been unable to give any insight on this issue. Until 1994, the number of theoretical models amounted up to 150 (see Nemiroff 1994), but once the cosmological nature of GRBs was established, only those models dealing with mechanisms able to produce 10^{51-54} erg in a few seconds were taken into consideration. Moreover, the observations of star-forming hosts link these events to the SFR rate in the Universe (Totani et al. 1998, Krumholz et al. 1998). Theoretical models can be classified into two groups:

i) coalescence of neutron stars in a binary system (Narayan et al. 1992): lifes of such systems are of the order of $\sim 10^9$ years, and large escape velocities are usual, putting them far away from the regions where their progenitors were born. The likely result is a Kerr black hole (BH), and the energy released during the merger process is $\sim 10^{54}$ erg. It is also possible that a $\sim 0.1\ M_\odot$ accretion disk forms around the BH and is accreted within a few dozen seconds, then producing internal shocks leading to the GRB (Katz 1997). There are variations of these models where one or two components are substituted for black holes (Paczyński 1991).

ii) a "failed" type-I SN (Bodenhaimer and Woosley 1983, Woosley 1993) or *hypernova* (as it has been called by Paczyński (1998) on the basis of the observational consequences): very massive stars (Wolf-Rayet) collapse forming a Keck BH and a 0.1–1 $M_\odot$ torus. The matter is accreted at a very high rate and the energy is extracted via the rotational energy of the BH or via the accretion energy from the disk. In any case, a "dirty fireball", is produced reaching a luminosity $\sim$ 300 times larger that than of a normal SN. This would happen every $\sim 10^6$ years. In this scenario, GRBs would be produced in dense enviroments near star forming regions (see also MacFadyen and Woosley 1999).

4. Summary

The existence of an X-ray afterglow in *most* bursts is confirmed. Out ot 15 BSAX pointings, 11 revealed a strong afterglow, and 4 displayed faint ones, leading to the detection of optical/IR/radio counterparts in 1997-98. In principle, it is possible to determine the redshift z of the host galaxies on the basis of absorption edges or emission lines present in the X-ray afterglow, by means of a prompt X-ray follow-up, as it has been shown for GRB 970508. In any case, only the population of bursts with durations of few seconds has been explored. Short bursts lasting less than 1 s, like GRB 980706, that follow the $-3/2$ slope in the log N–log S diagram (in contrast to the longer bursts) remain to be detected at longer wavelengths.

It is generally believed is that internal shocks produced by the "central engine" are responsible for the γ-ray emission, whereas the slowing down of a relativistic shell on the interstellar medium (an external shock) would cause the post-GRB emission at longer wavelengths (the afterglow).

BSAX and RXTE have opened a new window in the GRB field and it is widely accepted now that most GRBs, if not all, lie at cosmological distances. It is expected that BSAX, RXTE and CGRO will facilitate the discoveries of other counterparts and, together with the new high-energy observatories (AXAF, SPECTRUM X/Γ, XMM, INTEGRAL and HETE 2), will definitively solve the long-standing Gamma-Ray Burst mystery.

References

Antonelli, L., et al., 1997, IAU Circ, 6792
Atteia, J.-L., et al., 1987, ApJS, 64, 305
Bartolini, C., et al., 1998, in *Gamma-Ray Bursts*, eds. C.A. Meegan, R. Preece, & T. Koshut, AIP, 428, 540
Bloom, J.S., Kulkarni, S.R., 1998, GCN Circ, 161
Bloom, J.S., et al., 1998a, ApJ, 507, L25
Bloom, J.S., et al., 1998b, ApJ, 508, L21
Bodenhaimer, P., Woosley, S., 1983, ApJ, 269, 281
Bond, H., 1997, IAU Circ, 6655
Bonnell, J.T., Klebesadel, R.W., 1996, in *Gamma-Ray Bursts*, eds. C. Kouveliotou, M.F. Briggs, & G.J. Fishman, AIP, 384, 977
Bremer, M., et al., 1998, A&A, 332, L13
Castander, F., et al., 1997, IAU Circ, 6791
Castro-Tirado, A.J., 1998, in *Ultraviolet Astrophysics: beyond the IUE Final Archive*, eds: R. Gonzalez-Riestra, W. Wamsteker, & R. Harris, ESA Conf. Proc., SP-413, 659
Castro-Tirado, A.J., Gorosabel, J., Heidt, J., et al., 1997a, IAU Circ, 6598
Castro-Tirado, A.J., Gorosabel, J., Wolf, C., et al., 1997b, IAU Circ, 6657

Castro-Tirado, A.J., Gorosabel, J., Benitez, N., et al., 1998a, Sci, 279, 1011
Castro-Tirado, A.J., Metcalfe, L.P., Laureijs, R., et al., 1998b, A&A, 330, 14
Castro-Tirado, A.J., et al., 1998c, IAU Circ, 6848
Castro-Tirado, A.J., Zapatero-Osorio, M.R., Gorosabel, J., et al., 1999, ApJ, 511, L85
Celidonio, G., et al., 1998, IAU Circ, 6851
Chevalier, C., et al., 1997, IAU Circ, 6663
Costa, E., et al., 1997a, IAU Circ, 6533
Costa, E., Frontera, F., Heise, J., et al., 1997b, Nature, 387, 783
DiCiolo, L., et al., 1998, IAU Circ, 7074
Diercks, A., et al., 1997, IAU Circ, 6791
Djorgovski, S.G., et al., 1997, Nature, 387, 876
Djorgovski, S.G., et al., 1998, ApJ 508, L17
Djorgovski, S.G., et al., 1999, GCN Circ, 189
Feroci, M., Antonelli, L., Guainazzi, M., et al., 1998a, A&A, 332, L29
Feroci, M., et al., 1998b, IAU Circ, 6909
Fishman, G.J., Meegan, C.A., 1995, ARA&A, 33, 415
Frail, D., Kulkarni, Nicastro, L., Feroci, M., Taylor, G., 1997, Nature, 389, 261
Frail, D., et al., 1998, GCN Circ, 128
Frail, D., et al., 1999, in *Gamma-ray Bursts in the Afterglow Era*, A&AS, in press
Frontera, F., Costa, E., Piro, L., et al., 1998a, ApJ, 493, L67
Frontera, F., et al., 1998b, IAU Circ, 6853
Fruchter, A., Livio, M., Macchetto, D., et al., 1997, IAU Circ, 6747
Fruchter, A., 1999, ApJ, 512, L1
Fruchter, A., Pian, E., Thorsett, S., et al., 1999, ApJ, in press (astro-ph/9807295)
Galama, T., et al., 1997, Nature, 387, 479
Galama, T., et al., 1998a, ApJ, 497, L13
Galama, T., et al., 1998b, Nature, 395, 670
Garcia, M., et al., 1997, IAU Circ, 6661
Ginzburg, V.L., Syrovatskii, S.I., 1965, ARA&A, 3, 297
Goodman, J., 1986, ApJ, 308, L47
Gorosabel, J., Castro-Tirado, A.J., Willot, C., et al., 1998a, A&A, 335, L5
Gorosabel, J., Castro-Tirado, A.J., Wolf, C., et al., 1998b, A&A, 339, 719
Graziani, C., et al., 1999, in *Gamma-ray Bursts in the Afterglow Era*, A&AS, in press.
Greiner, J., Heise, J., 1997, IAU Circ, 6570
Greiner, J., Englhauser, J., Groot, P.J., Galama, T., 1997, IAU Circ, 6757
Groot, P.J., et al., 1997, IAU Circ, 6584
Groot, P.J., Galama, T.J., van Paradijs, J., et al., 1998, ApJ, 493, L27
Guarnieri, A., et al., 1997, A&A, 328, L13
Halpern, J., et al., 1997, IAU Circ, 6788
Hjorth, J., et al., 1999, *Gamma-ray Bursts in the Afterglow Era*, A&AS, in press
Hurley, K., et al., 1999, GCN Circ, 129
Funk, B., et al., 1996, in *Gamma-Ray Bursts*, eds. C. Kouveliotou, M.F. Briggs, & G.J. Fishman, AIP, 384
Hudec, R., et al., 1999, in *Gamma-ray Bursts in the Afterglow Era*, A&AS, in press
in ′t Zand, J., et al., 1998, IAU Circ, 6805
Katz, J.I., 1997, ApJ, 490, 633
Klebesadel, R., Olsen, R., Strong, I., 1973, ApJ, 182, L85
Klose, S., Meusinger, H., Lehmann, H., 1998, IAU Circ, 6864
Krumholz, M., Thorsett, S., Harrison, F., 1998, ApJ, 506, L81
Kulkarni, S., et al., 1998a, Nature, 393, 35
Kulkarni, S., et al., 1998b, Nature, 395, 663
Levine, A., et al., 1998, IAU Circ, 6966
MacFadyen, A., Woosley, S., 1999, ApJ, in press (astro-ph/9810274)
Marshall. F., et al., 1997, IAU Circ, 6683
Mazets, E.P., et al., 1981, Ap&SS, 80, 3

Meegan, C.A., Fishman, G.J., Wilson, R., et al., 1992, Nature, 355, 143
Mészáros, P., Rees, M.J., 1997, ApJ, 476, 232
Mészáros, P., Rees, M.J., 1998, MNRAS, 299, L10
Mészáros, P., et al., 1994, ApJ, 432, 181
Metzger, M.R., et al., 1997a, IAU Circ, 6655
Metzger, M.R., et al., 1997b, Nature, 387, 878
Mignoli, M., et al., 1997, IAU Circ, 6661
Muller, J., et al., 1998, IAU Circ, 6910
Murakami, T., Ueda, Y., Yoshida, A., et al., 1996, IAU Circ, 6732
Narayan, R., et al., 1992, ApJ, 395, L83
Nemiroff, R., 1994, Comments Astrophys, 17, 189
Palazzi, E., et al., 1998, A&A, 336, L95
Paczyński, B., 1986, ApJ, 308, L43
Paczyński, B., 1991, Acta Astron., 41, 257
Paczyński, B., 1998, ApJ, 494, L45
Pedersen, H., et al., 1998, ApJ, 496, 311
Pedichini, F., Di Paola, A., Stella, L., et al., 1998, A&A, 327, L36
Pian, E., et al., 1998a, ApJ, 492, L103
Pian, E., et al., 1998b, GCN Circ, 69
Pian, E., et al., 1998c, GCN Circ, 158
Piran, T., et al., 1999, Phys. Rep., in press (astro-ph/9810256)
Piro, L., et al., 1996, IAU Circ, 6467
Piro, L., et al., 1997a, IAU Circ, 6570
Piro, L., et al., 1997b, IAU Circ, 6617
Piro, L., et al., 1997c, IAU Circ, 6656
Piro, L., et al., 1998, GCN Circ, 155
Piro, L., et al., 1999, ApJ, in press
Pooley, G., Green, D., 1997, IAU Circ, 6670
Remillard, R., Wood, A., Smith, D., Levine, A., 1997, IAU Circ, 6726
Sahu, K., Livio, M., Petro, L., et al., 1997, Nature, 387, 476
Sari, R., Piran, T., Narayan, R., 1998, ApJ, 497, L17
Schaefer, B.E., et al., 1997, IAU Circ, 6658
Smith, D.A., et al., 1997, IAU Circ, 6718
Smith, M.J.S., et al., 1998a, IAU Circ, 6938
Smith, D.A., et al., 1998b, IAU Circ, 6938
Smith, I.A., Tilanus, R.P.J., van Paradijs, J., et al., 1999, A&A, in press (astro-ph/9811026)
Sofitta, P., et al., 1997, IAU Circ, 6797
Sofitta, P., et al., 1998, IAU Circ, 6884
Sokolov, V.V., Kopylov, A.I., Zharikov, S.V., et al., 1998a, A&A, 334, 117
Tavani, M., 1997, ApJ, 483, L87
Tanvir, N., et al., 1997, IAU Circ, 6796
Taylor, G.B., et al., 1997, Nature, 389, 263
Taylor, G.B., et al., 1998, ApJ, 502, L115
Totani, T., 1998, ApJ, in press (astro-ph/9805263)
van Paradijs, J., et al., 1997, Nature, 386, 686
Vietri, M., 1997, ApJ, 488, L105
Waxman, E., 1997, ApJ, 489, L33
Wijers, R.A.M.J., Galama, T.J., 1998, ApJ, in press (astro-ph/9805341)
Wijers, R.A.M.J., Rees, M.J., Mészáros, P., 1997, MNRAS, 288, L51
Woosley, S., 1993, ApJ, 405, 273
Yoshida, A., et al., 1999, in *Gamma-ray Bursts in the Afterglow Era*, A&AS, in press
Zapatero-Osorio, M.R., et al., 1998, IAU Circ, 6747
Zharikov, S.V., Sokolov, V.V., Barishev, Yu.V., 1998, A&A, 337, 356

MODELLING THE GROWTH OF DARK-MATTER HALOS

A. RAIG, E. SALVADOR-SOLÉ,
Departament d'Astronomia i Meteorologia
Universitat de Barcelona

AND

G. GONZÁLEZ-CASADO
Departament de Matemàtica Aplicada II
Universitat Politècnica de Catalunya

Abstract. A new model for the cosmological growth of dark matter halos is presented. We compare theoretical predictions with results of *N*-body simulations.

In the last years several semi-analytical models (SAMs) of galaxy formation and evolution in a cosmological frame have been developed (see Somerville 1998). In spite of the important simplifications made, these models describe fairly well the observed properties of galaxies at different redshifts.

The clustering of dark matter halos is one of the cornerstones for such SAMs. Current models are based on the extended Press-Schechter (1974; PS) formalism, which does not deal with a good estimate of the halo formation time (Manrique & Salvador-Solé, 1996) nor does it correctly account for accretion (Somerville & Kolatt, 1999). Here we report a new analytical model that improves these two aspects.

A capture that produces a rearrangement in the equilibrium state of the halo through violent relaxation is called merger, while if the capture barely affects the dynamical equilibrium, and just produces an increase in the mass of the halo it is called accretion. Salvador-Solé, Solanes, & Manrique (1998; SSM) introduced this distinction within the PS formalism by considering that the frontier between both processes depends on the relative mass: a halo of mass M accretes a mass ΔM if $\Delta M/M \leq \Delta_m$, otherways we say that it merges with a halo of mass ΔM and it is destroyed.

Astrophysics and Space Science is the original source of publication of his article. It is recommended that this article is cited as: *Astrophysics and Space Science* **263**: 27–30, 1999.

This distinction makes the merger rate introduced by Lacey & Cole (1993) in the PS formalism, $r_{LC}^{m}(M \rightarrow M', t)$, to split in two contributions (see Fig. 1): one giving rise to accretion ($\Delta M/M \leq \Delta_m$) and the second one to major mergers ($\Delta M/M > \Delta_m$). The *mass accretion rate* and *destruction rate* are therefore given by

$$r_{\rm mass}^{\rm a}(M,t) = \frac{{\rm d}M}{{\rm d}t} = \int_{M}^{M(\Delta_{\rm m}+1)} \Delta M \; r_{\rm LC}^{\rm m}(M \rightarrow M',t)\;{\rm d}M' \qquad (1)$$

and

$$r^{\rm d}(M,t) = \int_{M(\Delta_{\rm m}+1)}^{\infty} r_{\rm LC}^{\rm m}(M \rightarrow M',t)\;{\rm d}M', \qquad (2)$$

respectively.

From these quantities we can also obtain the *formation rate* by the continuity equation for the density of objects of mass M

$$r^{\rm f}[M(t),t] = \frac{{\rm d}\ln N[M(t),t]}{{\rm d}t} + r^{\rm d}[M(t),t] + \partial_M r_{\rm mass}^{\rm a}(M,t)|_{M=M(t)}. \qquad (3)$$

Furthermore, the model allows us to properly define the formation time of a halo, t_f, as the time when the object went through the last major merger. For the first time in these kind of models the formation time is not arbitrarly defined as the time some progenitor reached some fraction of the final mass, but it is related to a well-defined physical process: the rearrangement of the halo. SSM derived the *formation time distribution*

$$\Phi_{\rm f}(t) = r^{\rm f}[M(t),t] \exp\left\{-\int_{t}^{t_0} r^{\rm f}[M(t'),t']\,{\rm d}t'\right\}, \qquad (4)$$

the typical formation time being defined as the corresponding median value.

All the previous quantities above depend on the value of the phenomenological parameters Δ_m. This can be fixed by fitting the results of high resolution N-body cosmological simulations. From such simulations Navarro, Frenk, & White (1997) have shown that the density profile of dark matter halos has the universal form

$$\frac{\rho(x)}{\rho_{\rm crit}} = \delta_{\rm c} \frac{x_{\rm s}^3}{x(x+x_{\rm s})^2}. \qquad (5)$$

In equation (5), $\rho_{\rm crit}$ is the critical density of the universe, $x = r/R$ is the radius scaled to the so-called virial radius R, and $\delta_{\rm c} = \rho_{\rm c}/\rho_{\rm crit}$ and $x_{\rm s} = r_{\rm s}/R$ are two dimensionless parameters giving the characteristic density and scale radius, respectively, of the density profile. It is found that the less massive the halos, the denser they are. This mass-density relation seems to

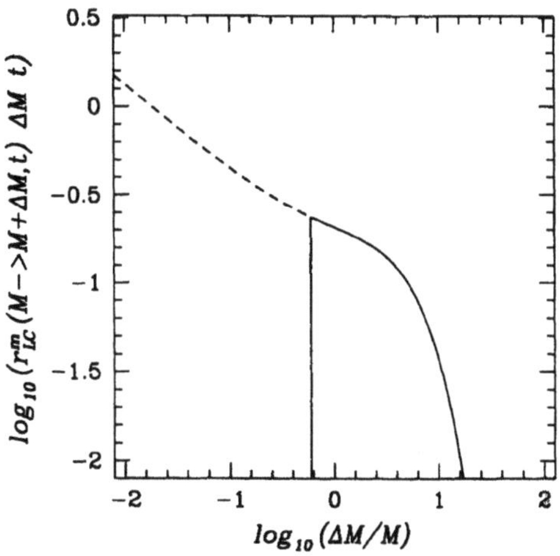

Figure 1. Contributions from major mergers (solid line) and accretions (dashed line) to the *merger rate* introduce by Lacey & Cole. The vertical line is the cut off at $\Delta M/M = \Delta_m$

reflect the fact that less massive objects form earlier, when the mean cosmic density was higher. Indeed, for $\Delta_m \sim 0.6$, SSM have found that a simple proportion between ρ_c and the value of ρ_{crit} at halo formation gives very good fits to the mass-density relation for masses in a range of four decades in any cosmogony analyzed. As pointed out by SSM, the proportion between ρ_c and the value of ρ_{crit} at halo formation would be a natural consequence of the following simple evolutionary scenario: the structure of halos would be fixed at their last major merger and would remain essentially unaltered since then until the next major merger, except for a progressive expansion of their envelope due to the accretion of surrounding layers. Through detailed energetic arguments Raig, González-Casado & Salvador-Solé (1998) have shown the consistency of this evolutionary scenario.

This interesting result relies, of course, on the analytical model of halo growth developed by SSM. Here we report a detailed comparison between the predictions of this model and N-body simulations carried out by Lacey and Cole (see Lacey & Cole, 1994).

In Figure 2 we plot the predicted and empirical *accretion rate*, *destruction rate*, and *formation rate* for a scale-free, $n = 0$, cosmology. As can be seen it is found a very good agreement between theory and simulations. A similar agreement is also obtained for other cosmologies. This therefore proves the reliability of our analytical model.

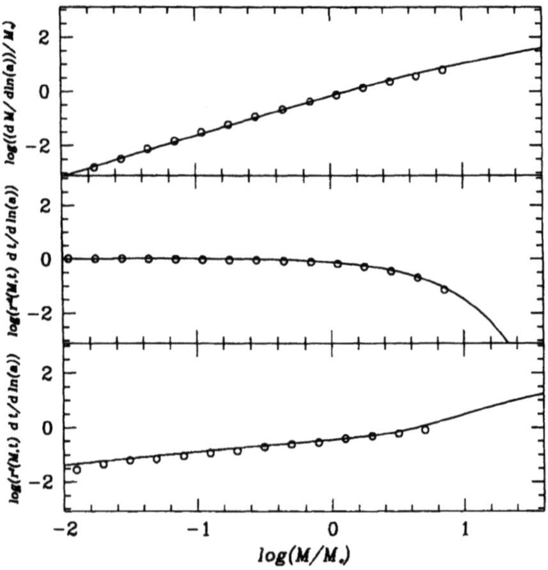

Figure 2. Comparisons between the theoretical predictions of the model (solid lines) and the results of *N*-body simulations, with FOF(0.2) groups for the case of a scale-free $n = 0$ universe (empty circles). Ordinates are in dimensionless units; in abcisae M_* is the characteristic mass of halos. *Upper panel*: mass accretion rate. *Middle panel*: destruction rate. *Lower panel*: formation rate

References

Lacey, C., Cole, S., 1993, MNRAS, 262, 627
Lacey, C., Cole, S., 1994, MNRAS, 271, 676
Manrique, A., Salvador-Solé, E., 1996, ApJ, 461, 504
Navarro,J.F., Frenk, C.S., White, S.D.M., 1995, ApJ, 490, 493
Press, W., Schechter, P., 1974, ApJ, 187, 425
Raig, A., González-Casado, G., Salvador-Solé, E., 1998, ApJ Lett., accepted (astro-ph/9810168)
Salvador-Solé, E., Solanes, J.M., Manrique, A., 1998, ApJ, 499, 542
Somerville R.S., 1998, Ph.D. Thesis
Somerville R.S., Kolatt, T.S., 1999, MNRAS, submitted (astro-ph/9711080)

STAR FORMATION AND COSMOLOGICAL SIMULATIONS

G. YEPES, D. ELIZONDO AND Y. ASCASIBAR
Grupo de Astrofísica, Dept. Física Teórica C-XI
Universidad Autónoma de Madrid, Cantoblanco 28049, Spain

Abstract. The inclusion of a detailed modeling of the short-scale baryonic physics in a large-scale cosmological simulation is crucial for a better comparison between observations and predictions from cosmological models. From a set of 3D hydrodynamical simulations which include a chemical model to account for the complex physics of the ISM at a sub-grid scale, we have been able to obtain a statistically significant sample of galaxy-type halos with observational properties, like colors and luminosities for different cosmological scenarios. From this data base, we have studied a number of different things, like Tully-Fisher relations, luminosity functions and environmental effects. Despite the progress made during the last few years in the modeling of the physics of ISM and star formation, more work is clearly needed.

1. Introduction

In our numerical model, matter is treated as a multi-phase fluid with different physical processes acting between them. The gas is treated as two separate phases, depending on its temperature: *Hot Gas* ($T_h > 10^4$K) and *Cold Gas Clouds* ($T_h < 10^4$K) from which stars are formed. The numerical code is a combination of a Particle-Mesh (PM) Poisson solver and an Eulerian PPM hydrodynamical code. The equations governing the star formation in each volume element in our simulations and detailed description of the N-body/hydro code were discussed in detail in Yepes et al (1997) (YK^3). Here we briefly summarize the main features. In each resolution element the amount of cold gas $m_{\rm cold}$ capable of producing stars is regulated by the mass of the hot gas $m_{\rm hot}$ that can cool on the time scale $t_{\rm cool}$, by the rate of forming new stable stars, and by supernovae, which heat and evaporate cold gas. The supernovae formation rate is assumed to be

Astrophysics and Space Science is the original source of publication of his article. It is recommended that this article is cited as: *Astrophysics and Space Science* **263**: 31–34, 1999.

proportional to the mass of cold gas: $\dot{m}_{SN} = \beta m_{cold}/t_*$, where $t_* = 10^8$yr is the time scale for star formation, and β is the fraction of mass that explode as supernovae ($\beta = 0.12$ for the Salpeter IMF). Each $1M_\odot$ of supernovae dumps 4.5×10^{49} ergs of heat into the interstellar medium *and* evaporates a mass $A \cdot M_\odot$ of cold gas. Small values of A imply large reheating of hot gas and small evaporation, which makes the gas to expand due to the large pressure gradients. The feedback parameter A is taken to be large ($A \sim 200$) resulting in low efficiency of converting cold gas into stars. Chemical enrichment due to supernovae is also taken into account in the following way: we assume solar composition for the gas in regions where star formation has taken place. In regions where no stars are present, we assume that the gas has primordial composition. The gas then cools with cooling rates, $\Lambda(T_h)$ corresponding to either primordial or solar plasmas. In order to mimic the effects of photoionization by quasars and AGNs, the gas with overdensity less than 2 was kept at a constant temperature of $10^{3.8}$K (Giroux & Shapiro 1986; Petitjean, Mücket, & Kates 1996).

2. Simulations

The purpose of the simulations was to obtain a sufficiently large catalog of "numerical galaxies", permitting reliable, statistically significant comparisons with observational quantities (Elizondo 1996). To this end, a set of 11 simulations were performed for each of the CDM, ΛCDM ($\Omega_\Lambda = 0.65$), and BSI models (Kates et al. 1995). COBE normalization was taken; baryon fractions were compatible with nucleosynthesis constraints (Smith et al. 1993) ($\Omega_B = 0.051$ for BSI and CDM, $\Omega_B = 0.026$ for ΛCDM). The box size was 5.0 Mpc, but with different Hubble constants given by $h = 0.7$ for the ΛCDM model and $h = 0.5$ for the CDM and BSI simulations. The simulations were performed at the CEPBA *(Centro Europeo de Paralelismo de Barcelona)* with 128^3 particles and cells (i.e. 39 kpc cell width). Effects of resolution have been checked by re-running 2 of the simulations with 256^3 cells and particles. It did not result in significant changes in global parameters (mass, luminosity) of galaxies (Elizondo et al. 1998a). To test the effects of supernovae feedbacks on the final observational properties of galactic halos, we have rerun 6 of the 11 simulations for each cosmological model, with different feedback parameter: $A = 50$ (strong gas reheating) and $A = \beta = 0$ (no reheating or mass transfer).

3. Results

The observational Tully-Fisher (Pierce and Tully 1992) relation involves line widths, from which one deduces rotational velocities. In simulations, the TF "rotational velocities" of galaxies were defined by the depth of the gravitational potential: $v_{grav} = \sqrt{GM/r}$, where M is the total mass of a galaxy within its assigned radius r. In Figure ?? we plot the magnitude-circular velocity relation for the galaxies in our simulations of the 3 cosmological models, for a feedback parameter $A = 200$ and no feedback at all. As can be seen, the main effect of feedback is to raise the slope of the TF, but it has a marginal effect in the scatter and the zero point of the relation. Large circular velocity halos ($\geq$ 150 km/s) are hardly affected by supernovae explosions. A detailed analysis of the results from these simulations can be found elsewhere (Elizondo et al. 1998a, 1998b).

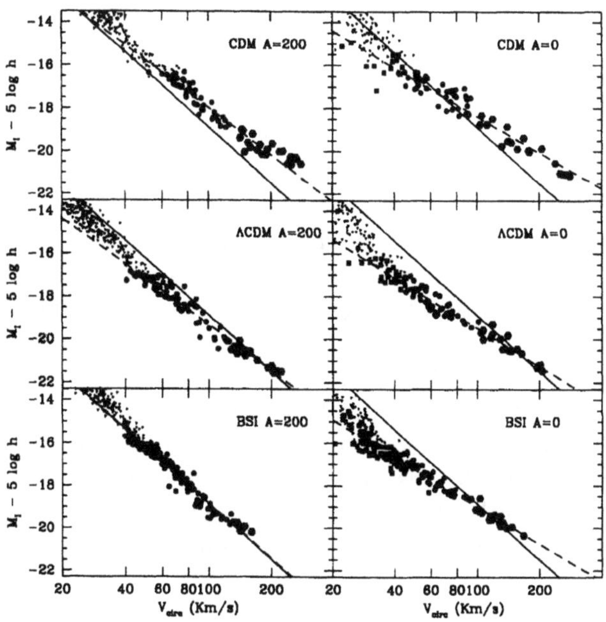

Figure 1. TF relation for the I band for halos found is simulations with $A = 200$ and no feedback ($A = \beta = 0$). Solid line is the observational relation from Pierce and Tully (1992).

4. Future developments and conclusions

Despite the success of our model to reproduce some of the properties of real galaxies, it is nevertheless a simplified model of the complex star-gas interactions. UV photoionization and chemical enrichment are treated in a phenomenological way. These effects could be very important, and could change the evolution of the gas component. There is a considerable work on the subject of chemical evolution from supernovas in "non-dynamical" galactic models (see e.g. Matteucci 1997). The advantage of using eulerian

PPM hydrodynamics is that it is simple to advect metals as a new "phase" of the gas density. In each of the cells of the simulations the hot gas will cool as $\Lambda(T_h, Z_h)$, with $Z_h = m_{\rm met}/m_{\rm hot}$ being the metallicity of the hot gas in each cell volume. Stars will then form with different metallicities and their luminosities can be computed by the new generation of spectral synthesis models. UV radiation from the stars is another important effect that has not yet been explored. Our intention is to make a self-consistent modeling of the photoionization heating of the gas from UV flux coming from the stars produced in the simulations. These effects can be very important in the early epoch of the Universe, when stars are produced in the first collapsing objects. The main goal of a cosmological simulation is to reach a dynamical range large enough so that internal structure of individual galaxies can be resolved in a simulated volume in which a reliable realization of the power spectrum can be obtained. The most promising way to achieve this goal is by using new numerical algorithms based on *Adaptive Mesh Refinement* (AMR) techniques. But, high resolution does not necessarily mean better results, unless the most important physical processes acting at the scales of interest have been included in the simulation. Our work goes in this direction. On one hand it is necessary to explore many approximations to properly model the complex physics of star formation and star-gas interactions (Ascasibar 1999). On the other hand, new AMR methods for gravity (e.g. Kravtsov et al. 1997) and hydrodynamics (Khokhlov 1998) have been developed. The next step is to put together the different pieces and build the next generation of numerical models to study galaxy formation from cosmological initial conditions. In this regard, the data base obtained from simulations performed with our present model (YK^3) will be very useful as a testbed for the new models we are developing.

References

Ascasibar, Y., 1999, Ph.D. Thesis, in preparation
Eisenstein, D.J., Loeb, A., 1996, ApJ, 459, 432
Elizondo, D., 1996, PhD Thesis, Universidad Autónoma de Madrid
Elizondo, D., Yepes, G., Kates, R., Müller, V., Klypin, A., 1998a, ApJ, in press
Elizondo, D., Yepes, G., Kates, R., Klypin, A., 1998b, New Astronomy, (in press)
Giroux, M., Shapiro, I., 1986, ApJS, 102, 191
Kates, R., Müller, V., Gottlöber, S., Mücket, J.P., Retzlaff, J., 1995, MNRAS, 277, 1254
Kravtsov, A., Klypin, A., Khokhlov, A., 1997, ApJS, 111, 73
Khokhlov, A., 1998, J. Compt. Phys., 143, 519
Matteucci, F., 1997, ASP Conf. Ser., 126, 495
Petitjean, P., Mücket, J, Kates, R., 1995, A&A, 295, L9
Pierce, M.J., Tully, R.B., 1992, ApJ, 387, 47
Smith, M.S., Kawano, L.H., Malaney, R.A., 1993, ApJS, 85, 219
Yepes, G., 1997, ASP Conf. Ser., 126, 279
Yepes, G., Kates, R., Khokhlov, A., Klypin, A., 1997, MNRAS, 284, 235 (YK^3)

GALAXY DISK FORMATION IN HIERARCHICAL HYDRODYNAMICAL SIMULATIONS

R. DOMÍNGUEZ-TENREIRO[1], P.B. TISSERA[1,2] AND A. SÁIZ[1]
[1] *Dpt. Física Teórica, UAM, Cantoblanco, 28049 Madrid, Spain*
[2] *IAFE, Casilla 67, Suc. 28, 1428 Buenos Aires, Argentina*

Abstract. We report some results on galaxy disk formation in hierarchical hydrodynamical simulations with characteristics compatible with observed spirals.

Understanding how galaxies form and evolve has been a long-standing ambition in Astronomy. Some recent developments in theoretical cosmology, in numerical computation technics and in the power and velocity of computers, have made it possible to attempt to numerically simulate the process of galaxy formation and evolution from primordial fluctuations. Even if still at its beginnings, this numerical approach to study galaxy formation seems promising, because the physics is introduced at a very general level (Newton laws, hydrodynamics equations, possibly a star formation algorithm), and then, the processes that play a role in galaxy assembly appear as a consequence. In particular, merger and interaction effects are naturally accounted for. However, so far, no hydrodynamical simulation of galaxy formation in fully consistent hierarchical cosmological scenarios had been able to produce extended disks similar to those observed in spirals. The problem was either the excessive loss of angular momentum by the gas clumps as they merge inside the dark haloes, when no star formation processes are considered, resulting in too concentrated disks (i.e., the so-called *disk angular momentum catastrophe* problem, hereafter DAMC [10] [11] [20] and references quoted therein), or the too early gas exhaustion into stars as it cools and collapses, leaving no gas to form disks at low z [17] [15]. As reported in [13] [8], DAMCs can be avoided by stabilizing disks with bulges, that are unavoidably formed when star-forming processes are considered. So, even a simple implementation of these processes, provided that it does not lead to an excessive gas depletion at high z, could be enough to ensure the formation of disks and their stability.

In this paper we report some results on the formation of disk-like structures, with characteristics similar to those observed in spirals, in a hierarchi-

Astrophysics and Space Science is the original source of publication of his article. It is recommended that this article is cited as: *Astrophysics and Space Science* **263**: 35–38, 1999.

TABLE 1. SOME CHARACTERISTICS OF DLOS[a]

DLO	1	2	3	4	5	6	7	8	9	10	HRS
N_{gas}	348	359	307	311	210	151	227	189	108	109	1713
N_{star}	278	240	211	215	95	69	79	157	99	47	1380
n	1	1	2	1	2	1	1	2	1	2	2
B/D	1.19	1.19	1.55	1.60	1.15	1.22	2.02	1.31	2.95	1.30	0.86
r_b	0.74	0.74	0.85	0.74	0.54	0.53	0.99	0.49	0.54	0.40	1.13
r_d	7.33	5.66	10.90	9.98	6.50	5.61	6.56	5.29	7.07	9.75	9.58
LS	-0.17	-0.16	-0.11	-0.09	-0.14	-0.09	-0.19	-0.18	-0.15	-0.18	0.04
$\log \kappa$	3.33	3.48	3.02	3.08	3.15	3.11	3.19	3.30	2.97	2.72	2.90

[a]Distances are given in kpc and κ is in $M_\odot$ pc^{-2}

cal hydrodynamical simulation [18] [17] [8]. We have followed the evolution of 64^3 particles in a periodic box of 10 Mpc (H_0 =50 km s^{-1} Mpc^{-1}) using a SPH code coupled to the high resolution AP3M code [16], *including a star formation algorithm* with star formation efficiency $c = 0.01$ (S1 simulation). The initial distribution of positions and velocities, set using COSMICS, is consistent with a standard flat CDM cosmology, with $\Omega_b = 0.1, \Lambda = 0$ and $b = 2.5$. All, dark, gas and star particles have the same mass, $M = 2.6 \times 10^8$ $M_\odot$. The gravitational softening length is 3 kpc and the minimum allowed smoothing length is 1.5 kpc. The low c value used allowed us to have available gas to form disks at low z. To some extent, this condition could mimic the effects of energy injection from supernovae explosions.

Baryonic objects forming disk-like structures (DLOs) identified in S1 have stellar bulge-like cores and extended, populated disks. Some characteristics of those with $N_{baryon} > 150$ are given in Table 1. To ascertain how well they compare with observed spirals, we have measured their bulge and disk scales by fitting a Sérsic law [14] for the bulge (with free shape parameter n), and a pure exponential law (i.e., with $n = 1$) for the disk, to their *mass* density profiles projected on the disk plane. The optimal bulge shape parameters, n, for different DLOs are given in Table 1. Like for observed spiral bulges [1] [7] [5] [4], in most cases $n = 1$ is the best fit. As noted by [1], the shape parameter is correlated with the spiral type. The objects in our sample would be $T \geq 2$ in the RC2 classification scheme. The scale parameters, r_b and r_d, corresponding to a double exponential fit, are given in Table 1 and plotted in Fig. 1. From this Figure it can be seen that the scales are consistent with the values found by [4] and [5] from 1D surface brightness decompositions of Sb - Sc spirals. The corresponding *mass* B/D ratios are given in Table 1. An estimation of the disk and bulge mass-to-light ratios, $\Upsilon_d = 2.1h$ and $\Upsilon_b = 7.7h$ [19], makes them compatible with

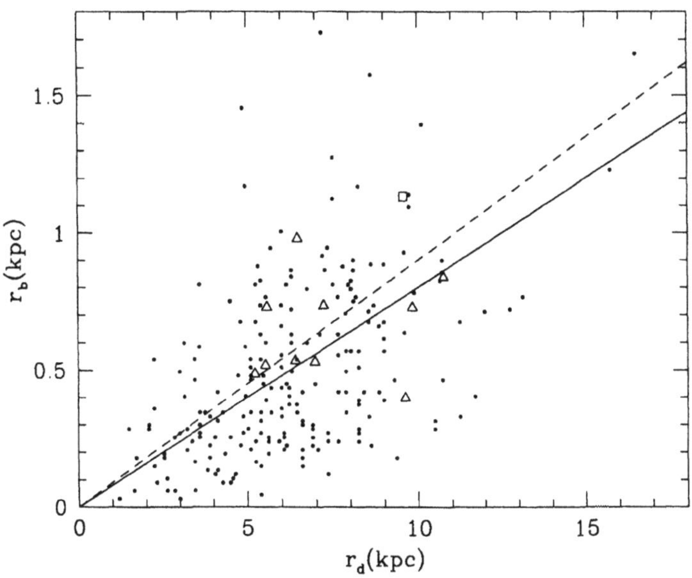

Figure 1. The scale parameters r_b versus r_d for DLOs in S1 (*open triangles*), and HRS (*open square*). *Points* are from 1D decompositions of surface brightness profiles given by [4]. The *dashed (solid)* line is a fit to the S1 values (data).

observed *luminosity B/D* ratios.

The rotation curves (RCs) of S1 DLOs fit very well the tangential velocity of the baryonic particles, suggesting that they are in rotational supported equilibrium within the potential well produced by both the dark haloes and the baryons themselves [17]. The values we have obtained for some parameters characterizing these RCs are given in Table 1 (logarithmic slope, LS, and compactness parameter, $\kappa = V^2(2.2r_d)/Gr_d$. Their comparison to observed values [3] [2] [12] [6], indicates that S1 DLOs *have observational counterparts*, even if we miss some variability.

Concerning the angular momentum content, in Fig. 2 we plot the specific total angular momentum at $z = 0$ versus mass for dark haloes, for the gas masses inside r_{opt}, and for the stellar components. We see that DLOs are inside the box defined by observed spiral disks in this plot [9].

To make sure that disks in S1 do not result from unphysical gas heating [11], we have run a higher resolution simulation (64^3 particles in a periodic box of 5 Mpc, with cosmological and star formation parameters similar to those in S1; hereafter HRS). Only one disk with mass comparable to those in S1 forms. Its analysis has shown that its structural and dynamical characteristics are compatible with observations (see Table 1 and Fig. 1) and that the physical processes leading to its formation are essentially the same as those at work in S1 (see Fig. 2). We conclude that once DAMCs

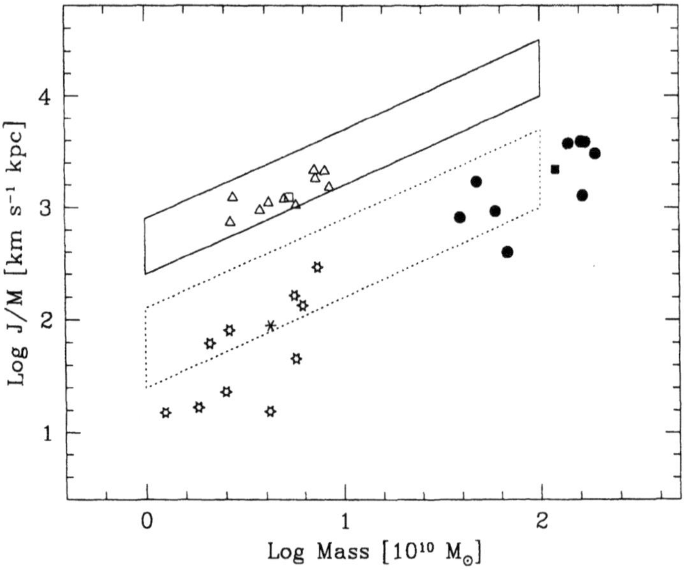

Figure 2. The specific angular momentum at $z = 0$ versus the mass, for halos in S1 (*filled circles*), and HRS (*filled square*); the gas component inside $r_{\rm opt}$ in S1 (*open triangles*) and HRS (*open square*); and the stellar component in S1 (*open stars*) and HRS (*asterisk*). The *solid (dotted)* box shows the region occupied by the spiral disks (ellipticals).

are avoided by stellar bulges, gas collapse proceeds with specific angular momentum conservation, producing disks similar to observed ones.

References

1. Andredakis, Y.C., Peletier, R.F., Balcells, M., 1995, MNRAS, 275, 874
2. Broeils, A.H., 1992, Ph.D. thesis, Univ. Groningen
3. Casertano, S., van Gorkom, J.H., 1991, AJ, 101, 1231
4. Courteau, S. 1997, in *Morphology & Dust Content in Spiral Galaxies* eds. D. Block & M. Greenberg, (Dordrecht: Kluwer)
5. Courteau, S., de Jong, R.S., Broeils, A.H., 1996, ApJ, 457, L73
6. de Blok, W.J.G., McGaugh, S.S., van der Hulst, J.M., 1997, MNRAS, 283, 18
7. de Jong, R.S., 1995, Ph.D. thesis, Univ. Groningen
8. Domínguez-Tenreiro, R., Tissera, P.B., & Sáiz, A. 1998, ApJ Letters, in press
9. Fall, S.M. 1983, IAU Symp. 100 *Internal Kinematics and Dynamics of Galaxies*, ed. E. Athanassoula (Dordrecht: Reidel)
10. Navarro, J.F., Frenk, C.S., White, S.D.M., 1995, MNRAS, 275, 56
11. Navarro, J.F., Steinmetz, M., 1997, ApJ, 478, 13
12. Rhee, M.H., 1996, Ph.D. thesis, Univ. of Groningen
13. Sáiz, A., Tissera, P.B., Domínguez-Tenreiro, R., 1998, these proceedings
14. Sérsic, J.L., 1968, *Atlas de Galaxias Australes* (Obs. Astronómico de Córdoba)
15. Steinmetz, M., Navarro, J.F., 1998, astro-ph/9808076 preprint
16. Thomas, P.A., Couchman, H.M.P., 1992, MNRAS, 257, 11
17. Tissera, P.B., Domínguez-Tenreiro, R., 1998, MNRAS, 297, 177
18. Tissera, P.B., Lambas, D.G., Abadi, M.G., 1997, MNRAS, 286, 384
19. van den Bosch, F.C., 1998, astro-ph/9805113 preprint
20. Weil, M.L., Eke, V.R., Efstathiou, G., 1998, astro-ph/9802311 preprint

ANALYTICAL MODELLING OF GALAXY FORMATION

ALBERTO MANRIQUE AND EDUARD SALVADOR-SOLÉ
Dpt. d'Astronomia i Meteorologia, Universitat de Barcelona
Martí i Franquès 1, E-08028 Barcelona, Spain

Abstract. We present an analytical model of galaxy formation in hierarchical cosmogonies with new prescriptions for the clustering of dark matter halos, their internal structure, and the modelling of galaxies. These improvements with respect to current semi-analytical models allows us, respectively, to follow more accurately the evolution of halos, to study the spatial segregation of galactic properties within clusters, and to obtain more confident predictions on the Tully-Fisher and the Fundamental Plane relations.

1. Introduction

Although recent observational data support the evolution of galactic properties along time, it is not yet fully understood the processes involved in galaxy formation and evolution. From the theoretical viewpoint, a great deal of effort is being put to shed light on these issues. There are two ways to study them: by means of numerical simulations using N-body plus hydrodynamical codes (see Domínguez-Tenreiro, Tissera & Sáiz and Yepes, Elizondo & Ascasibar in these volume), and by means of semi-analytical models (SAMs). The first method is well suited to study the detailed physical processes taking place in the formation and evolution of individual galaxies, while the second one is better suited to study the statistical properties of galactic populations in different environments (i.e., over a wide dynamical range) and for a broad region of the parameter space.

The seminal work of White & Frenk (1991) put the foundations of SAMs. These authors gave a set of simple although physically motivated recipes (see § 2) to model the formation and evolution of galaxies in hierarchical cosmogonies. Based on this work and almost at the same time, Kauffmann, White & Guiderdoni (1993) and Cole et al. (1994) presented

Astrophysics and Space Science is the original source of publication of his article. It is recommended that this article is cited as: *Astrophysics and Space Science* **263**: 39–42, 1999.

improved SAMs capable of fitting observed properties such us the luminosity function or the Tully-Fisher relation, although not simultaneously. More recently, Somerville & Primack (1998) have attempted to solve this discrepancy by introducing the effects of extinction by dust.

Here we present a fully analytical model of galaxy formation which shares the basic recipes with SAMs, but introduces several changes leading to a more realistic modelling.

2. Ingredients to model galaxy formation

In hierarchical scenarios galaxies form by cooling of baryonic gas within dark matter halos. Therefore, to study the evolution of galaxies, one needs to follow the growth history of halos. For this purpose SAMs adopt the extended Press & Schechter (1974) clustering model developed by Lacey & Cole (1993) (hereafter the LC model) which is used to build merger trees of individual halos by means of Monte Carlo simulations.

Dark matter halos are assumed to be spherical with a simple density profile. SAMs usually adopt a singular isothermal profile $\rho(r) \propto r^{-2}$, characterized by the lack of any scale for halos of a given mass at any given epoch. In this way the consistency with the LC clustering model describing the mass growth of halos is automatically guaranteed.

Contrary to dark matter, the realistic modelling of colliding baryons is very complicated and, in many aspects, poorly known. Therefore, some simplifications are needed to make it tractable. SAMs take the viewpoint that after a merger, the hot gas of the progenitors is shock heated to the virial temperature of the new halo, while the most massive galaxy falls rapidly to the center by dynamical friction. Then the hot baryonic gas experiences radiative cooling, progressively falls to the halo center and is deposited in a centrifugally supported disk where star formation takes place. The rate at which stars form is proportional to the available amount of cold gas. Feedback mechanisms, such as SN explosions, also operate by reheating part of the cold gas, which affects star formation, and blowing it backwards to the outer halo. Satellites (either brought by the progenitors at halo formation or accreted since then) may also sink to the halo center owing to dynamical friction. The capture of a sufficiently massive satellite by the central galaxy is able to destroy the disk giving rise to a stellar bulge through a violent starburst.

3. Innovations with respect to SAMs

SAMs are successful in reproducing many observable properties of galaxies (e.g., luminosity functions, counts, morphological populations) and have been applied to try to explain a number of topics related to galaxy evolution (e.g., the Butcher-Oemler effect, the stellar formation history). Yet, they have important limitations due to the simple rules involved to describe the physical mechanisms at work, and still need to be greatly improved. Our model introduces some novelties with the aim of circumventing some of these shortcomings. In this section we briefly comment on three main improvements related to the monitoring of the hierarchical growth of dark matter halos, the internal structure of these systems, and the detailed modelling of galaxies.

A proper characterization of mergers and accretion is crucial to obtain the properties of dark matter halos at a given time. The LC clustering model gives very good fits to the results of N-body simulations. However, in this formalism no distinction is made between accretion and major mergers: any halo mass increase is treated as a merger. As a consequence, the times of formation or destruction of a halo, which play an important role in SAMs, are rather ambiguous concepts. The distinction between accretion and merger is a good criterion for deciding whether a halo forms or simply evolves in a quasi-steady way. In our model we use the gravitational clustering model developed in Salvador-Solé, Solanes & Manrique (1998) (see also Raig in these volume). This enables one to properly follow the formation of halos through (binary) major mergers of similar massive progenitors, and their gentle mass increase by accretion of other tiny halos. On the other hand, instead of building merger trees for individual halos, we obtain ensemble average properties for a wide range of masses. This procedure has the advantage of allowing us to calculate the properties for the whole family of halos at a given epoch. In this way, quantities such as the evolution of the density, temperature, and metallicity of the inter cluster gas or the effects of a previous ionizing population of objects, such as quasars or Population III stars, can be readily computed.

Another great success of this gravitational clustering model is that it consistently describes the time evolution of halos with an internal structure in agreement with the results of recent high resolution N-body simulations (Navarro, Frenk & White (1997), NFW). These indicate, indeed, the existence of a universal density profile of the form $\rho(r) \propto [r(r+r_s)^2)]^{-1}$, where r_s is a characteristic scale over which the logarithmic slope changes from -1 ($r \ll r_s$) and -3 ($r \gg r_s$). This scale length correlates in a specific manner with halo mass. In Raig, González-Casado & Salvador-Solé (1998), it has been shown that the NFW halo density profile is closely linked to the processes of merging and accretion. The inability of SAMs to handle with such a distinction hinder them to properly implement the NFW density profile in that prescription. In contrast, in our analytical model of galaxy formation this can be fully accounted for. In particular, this description allows one to specify, in broad lines, the spatial location of the material which is assembled in a major merger or which is gently accreted since then. This has the advantage of making possible to follow the spatial segregation of galactic properties inside big halos, i.e., galaxy clusters.

Finally, SAMs model galaxies in a rather poor way. They only distinguish between bulge and disk, and use the luminosity ratio between them to classify galaxies morphologically. For some purposes this is enough. But if fine galactic properties such as rotation curves or the Fundamental Plane relation, or the distinction between high and low surface brightness disks are intended to be accounted for, a more thorough description is required. For these reasons we have introduced a simple but physically motivated modelling of the internal structure of galaxies, taking special care on the way in which their structure is truncated when their host halo merges with or is accreted by another halo. This procedure, in addition with the more accurate description of the evolution of halos and central galaxies during accretion, tends to alleviate the problem of the overmassive central galaxies common to all SAMs.

4. Conclusions

Not only can our model of galaxy formation be applied to the usual topics studied by means of SAMs, but also to new ones not yet examined. Concerning the former issue, we hope to obtain more reliable Tully-Fisher and Fundamental Plane relations owing to the better description of morphological types given by the galaxy modelling. But the main interest of the new model is that it is well suited to investigate the origin of the spatial segregation of galactic properties inside clusters, or to study the effects of galaxies and related objects (quasars, population III stars) on the inter cluster medium and the consequences of such a large scale feedback on the physical properties of galaxies and clusters.

References

Cole, S., Aragón-Salamanca, A., Frenk, C.S., Navarro, J.F., Zepf, E., 1994, MNRAS, 271, 781
Kauffmann, G., White, S.D.M., Guiderdoni, B., 1993, MNRAS, 264, 201
Lacey, C.G., Cole, S., 1993, MNRAS, 262, 627
Navarro, J.F., Frenk, C.S., White, S.D.M., 1997, ApJ, 490, 493
Press, W.H. and Schechter, P., 1974, ApJ, 187, 425
Raig, A., González-Casado,G., Salvador-Solé, E., 1998, ApJ, 508, L129
Salvador-Solé, E., Solanes, J.M., Manrique A., 1998, ApJ, 499, 542
Somerville, R.S., Primack J.R., 1998, submitted to MNRAS (astro-ph/9802268)
White, S.D.M., Frenk, C.S., 1991, ApJ, 379, 52

THE STABILIZING ROLE OF STELLAR BULGES IN GALAXY DISK FORMATION

A. SÁIZ[1], P.B. TISSERA[1,2] AND R. DOMÍNGUEZ-TENREIRO[1]
[1] *Dpt. Física Teórica, UAM, Cantoblanco, 28049 Madrid, Spain*
[2] *IAFE, Casilla 67, Suc. 28, 1428 Buenos Aires, Argentina*

The assembly of disk-like objects (DLOs) in hierarchical hydrodynamical simulations proceeds through different phases: a) *Quiescent* phases, where the diffuse halo gas component cools and collapses with specific angular momentum (j) conservation. In these phases, extended, thin disks are formed, similar to those observed in spirals (Fall & Efstathiou's model of disk formation [9]). It has been shown that such disks will have an *exponential* projected mass density profile [6]. b) *Violent* phases, when the system undergoes interaction and merger events. The output of these violent phases critically depends on the availability of gas to regenerate a disk structure, on the one hand, and on the structural stability of the objects involved in the event, on the other. Concerning stability, cold thin disks are known to be violently unstable against the bar mode ([4] [11] and references therein). Massive haloes can stabilize disks, but not every halo is able to stabilize the *exponential* disk that would form in its last *quiescent* phase, whose scale would be determined by j conservation [13] [20]. In these cases, a central compact bulge is needed to ensure stability [5] [20]. Otherwise, a bar instability could be easily triggered by interactions and mergers, followed by strong gas inflow due to important j loss (this is the so-called disk angular momentum catastrophe, DAMC [12]).

Mergers and interactions play a fundamental role in galaxy assembly. Unlike semianalytical models [6] [10] [13], hydrodynamical simulations of galaxy formation in fully-consistent cosmological scenarios naturally take into account interaction and merger effects, so that they are a promising working tool. However, so far, hydrodynamical simulations in hierarchical scenarios have been unable to produce disks similar to those observed in spirals. The objects formed are either gaseous and too concentrated, due to an excessive j loss (i.e., to DAMCs), when no star formation processes are considered at all ([14] [15] [21] and references therein), or rather stellar and without disks, when these processes have been included [16] (in this

Astrophysics and Space Science is the original source of publication of his article. It is recommended that this article is cited as: *Astrophysics and Space Science* **263**: 43–46, 1999.

case, the particular star formation implementation used leads to too early gas transformation into stars). This unability to reproduce observed disks has been interpreted by many authors in the field of galaxy formation as an evidence that some of the essential pieces in the art of galaxy building are not accounted for in present-day numerical simulations. However, in [7] and [8] we present results of a hydrodynamical simulation (S1) in a hierarchical scenario [18], where a simple implementation of star formation that prevents early gas exhaustion, but allows *compact stellar bulges* to form [17], produces galaxy-like objects with *disks similar to those observed in spirals.*

The stabilizing character of bulges against the bar mode hints at the possibility that the process at work in the DAMCs observed in hydrodynamical simulations so far, is related to disk instabilities. This point can be made clearer by comparing the assembly of DLOs in S1 to that in another simulation (S2), sharing the same initial conditions with S1 (so that haloes and their merger trees are identical in both of them), but where the star formation has been turned off. In S1, the process is as follows: i) First, dark matter haloes collapse at high z forming a first generation of (very small) disks and stars. ii) Then, the first unstabilizing mergers at high z happen, resulting in disk disruption and rapid mass inflow to the central regions with angular momentum loss and violent star formation, mainly at the central regions. Also, most preexisting stars will concentrate at the center of the new object through violent relaxation. These two processes help build up a central stellar bulge-like structure. iii) After the first mergers, a disk is regenerated through an infall of gas particles, either belonging to the baryonic merging clumps or diffuse. For example, a compact stellar bulge and an almost cold disk in a particular object, DLO #1, of S1 at $z = 0.57$ are apparent in Fig. 1a. iv) After disk regeneration, the system can undergo new major merger events at lower z. During the orbital decay phase, previous to the actual fusion of the DLOs, most of their orbital angular momentum is transported to (the particle components of) each host halo, spinning it up (as in [1] [2] [3]). Because, now, the disks involved in the merger are stabilized by their bulges, no strong gas inflow occurs in this phase (as in [12]). As the disks approach one another, they are heated and finally disrupted, but the high efficiency of gas shocking and cooling, and the symmetry of the central potential, quickly put those of their gas particles with high angular momentum into a new intermediate disk, while their low angular momentum particles sink to the center where most of them are transformed into stars, feeding the bulge. The stellar bulge of the smaller DLO is eventually destroyed and incomplete orbital angular momentum loss puts most of its stars on the remnant disk (Fig. 1b, note incomplete relaxation). v) Relaxation and disk regeneration are completed. Most of disk external particles are supplied by infall, as in iii) (Figs. 1c,1d).

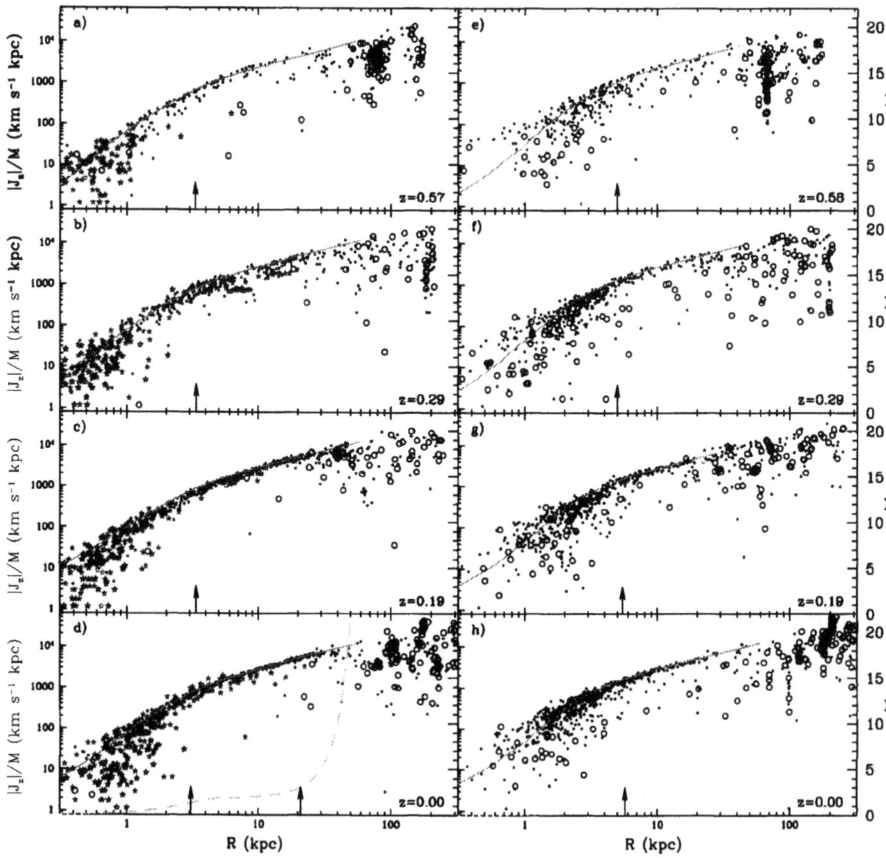

Figure 1. Specific angular momentum component along $\vec{J}_{\rm dis}$ for each baryon particle of halo #1, versus their positions, at different z around its last major merger. Left panels: S1 version. Right panels: S2 version. *Circles*: gas particles, *stars*: stellar particles; *open symbols*: counterrotating particles. *Solid lines*: $v_c(R)R$, a small dispersion about this curve indicates a disk configuration; a large dispersion indicates that particles have been involved in DAMCs. *Dotted lines*: $X_2(R)$ for actual disks at each z; *dashed line*: $X_2(R)$ for the pure exponential version. *Arrows* mark $R_{\rm st}^{\rm ad}$ and $R_{\rm st}^{\rm ped}$, the stability thresholds where $X_2(R) = 3$.

The assembly of galactic-like objects in S2 follows the same stages. The main difference is that in S2, the i) and ii) stages do not result in a stellar core, and, consequently, in iii) stage an unstable gas disk is formed, susceptible to grow bars. In particular, during the orbital decay phase in iv), strong gas inflow and j loss are induced (i.e., a DAMC, see Fig. 1e, and [2] [12]). The actual fusion completes the gas inflow (Figs. 1f, 1g), involving most of the gas particles originally in the merging disks. Few of them are left for disk regeneration, so that, in phase v), new disks are formed almost only from halo gas particles (Fig. 1h), and hence their low population.

In Fig. 1d we see the end product in S1: most gas particles placed at $R_i \lesssim 30$ kpc form a cold thin disk. By contrast, in S2 (Fig. 1h), most gas particles are at $R_i \lesssim 6$ kpc and have been involved in a DAMC. Gas particles at $R_i \gtrsim 30$ kpc are disordered in both S1 and S2, and they roughly follow the velocity distribution of dark matter.

Disk stability is usually analyzed through the $X_2(R)$ parameter [4] [19]. It has been calculated for the disk component of the DLOs at different z, and, also, for their *pure exponential disk (ped)* version (i.e., putting all their respective baryonic masses distributed as a *ped*). It is apparent from Fig. 1 that disks in S1 or S2 appear *only* where they are stable, i.e., at $R > R_{\rm st}^{\rm ad}$ (and provided that they have had enough time to form after the last merger). By contrast, the *ped* version of DLO #1 at $z = 0$ would be stable only at larger R ($R > R_{\rm st}^{\rm ped} \simeq 21$ kpc). This behaviour is common to the other DLOs identified either in S1 or S2 and it further suggests that DAMCs observed in S2 (and, presumably, in other hierarchical simulations lacking star formation) result from the unability of haloes to stabilize as *pure exponential disks* the main gaseous clump they host, so that interaction and mergers can trigger strong gas inflows. Moreover, DAMCs can be easily avoided by stabilizing the disks with stellar bulges.

References

1. Barnes, J.E., 1992, ApJ, 341, 425
2. Barnes, J.E., Hernquist, L., 1991, ApJ, 370, L65
3. Barnes, J.E., Hernquist, L., 1996, ApJ, 471, 115
4. Binney, J., Tremaine, S., 1987, *Galactic Dynamics*, Princeton Univ. Press
5. Christodoulou, D.M., Shlosman, I., Tohline, J.E., 1995, ApJ, 443, 551
6. Dalcanton, J.J., Spergel, D.N., Summers, F.J., 1997, ApJ, 482, 659
7. Domínguez-Tenreiro, R., Tissera, P.B., Sáiz, A., 1999, this volume
8. Domínguez-Tenreiro, R., Tissera, P.B., Sáiz, A., 1999, ApJL, in press
9. Fall, S.M., Efstathiou, G., 1980, MNRAS, 193, 189
10. Kauffmann, G., 1996, MNRAS, 281, 475
11. Martinet, L., 1995, Fund. Cosmic Phys., 15, 341
12. Mihos, J.C., Hernquist, L., 1994, ApJ, 425, L13
13. Mo, H.J., Mao, S., White, S.D.M., 1998, MNRAS, 295, 319
14. Navarro, J.F., Steinmetz, M., 1997, ApJ, 478, 13
15. Navarro, J.F., Frenk, C.S., White, S.D.M., 1995, MNRAS, 275, 56
16. Steinmetz, M., Navarro, J.F., 1999, ApJL, submitted (astro-ph/9808076)
17. Tissera, P.B., Domínguez-Tenreiro, R., 1998, MNRAS, 297, 177
18. Tissera, P.B., Lambas, D.G., Abadi, M.G., 1997, MNRAS, 286, 384
19. Toomre, A., 1981, in *The Structure and Evolution of Normal Galaxies*, eds. S.M. Fall & D. Lynden-Bell, Cambridge: Cambridge Univ. Press, p. 111
20. van den Bosch, F.C., 1999, ApJ, 507, 601
21. Weil, M.L., Eke, V.R., Efstathiou, G., 1999, MNRAS, 300, 773

ANALYSIS OF THE DIFFERENCE LIGHT CURVE OF THE GRAVITATIONAL MIRAGE QSO 0957+561

R. GIL-MERINO AND L.J. GOICOECHEA
Departamento de Fisica Moderna, Universidad de Cantabria
Avda. de los Castros, E-39005 Santander, Cantabria (Spain)
E-mail: goicol@besaya.unican.es

AND

M. SERRA, A. OSCOZ, E. MEDIAVILLA AND J. BUITRAGO
Instituto de Astrofisica de Canarias (IAC)
Via Lactea, E-38200 La Laguna, Tenerife (Spain)
E-mail: mserra@iac.es,aoscoz@iac.es,emg@iac.es,jgb@iac.es

Abstract. From optical data of Q0957+561 for 1995/1996 seasons (in the g band) and 1996/1997 seasons (in the R band) they are derived upper limits on the amplitude of the rapid microlensing fluctuations of 0.07 mag and 0.05 mag, respectively. The photometry in the R band (1996/1997 seasons) was taken at Teide Observatory (IAC-80 Telescope), Instituto de Astrofisica de Canarias, Spain. In this contribution, we also study the microlensing history (from 1981 to 1996) of the Twin QSO, which is a curious system with two optical images of similar brightness.

1. Introduction

Since the discovery of the first lens system, the Twin QSO (0957+561A,B), two decades ago (Walsh el al. 1979), there is no doubt about the importance of this gravitational mirage. A giant elliptical (cD) galaxy is partially responsible for the phenomenon, and so, the distribution of the luminous stars and compact dark objects (MACHOs) can induce variations in one or both of the images. These variations are caused by microlensing and, in principle, can be related to the masses of the MACHOs (e.g., for a small source, with 100% of the dark matter in compact objects, the probability of an encounter with a MACHO must be much greater than the probability

Astrophysics and Space Science is the original source of publication of his article. It is recommended that this article is cited as: *Astrophysics and Space Science* **263**: 47–50, 1999.

of an encounter with a luminous star). The fluctuations on a timesacale of weeks or months are related to the so-called short-timescale microlensing, whereas the fluctuations on a timescale of years are associated with the so-called long-timescale microlensing.

If an intrinsic variation (which is originated in the QSO) is observed in the light curve A (image A) at time t, then it should be seen in the light curve B (image B) at time $t + \Delta\tau_{BA}$, being $\Delta\tau_{BA}$ the time delay due to lensing by the deflector (galaxy + cluster). Therefore, the relationship between both light curves, microlensing and time delay is $B(t) = A(t - \Delta\tau_{BA}) + l_{BA} + m(t)$, where l_{BA} is the magnification ratio in absence of microlensing and $m(t)$ is the microlensing variability (which is caused by the granularity of the lens galaxy). Using a time delay of about 417 days, the behaviour of $m(t)$ from 1981 to 1996 (the microlensing history) was discussed by Pelt et al. (1998, hereafter P98). They suggested the existence of a MACHOs-event with a timescale of 5 years (1982-1986) and an amplitude of 0.25 mag. On the other hand, there are also evidences in favor of a significant short-timescale microlensing (Schild 1996, Goicoechea et al. 1998a). Recently, Schmidt & Wambsganss (1998, hereafter SW98) presented an interesting study of the short-timescale microlensing for 1995/1996 seasons (in the g band). They adopted a time delay of 417 days and found rapid microlensing variations no greater than 0.05 mag. For small source size, the authors claimed a lower limit for MACHOs-mass of $10^{-2}\ M_{\odot}$.

In this contribution, using two time delays: 417 days and 424 days (a new self-consistent test suggests that the photometry by Kundic et al. 1997, hereafter K97, is in excellent agreement with a delay of 423–424 days; see Goicoechea et al. 1998b), we analyze the microlensing signal in the Twin QSO. In Sect. 2, we discuss on the long-timescale microlensing and reinterprete the observed trend. Sect. 3 is dedicated to the short-timescale microlensing, using photometry made by K97 in the g band and new data taken at Teide Observatory for 1996/1997 seasons (in the R band). Finally, our conclusions are presented in Sect. 4.

2. Long-timescale microlensing

The radio emission region is much larger than the optical one, and therefore, it is likely magnified only by the macrolens ratio. In this manner, assuming $l_{BA} = (B - A)_{radio} = +0.31$ mag, the optical microlensing will be $m = (B - A)_{optical} - (B - A)_{radio}$. In order to infer the whole difference light curve $(A - B)_{optical}$, P98 used a time delay of about 417 days and optical data from 1981 to 1996. They saw a microlensing "event" from 1982 to 1986 with an amplitude (relative to another microlensing feature) of 0.25 mag and interpreted it as microlensing caused by MACHOs in the lensing galaxy. This conclusion is based on a large dataset with 1068 observations of the Twin QSO in the R band (which was kindly provided to us by Rudy Schild).

We however disagree with the point of view by P98 about the variability from 1981 to 1985 (it was interpreted as a MACHOs-event), at least for two reasons: (*i*) if one takes the delay of 424 days, the "event" leads to $m \approx 0$, which should be a very surprising coincidence; in fact, the optical microlensing history suggests that around 1983 ends an event and begins another long event, and (*ii*) the fluctuation represents a single demagnification of image B, but it is not a very natural variability (either several magnifications and demagnifications or a single magnification should be more reasonable physical processes). Considering the global trend as basically an event with timescale ≥ 15 years and amplitude of 1/3 mag (our point of view), we can infer some limits on the MACHOs-mass. Assuming a flat universe and a Hubble constant of $H_0 = 66$ km s^{-1} Mpc^{-1} (see, e.g., Oscoz et al. 1997), we differentiate two cases: for a small source, the MACHOs-mass would be $M \geq 0.2\ M_{\odot}$, while for an extended source, we find the constraint $M > 10^{-3}\ M_{\odot}$. Here, a source is assumed as a small object when its angular radius is less or equal than the angular radius of the Einstein ring associated with a MACHO of mass M.

3. Short-timescale microlensing

With respect to the photometry made by K97 (1995/1996 seasons) in the g and r bands, we will focus our attention on the data in the g band, because the presumably weak contamination in this band. As in the previous section, we also consider the two time delays (see Introduction). The microlensing curves are computed from $m(t) - < m > = B(t) - < B > -[A(t - \Delta\tau_{BA}) - < A >]$. For a time delay of 417 days, we basically agree with the results by SW98 (a small discrepancy is raised in the data selection, but it can be seen as an anecdotical fact). On the other hand, for a delay of 424 days, they appear a microlensing peak with an amplitude of about 0.05 mag and a bound (upper limit) of 0.07 mag.

We analyse now the data, in the R band, obtained with the IAC-80 Telescope at Teide Observatory (Spain) from February 1996 to July 1997. We use 33 photometric measurements of the image A and 36 photometric data of the image B. The errors in this dataset are very large (the mean values are about 0.023–0.026 mag), and so, we re-reduced the photometry, making groups with individual measurements. The raw and re-reduced datasets are available in: `http://besaya.unican.es/~goicol`. For the two delays (417 and 424 days), we find an upper limit (on $|m - < m > |$) of 0.05 mag.

4. Conclusions and expectances

1. The microlensing history of the Twin QSO (from 1981 to 1996) is in agreement with the existence of an event beginning at 1983 or before and remaining up to the present. This event can be related to the MACHOs-mass in the main lens galaxy (a cD galaxy). Assuming a granular distribution of the matter, only masses $> 10^{-3}\ M_{\odot}$ may cause the phenomenon.

2. The adopted time delay plays a crucial role in rapid microlensing studies. From data for three seasons (1995, 1996 and 1997), we derived a robust global upper limit on the amplitude of the rapid microlensing fluctuations of 0.07 mag. For 1996/1997 seasons, even fluctuations with an amplitude greater than 0.05 mag can be ruled out.

3. Assuming $\Delta\tau_{BA} = 424$ days, the microlensing curve for 1995/1996 seasons shows a prominent event. On the contrary, for 1996/1997 seasons, the microlensing curve is consistent with $m- < m > = 0$. There is an apparent correlation between existence (absence) of rapid microlensing variability and activity (calmness) of the source. This point merits more attention.

4. Future work must confirm (or reject) the correlation pointed out in the previous item. Also, new annual differences $(B - A)_{optical}$ are fundamental for tracing the evolution of the long-timescale event. Finally, we remark that the attainment of new upper limits on the short-timescale microlensing and the characterization (amplitude, duration, etc) of possible rapid microlensing events could be basic works for deciding on the nature of the dark matter in the cD galaxy as well as the physical properties of the emitter (QSO).

5. Acknowledgements

This project was supported by Instituto de Astrofisica de Canarias and Universidad de Cantabria funds and DGESIC (Spain) grant PB97-0220-C02.

References

Goicoechea et al., 1998a, ApJ 492, 74
Goicoechea et al., 1998b, preprint (astro-ph/9801315)
Kundic et al., 1997, ApJ, 482, 75 (K97)
Oscoz et al., 1997, ApJ, 479, L89
Pelt et al., 1998, A&A, 336, 829 (P98)
Schild, R., 1996, ApJ, 464, 125
Schmidt, R., Wambsganss, J., 1998, A&A, 335, 379 (SW98)
Walsh, D., Carswell, R. F., Weymann, R. J., 1979, Nature, 279, 381

THE CASTLES PROJECT

J. A. MUÑOZ, E. E. FALCO, C. S. KOCHANEK, J. LEHÁR AND
B. A. MCLEOD
Harvard-Smithsonian Center for Astrophysics
60 Garden St., Cambridge, MA 02138

AND

C. D. IMPEY, H.-W. RIX AND C. Y. PENG
Steward Observatory, University of Arizona
Tucson, AZ 85721

Abstract. We describe the goals of the CASTLES (CfA–Arizona–Space–Telescope–LEns–Survey)[1] project including a sample of NICMOS images of gravitational lenses and a brief list of the preliminary findings.

1. Introduction

Gravitational lenses are powerful astrophysical tools to investigate cosmology, the Hubble constant, galactic structure and evolution, dust extinction, and AGN host galaxies. There are now 47 known galaxy lens systems, and their broad astrophysical utility relies on accurate photometry, astrometry and redshifts for as many systems as possible. These lenses consist of 2-4 source images (AGNs, quasars, hosts), superposed on a foreground lens galaxy within a diameter of $\sim 1''$. Thus, precise and quantitative studies in the optical and IR require Hubble Space Telescope (HST) observations.

2. CASTLES Goals and Preliminary Results

CASTLES is an ongoing HST survey of all the known lensed systems and lens candidates. We are obtaining images with NICMOS and WFPC2 in the H, I and V bands. Some of its goals are: to create a complete, uniform high-resolution photometric sample of the known galaxy-mass lenses; to obtain redshift estimates for all lens galaxies which lack spectroscopic redshifts;

[1]see http://cfa-www.harvard.edu/castles

Astrophysics and Space Science is the original source of publication of his article. It is recommended that this article is cited as: *Astrophysics and Space Science* **263**: 51–54, 1999.

to find any source or lens components that have escaped detection and determine their photometric properties; to obtain precise astrometric data for all source components to improve lens models and estimates of H_0; and to investigate the wide field environments of the lens galaxies and their role in lensing.

In this contribution we present examples of CASTLES images (see Figure 1) and a brief list of the preliminary findings:

Missing Lenses: As expected, the deeper CASTLES optical and IR observations are detecting most "missing" lens galaxies, even in cases where bright quasar images and small separations make this difficult. Among the 10 doubles in Lehár et al. (1999), we detected 9 lens galaxies (4 of which are discoveries). We failed only in the case of Q1208+1011, due to contrast problems. We never detect lenses in the wider quasar pairs (UM425, Q1429–008, Q1634+267, MGC2214+3550, and Q2345+007) whose extreme limits on the lens M/L make lensing improbable and point to the binary quasar interpretation (see Kochanek, Falco & Muñoz, 1999).

Host Galaxies: CASTLES has detected the lensed AGN host galaxies for 7 quasars (MG0414+0534, Q0957+561, PG1115+080, H1413+117, CTQ414, BRI0952–0115, HE1104–1805), and 4 radio galaxies (MG0751 +2716, MG1131+0456, B1600+434, B1608+656). Lensing provides a unique opportunity to study $z > 1$ host galaxies, because the hosts are magnified to detectable angular sizes. A preliminary analysis (i.e. demagnification) finds that $L_{\rm host} \lesssim L_*$ and indicates that even the most luminous quasars (e.g. PG1115+080) need not reside in particularly luminous hosts. For 2 of the QSO pairs (MGC2214+3550 and Q2345+007), the host morphology proves the systems are quasar binaries and not "dark" lenses.

Photometric Lens Redshifts: We have found that we can accurately estimate photometric lens redshifts by mapping the lenses onto the local "fundamental plane" using passive evolution models for E- and K-corrections. These Bayesian estimates simultaneously use the available color, luminosity, mass, and structural information. Tests on ten double-image systems (Lehár et al. 1999) demonstrate agreement within $\Delta z_L \lesssim 0.1$ with spectroscopic redshifts.

Galaxy Structure: The CASTLES observations provide detailed constraints on lens galaxy structure, through shape and profile fitting. We continue to find that most lens galaxies have de Vaucouleurs profiles, as well as shapes, and colors consistent with passively evolving early-type galaxies; e.g. Q0142–100, BRI0952–0115, Q1017–207, B1030+071, HE1104–1805 (Lehár et al. 1999); MG1131+0456 (Kochanek et al. 1999) and PG1115+080 (Impey et al. 1998). B0218+357 and PKS1830–211 (Lehár et al. 1999) are spirals but present a more complex photometric picture due to the high molecular gas column densities and implied dust extinction.

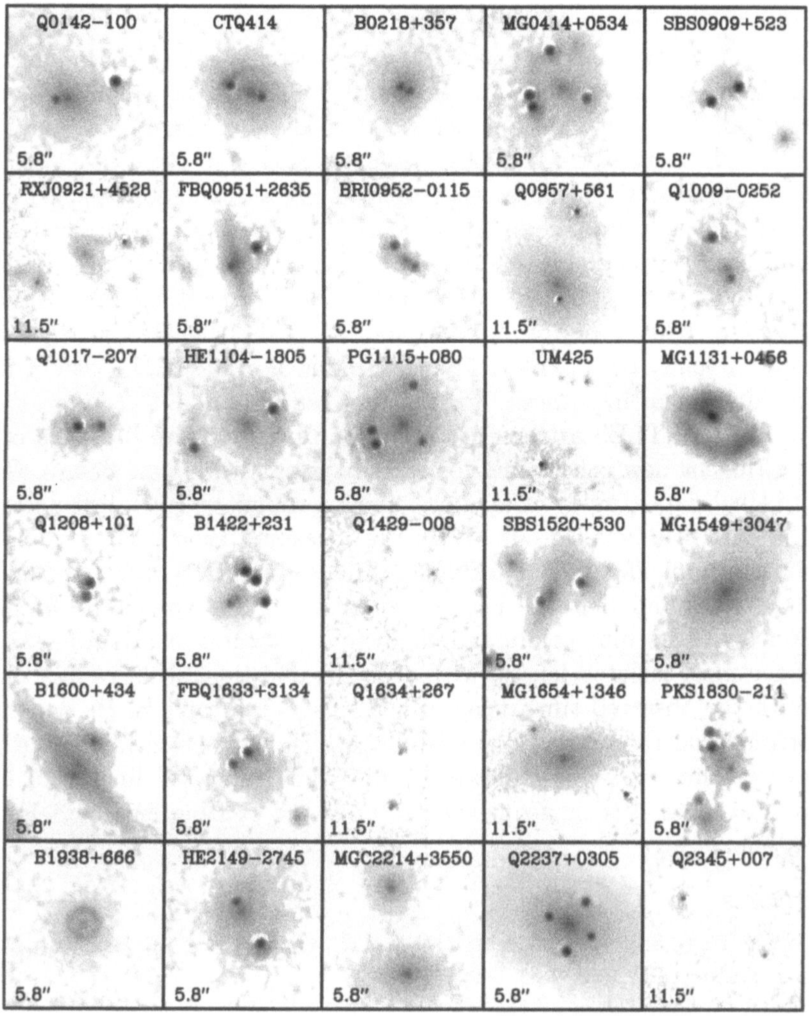

Figure 1. NICMOS images of CASTLES targets, showing the H-band images after deconvolving the HST H-band point-spread function. The target name and field size are displayed for each panel.

Dark Matter: We constrain and study the dark matter distribution of lenses by comparing their luminosity distribution to our improved lens mass models (Lehár et al. 1999). While most of the 2-image lenses can be fitted by a single ellipsoidal lens, some require substantial external shear, and most require such shear to explain misalignments between the lens mass and its light. Our observations reveal neighboring galaxies that produce the required shear (e.g. MG1131+0456, Kochanek et al. 1999). In some cases we can study the mass profile. We can use the mass determinations to estimate M/L and explore its value and evolution with redshift.

Lens Galaxy Extinction: We use lensed image flux ratios as a function of wavelength to measure the differential extinction in the lens galaxies. Of the 10 doubles in Lehár et al. (1999), 2 are spirals (B0218+357 and PKS1830–211) and are heavily attenuated. The rest are ellipticals; 4 contain modest but measurable amounts of dust (SBS0909+523, Q1009−0252, HE1104−1805, and Q1208+101), and 4 show little or no differential extinction. PG1115+080 (Impey et al. 1998) and MG1131+0456 (Kochanek et al. 1999) are virtually transparent. The transparency of MG1131+ 0456 rules out the "dusty lens hypothesis" for the very red lenses.

H_0: The CASTLES astrometric precision of better than 3 mas is providing stringent new constraints for lensed systems with time delays. Our PG1115+080 data (Impey et al. 1998) show a round, $r^{1/4}$-law lens galaxy, which is part of a small group. With the measured time delay our models robustly limit $H_0 \lesssim 65 \pm 5\,\mathrm{km\ s^{-1}\ Mpc^{-1}}$. In B0218+357 (Lehár et al. 1999), we found that the lens was best fit by an exponential profile, consistent with its blue colors and small image separation. B0218+357 is one of the most isolated lenses, with an estimated tidal shear of less than 1%. With the observed time delay of $\sim$ 12 days, the value of H_0 depends very strongly on the precise lens position, which we hope to improve with new observations. Our new image of Q0957+561 shows 2 images of the quasar host galaxy; their geometry rules out the existing lens models for the system.

References

Impey, C.D., Falco. E.E., Kochanek, C.S., Lehár, J., McLeod, B.A., Rix, W.-H., Peng, C.Y., Keeton, C.R., 1998, ApJ, 509, 551

Kochanek, C.S., Falco E.E., Muñoz, J.A., 1999, ApJ, 510, 590

Kochanek, C.S., Falco. E.E., Impey, C.D., Lehár, J., McLeod, B.A., Rix, W.-H., Keeton, C.R., Muñoz, J.A., Peng, C.Y., 1999, ApJ, in press

Lehár, J., Falco. E.E., Impey, C.D., Kochanek, C.S., McLeod, B.A., Rix, W.-H., Keeton, C.R., Muñoz, J.A., Peng, C.Y., 1999, ApJ, to be submitted

IDENTIFICATION AND STUDY OF DISTANT GALAXIES THROUGH CLUSTER LENSES

R. PELLÓ AND J.P. KNEIB
Observatoire Midi-Pyrénées, UMR 5572
14, Avenue E. Belin, F-31400 Toulouse, France

AND

G. BRUZUAL
Centro de Investigaciones de Astronomía (CIDA)
Apartado Postal 264, Mérida, Venezuela

Abstract. We present new results obtained when using clusters of galaxies as gravitational lenses to build up an independent sample of galaxies with $z \gtrsim 2$. The amplification effect provides a tool to study the spectrophotometric and morphological properties of such galaxies 1 to 3 mag. deeper with respect to field surveys. Distant sources are selected close to the critical lines in clusters where the mass distribution is well known, using photometric redshifts computed on a large wavelength interval. This method allows to reduce the biases towards luminous and active star–forming objects.

1. Introduction

The identification and study of extremely high-z galaxies remains the most direct way to constrain models of galaxy formation and evolution. The amplification close to the critical lines in clusters-lenses is typically $\Delta m \sim 2$ to 3 mags, but it is still ~ 1 mag at $1'$ from the center of the lens. Thus, clusters of galaxies can be used to construct *independent* and complementary samples of high-z galaxies with respect to field surveys. The signal/noise ratio in spectra of amplified sources is improved beyond the limits of conventional spectroscopy, allowing a detailed study of the stellar content and evolutionary state of such galaxies. Using photometric redshift estimates (hereafter $z_{\rm phot}$), it is possible to access the statistical properties of galaxies on an enlarged magnitude domain. The aim of our program is to produce

Astrophysics and Space Science is the original source of publication of his article. It is recommended that this article is cited as: *Astrophysics and Space Science* **263**: 55–58, 1999.

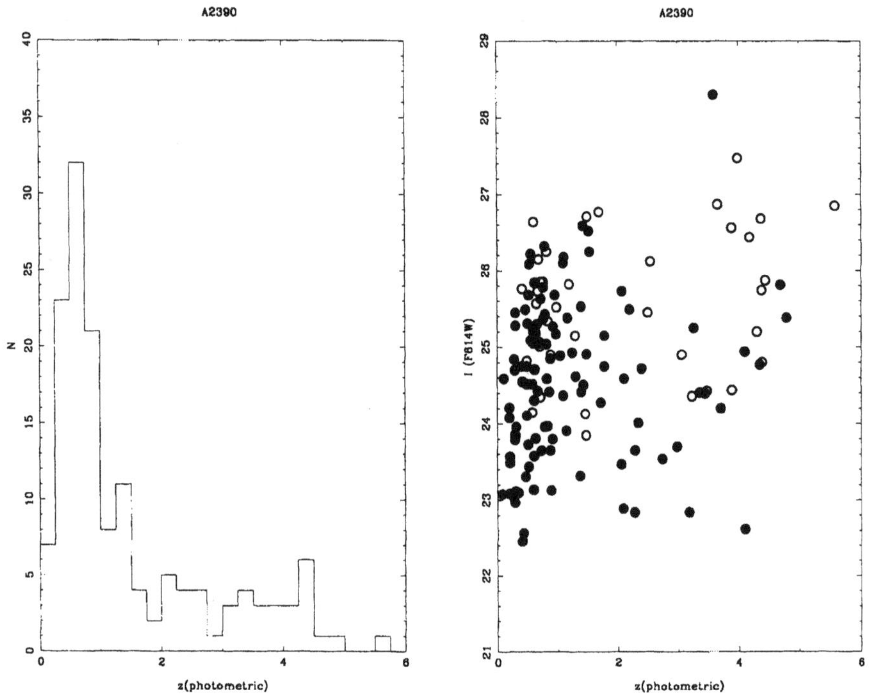

Figure 1. (left) $z_{\rm phot}$ distribution for the most secured candidate sources in A2390. (right) I (F814W) magnitude, uncorrected for amplification, versus photometric redshift. Black dots refer to the most reliable estimates.

a new independent sample of sources, in order to improve our knowledge of the faint population of galaxies at $2 \lesssim z \lesssim 7$. Recent examples of highly-magnified sources behind clusters lenses strongly encourages this approach: the star-forming source #384 in A2218 at z=2.51 (Ebbels et al. 1996), the luminous z=2.7 galaxy behind the EMSS cluster MS1512+36 (Yee et al. 1996), the three z$\sim$ 4 galaxies in Cl0939+47 (Trager et al. 1997), a z=4.92 system in Cl1358+62 (Franx et al. 1997, Soifer et al. 1998) and the two red galaxies at $z \sim 4$ in A2390 (Frye & Broadhurst 1998; Pelló et al. 1998).

2. Selection criteria

High-z candidates are selected according to two criteria: they lie close to the appropriate critical lines, and they have a $z_{\rm phot} \geq 2$. In order to keep the amplification uncertainties below $\Delta m < 0.3$ mags, only those clusters whose mass distribution is highly constrained by multiple images are useful for this program. A detailed description of the expected 2D distribution of high-z galaxies behind clusters can be found in Bézecourt et al. (1998). $z_{\rm phot}$'s are computed using the standard minimization method by Miralles

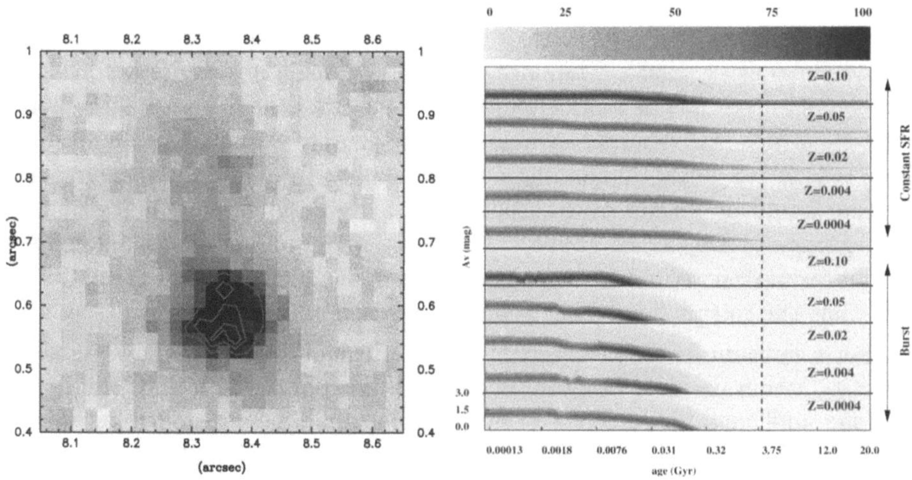

Figure 2. (right) Likelihood map for the $z = 4.05$ source H5 in A2390, where dark shading represents the most probable regions in the parameter space defined by the type of SFR, metallicity, reddening and age of de dominant stellar population. The dotted line is the cosmic age ($H_0 = 50$ km s^{-1} Mpc^{-1}, $q_0 = 0$). (left) Reconstruction of H5 on the $z = 4.05$ source plane, according to the modelling by Kneib et al. (1998). $0.1''$ corresponds to 1.6 kpc ($H_0 = 50$ km s^{-1} Mpc^{-1}, $q_0 = 0.1$).

& Pelló (1998). The spectral energy distribution (SED) of each source is compared to a set of 255 templates obtained using the Bruzual & Charlot evolutionary code (GISSEL98, Bruzual & Charlot 1993). The photometric SED is obtained on a large wavelength interval, including near–IR photometry. It is worth noting that such a selection procedure allows to reduce the biases towards intrinsically luminous and/or active star–forming sources.

3. Results and discussion

This method has been applied to 3 clusters (A2390, A370 and A2218), using 4m telescopes (CFHT, WHT, NTT and 3.6m at ESO), on a previous extended Toulouse–Cambridge–Barcelona collaboration. Figure 1 displays the $z_{\rm phot}$ distribution of arclets in the field of A2390, one of the best studied lenses. Arclets were selected in this case according to morphological criteria (surface brightness, elongation, orientation, position) on deep HST/WFPC2 images. 35 sources are identified at $z \geq 2.0$ up to $z \sim 5.5$, and 4 additional candidates are found up to $z \sim 6-7$, with a lower probability. Details on the mass modelling can be found in Kneib et al. (1998).

The SFR and the stellar masses involved in these sources can be derived when a good estimate of z is available. Although there is a degeneracy in the SFR–age–metallicity–reddening space, it can be constrained using the

GISSEL98 code when the IMF and the upper mass limit for star–formation are fixed. The two $z = 4.05$ sources behind A2390 illustrate this point (Fig. 2, and Pelló et al. 1998). They are dominated by massive OB stars at the wavelengths seen in the visible bands, and the presence of Lyα in emission points towards star–forming systems. These sources are intrinsically bright ($M_B^* - 2$ to -1 mags, depending on A_v), slightly brighter than the $z \sim 4$ objects found by Trager et al. (1997), and also brighter than the $z = 5.34$ field galaxy of Dey et al. (1998). The morphology of such lensed sources can be reconstructed advantageously through the lens modelling (Fig. 2b), and thus the morphology–SED relation can be studied in details.

For a subsample of spectroscopically confirmed objects, we have tested the $z_{\rm phot}$ accuracy as a function of the relevant parameters: SFR, reddening, age and metallicity of the stellar population. The agreement between the $z_{\rm phot}$, the spectroscopic and the lensing redshift obtained from inversion methods (Ebbels et al. 1998) is good up to at least $z \lesssim 1.5$. For higher z, the results on the most amplified sources are promising (Pelló et al. 1998), but an enlarged spectroscopic sample is urgently needed. A large VLT Program is presently going on, involving different european institutions (ESO/IAP–DEMIRN/Toulouse/Roma/MPE–Heidelberg), and the U. of Chile, aiming to perform the spectroscopic follow up of high–z candidates selected from the preparatory visible and near-IR photometry at ESO (NTT, 3.6m). Spectroscopic surveys with 10m telescopes should go further on this study, especially to estimate more precisely the metallicities using the UV absorption lines seen in the visible, or the usual visible indicators observed in the near–IR bands. All multi–wavelength studies on cluster–lenses should benefit from the amplification factor to improve the detection levels with respect to the field. A spectacular example is the deep ISOCAM survey of A2390 (Altieri et al. 1998). For most statistical purposes, $z_{\rm phot}$'s should be accurate enough to discuss the properties of these extremely distant galaxies. Inversely, this sample of amplified sources would allow to calibrate $z_{\rm phot}$ techniques up to much fainter limits compared to field surveys.

References

Bézecourt, J., Pelló, R., Soucail, G., 1998, A&A, 330, 399
Bruzual, G., Charlot, S., 1993, ApJ, 405, 538
Ebbels, T.M.D., et al., 1996, MNRAS, 281, L75
Ebbels, T.M.D., et al., 1998, MNRAS, 295, 75
Franx, M., et al., 1997, ApJ, 486, 75
Frye, B., Broadhurst T., 1998, ApJ, 499, 115
Miralles, J.M., Pelló, R., 1998, ApJ, submitted (astro-ph/9801062)
Pelló, R., et al., 1998, A&A, submitted (astro-ph/9809151)
Soifer, B.T., et al., 1998, ApJ, 501, 171
Trager, S.C., Faber, S.M., Dressler, A., Oemler, A., 1997, ApJ, 485,92
Yee, H.K.C., et al., 1997, AJ, 111, 1783

VLBA MONITORING OF THE QSO 3C 345

E. ROS, J.A. ZENSUS AND A.P. LOBANOV
Max-Planck-Institut für Radioastronomie, Bonn, Germany

1. Introduction.

The QSO 3C 345 (V=16, z=0.595) is a core-dominated radio source that displays apparent superluminal motions, with components traveling in the parsec-scale jet along curved trajectories and speeds up to $10c$ (Zensus et al., 1995). We have observed this QSO with the very long baseline interferometry (VLBI) technique using the NRAO Very Long Baseline Array (VLBA) at three epochs and four frequencies, in order to monitor it in total and linear polarization intensity.

2. Observations and Imaging.

We observed 3C 345 in 1995.84, 1996.41, and 1996.81, using the VLBA at 22, 15, 8.4, and 5 GHz, and recording with a 16 MHz bandwidth at all frequencies. At each frequency, the source was observed for about 14 hrs, using 5-minute scans and interleaving all observing frequencies. Some calibrator scans (on 3C 279, 3C 84, NRAO 91, OQ 208 and 3C 286) were inserted during the observations.

2.1. TOTAL INTENSITY.

After the fringe-fitting process, we exported the data into the differential mapping program DIFMAP (Shepherd et al., 1994), and we obtained total intensity images using the hybrid mapping technique. The components C7 and C8 can be identified close to the core D, in the images at higher frequencies. At the lower frequencies, the jet extends to the NW direction, turning to the N at $\geq$20 mas distances from the core. In Fig. 1, we show the central region of the source at 22 GHz for the three sampled epochs.

Astrophysics and Space Science is the original source of publication of his article. It is recommended that this article is cited as: *Astrophysics and Space Science* **263**: 59–62, 1999.

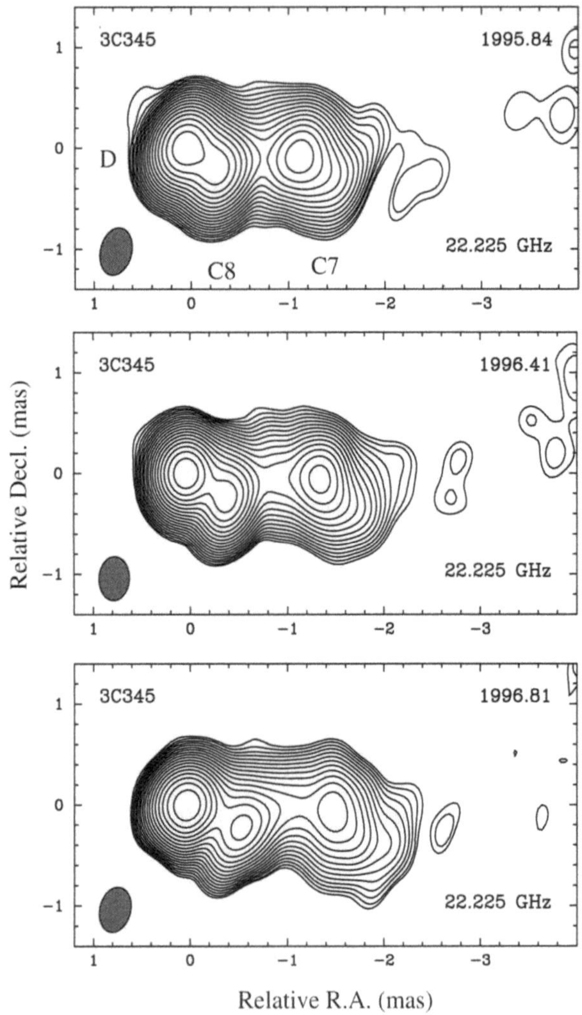

Figure 1. Total intensity images of the QSO 3C 345 at 22 GHz for the epochs 1995.84, 1996.41, and 1996.81. The synthesized beams are, respectively, 0.480×0.326, P.A. -15°.0, 0.433×0.316, P.A. -3°.1, and 0.451×0.322, P.A. -15°.5. The total fluxes recovered in the maps are 6.98, 4.51 and 4.61 Jy. The peaks of brightness are 2.56, 1.80, and 2.09 Jy/beam. The contours displayed are, 5.1, 3.7, and 4.2 mJy ×(-1,1.4,...,256,362).

To describe the structures observed within ~3 mas distance from the core, we fit elliptical components with Gaussian brightness profiles to the visibility data at 15 and 22 GHz. The change measured in the fitted positions for the component C8 indicates that the component was moving away from the core, at an angular speed of 0.26±0.08 mas/yr at 22 GHz, which corresponds to an apparent speed of $5.1 \pm 1.8h^{-1}$ c (assuming H_0=100h km s^{-1} Mpc^{-1} and q_0=0.5). The trajectory of C7 at the same frequency shows a proper motion of 0.29±0.08 mas/yr, or apparent speed of $5.7 \pm 1.8h^{-1}$ c. At 15 GHz the respective values are C8: 0.23 ± 0.11 mas/yr ($4.5 \pm 2.2\,h^{-1}c$) and C7: 0.30 ± 0.11 mas/yr ($5.9 \pm 2.2\,h^{-1}c$).

2.2. POLARIZED INTENSITY.

Since the early works of Cotton et al. (1984) up to now, important progress has been made in polarimetric VLBI observations. The VLBA has standardized feeds with low instrumental polarization, and the relatively small size of the antennas is compensated by the excellent performance of the receivers. In addition, a new method for self-calibrating the polarimetric data has been introduced (Leppänen et al., 1995), enabling D-term determination by using the program source itself. Using this method, we have calibrated the instrumental polarization, determining the feed solutions for the receivers of all the antennas. This has allowed us to obtain maps of the linearly polarized emission from the source ($P = Q + iU = pe^{2i\chi} = mIe^{2i\chi}$, where Q and U are the Stokes parameters, $p = mI$ is the polarized intensity, m is the fractional linear polarization, and χ is the position angle of the electric vector in the sky). In Fig. 2, we show a composition of the total intensity I (grey scale), the polarized intensity p (contours) and the electric vector orientation angle χ (segments, length proportional to p) for the 22 GHz observations in 1995.84.

In Fig. 2 the electric vector is aligned with the extremely curved jet direction at the inner 3 mas both for the core and the C7 component. The component C8 is boosted in polarized emission, showing an electric vector apparently perpendicular to this jet direction. Gómez et al. (1994a; 1994b) have modeled features similar to the ones presented here. They reported anticorrelation between the polarized and the total intensity flux close to a shock wave along a jet in its early evolution (near to the core). We should point out that the possible superposition between components with different electric vector orientations can lead to a cancellation of flux and produce apparent separations between polarized emission components. The brighter polarized emission in C8 might be explained by a shock wave in a curved jet geometry. More generally, the features of Fig. 2 can be explained in terms of a comprehensive shock model (Wardle et al., 1994) in the framework of a helical geometry for the motion of the components (Steffen et al., 1995).

3. Conclusions.

We have monitored the superluminal QSO 3C 345 at three epochs within one year, observing with the VLBA at four frequencies. We have presented

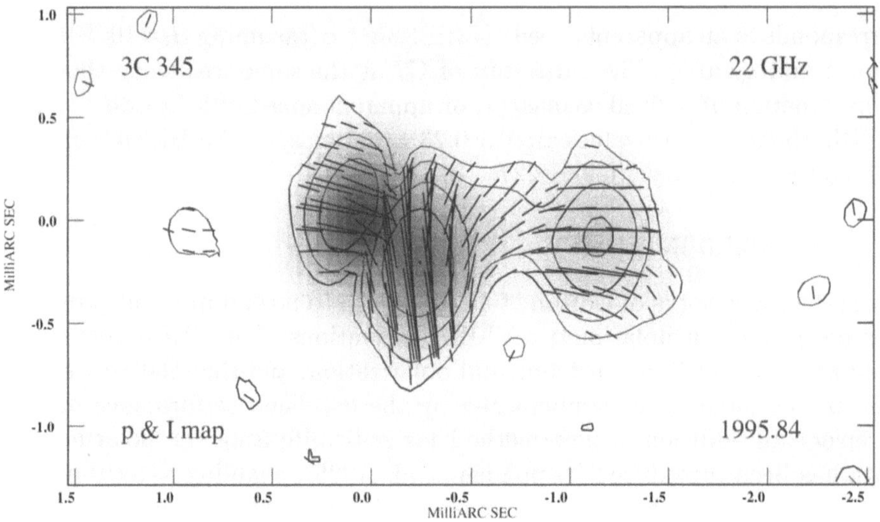

Figure 2. VLBA p, I and χ images of 3C 345 at 22 GHz, epoch 1995.84. The polarized intensity p is represented with contours (value of 7 mJy/beam×(1,2,4,8,16), brightness peak of 112 mJy/beam) over a grey scale total intensity map (grey scale above 7 mJy/beam, check Fig. 1, top for parameters), and the superimposed electric vectors (χ, length $\propto p$).

some results of these studies at the higher frequencies, showing the superluminal motions of components C8 and C7 with respect to the core component D, and the remarkably complex polarization structure near the core, which provides evidence for emerging components and changing projected jet direction within 3 mas from the core. The twist in the orientation of the electric vector along the jet can be explained in terms of an extremely curved helical geometry, following Steffen et al. (1995). The electric field is parallel to the jet direction, and the boosting in the polarized emission at the component C8 and the change in the vector orientation can be the result of the presence of a shock wave in a bent jet.

References

Cotton, W.D., Geldzahler, B.J., Marcaide, J.M., Shapiro, I.I., Sanromà, M., Rius, A., 1984, ApJ, 286, 503
Gómez, J.L., Alberdi, A., Marcaide, J.M., 1994a, A&A, 284, 51
Gómez, J.L., Alberdi, A., Marcaide, J.M., Marscher, A.P., Travis, J.P., 1994b, A&A, 292, 33
Leppänen, K.J., Zensus, J.A., Diamond, P.J., 1995, AJ, 110, 2479
Shepherd, M.C., Pearson, T.J., Taylor, G.B., 1994, BAAS, 26, 987
Steffen, W., Zensus, J.A., Krichbaum, T.P., Witzel, A., Qian, S.J., 1995, A&A 302, 335
Wardle, J.F.C., Cawthorne, T.V., Roberts, D.H., Brown, L.F., 1994, ApJ, 437, 122
Zensus, J.A., Cohen, M.H., Unwin, S.C., 1995, AJ, 443, 35

THE SHAPE OF THE BLUE/UV CONTINUUM OF B3-VLA RADIO QUASARS

Dependence on redshift, blue/UV luminosity and radio power

R. CARBALLO AND J.I. GONZALEZ-SERRANO, S. SANCHEZ

Instituto de Física de Cantabria y Dpto. de Física Moderna CSIC y Univ. de Cantabria, 39005 Santander,Spain

C.R. BENN

Isaac Newton Group of Telescopes, Royal Greenwich Observatory 38780 Santa Cruz de La Palma, Tenerife,Spain

AND

M. VIGOTTI

Istituto di Radioastronomia, Gobetti 101, 40129 Bologna, Italy

Abstract. We present some results based on optical photometry of an unbiased subsample of 64 sources, from a sample of radio quasars 80 per cent complete. The quasars (z=0.4–2.8) show approximately a similar spectral energy distribution (SED). Only two quasars appear clearly differenciated, exhibiting redder colours than the rest, and they have low or moderate redshifts (z=0.50 and 1.12). Composite broad-band spectra for the remaining 62 quasars reveal the presence of a break in the SED at $\sim$ 3000 Å, where the average slope α (defined by $S_\nu \propto \nu^\alpha$) changes from $\alpha_{\rm blue}$=0.11±0.16 at λ>3000 Å to $\alpha_{\rm UV}$=–0.66±0.15 at λ<3000 Å. Composite spectra from a low and a high redshift subsample, with the redshift separation at the median value z=1.2, yields $\alpha_{\rm UV}$=–0.87±0.20, for z<1.2 and $\alpha_{\rm UV}$=–0.48±0.12 for z>1.2 and the difference is 98 per cent significant. This trend could be intrinsic or could arise from a dependence between $\alpha_{\rm UV}$ and the blue/UV luminosity, since the three quantities $\alpha_{\rm UV} - L_{\rm blue/UV} - z$ are correlated. Finally, an intrinsic correlation is found between blue/UV and radio luminosity.

Astrophysics and Space Science is the original source of publication of his article. It is recommended that this article is cited as: *Astrophysics and Space Science* **263**: 63–66, 1999.

1. Introduction

The presence of dust reddening towards quasars has been claimed in the literature for the interpretation of different results, like the excess of the metagalactic UV flux J_ν compared to the predictions from QSO luminosity functions, the spectrum of the X-ray background, etc. Webster et al. (1995) found a broad range of B–K colours for flat-spectrum radio quasars and interpreted this scatter in terms of dust reddening, implying extinctions $A_{\rm B}\sim5$ mag for a substantial fraction of quasars. Serjeant & Rawlings (1996) argue that the red colours of flat-spectrum quasars are likely related to the core-synchrotron component enhanced by relativistic beaming, characteristic of flat-spectrum sources, which extends to the optical-infrared. In fact the B–K colours of radio quasars selected at low frequency are not as red as measured by Webster et al. for flat-spectrum quasars, and the reddest colours are generally related to a K excess due to the contribution of the host galaxy or a flat-spectrum component (Benn et al. 1998, Carballo et al. 1998).

As an additional step to quantify the amount of reddening in steep-spectrum radio quasars, we have performed a study of their optical SED, which would be significantly affected by obscuration. We used the B3-VLA Quasar Sample (Vigotti et al. 1997), which being selected at 408 MHz, contains predominantly steep-spectrum quasars.

2. Observations and results

UBVR photometry of the quasars was obtained, through CCD images, on 1997 at the 1.0-m JKT (La Palma) and at the 2.2-m telescope on Calar Alto. Magnitude errors were typically below 0.15 mag.

The R-band magnitude distribution of the quasars in the 80 per cent complete sample has a peak at R$\simeq$18.5 and decreases abruptly beyond R$\simeq$19 mag. The mean colours derived for the quasars are $\langle$U–R$\rangle$=–0.08, $\langle$B–R$\rangle$=0.64 and $\langle$U–V$\rangle$=–0.38 with dispersions 0.45, 0.27, and 0.42, respectively. Two of the quasars stand out for being particularly red, with U–R>1. Fitting power laws to the individual SEDs we find a large range of slopes, from 0.4 to –1.7. The mean slope and dispersion are –0.39 and 0.38 respectively. Excluding a tail to steep indices, that makes the distribution not-gaussian, we find –0.21 and 0.34 for the mean and dispersion. The average slopes are very similar to the value of –0.32 obtained by Francis et al. (1991) for optically selected quasars.

Composite SEDs normalized at various wavelengths $\lambda_{\rm n}$ were constructed from the sample (excluding the two red quasars) and the overall shape is very similar for the different normalizations. A clear abrupt change occurs at around 3000 Å where the SED starts to steepen towards high frequen-

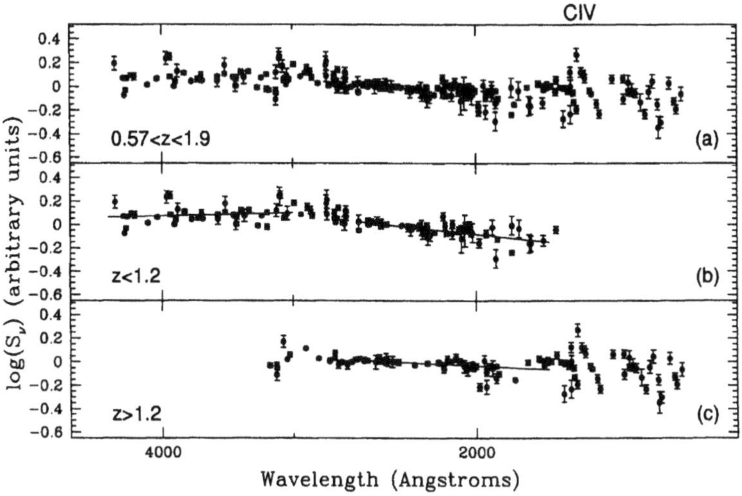

Figure 1. Composite SED obtained normalizing the individual SEDs at 2400 Å. The straight lines show the power law fits to the blue and UV ranges

cies. The only emission feature above the continuum revealed from the composites is the CIVλ1549 emission line. Parameterizing with power laws the composite SEDs at either side of the 3000 Å break we obtain an average $\langle\alpha_{\rm blue}\rangle$=0.11±0.16 for the range 4500–3000 Å and $\langle\alpha_{\rm UV}\rangle$=–0.66±0.15 for 2600–1700 Å. These values are in good agreement with those obtained by Natali et al. (1998) for optically selected quasars in similar redshift range, with α=0.15 for λ>3000 Å and α=–0.65 for λ<3000 Å. Therefore the average reddening for radio selected quasars is expected to be similar to that found for optically selected ones. Figure 1a shows the composite SED for $\lambda_{\rm n}$=2400 Å.

Separating the quasar sample in two redshift bins, with the cut at z=1.2, we find a better agreement between the values of $\alpha_{\rm blue}$ and $\alpha_{\rm UV}$ obtained for the different normalizations, and a hardening of $\alpha_{\rm UV}$ with redshift, with $\langle\alpha_{\rm UV}\rangle$=–0.87±0.20 for z<1.2 and $\langle\alpha_{\rm UV}\rangle$=–0.48±0.12 for z>1.2 (98 per cent significance). Figures 1b and 1c show the same composite as for Fig. 1a, for the low and the high redshift subsamples. A similar hardening at z>1 in this range was previously reported by O'Brien et al. (1988) for quasars with available UV(*IUE*) and X-ray data. The average spectral index $\alpha_{\rm blue}$ for the low redshift B3-VLA quasars is $\langle\alpha_{\rm blue}\rangle$=0.21±0.16, therefore these quasars show a steepening from the blue to the UV range $\langle\alpha_{\rm blue}\rangle$–$\langle\alpha_{\rm UV}\rangle$=1.08±0.26. The composite SEDs for high redshift quasars do not cover the low frequency range, therefore the presence of a break at 3000 Å cannot be analyzed for these quasars.

Separating the quasar sample in two bins of low and high luminosity at 2400 Å and using the composites for both subsamples, we also find a $\alpha_{\rm UV}$–L_{2400} trend (89 per cent significance), in the sense of a hardening of the slope

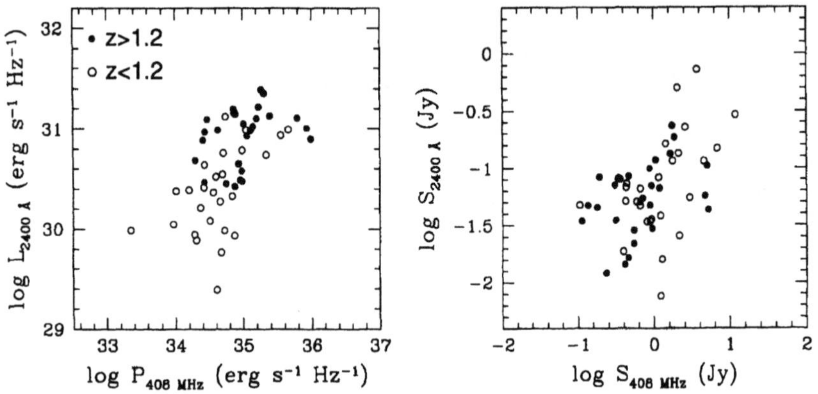

Figure 2. L_{2400} versus $P_{408\ \mathrm{MHz}}$ in luminosity-luminosity and flux-flux planes for 54 B3-VLA quasars.

with increasing luminosity. There is in addition a $L_{2400} - z$ trend, therefore the three parameters L_{2400}, α_{UV} and z appear correlated. The observed trends are consistent with the predictions from acretion disk models for the case of constant black hole mass and accretion rate increasing with redshift (Wandel & Petrosian 1988), since both α_{UV} and L_{2400} increase with $\dot{m}$ in these models. If the $L_{2400} - z$ correlation arises from a selection effect (from the flux-limits of the radio sample) only one of the correlations α_{UV}–z or α_{UV}–L_{2400} needs to be intrinsic.

Our data show an intrinsic correlation between blue/UV luminosity and radio power, confirmed in the flux-flux plane (see Figure 2). In fact, since the same optical flux limit was used in the identification of all quasars, the correlation in fluxes is most likely intrinsic. The fact that high and low redshift quasars show similar overall trends indicates that the correlations are not induced by independent evolution of the radio and optical luminosity with redshift. A linear fit to the $L_{2400} - P_{408}$ trend yields $L_{2400} \propto P_{408}^{0.52\pm0.10}$

References

Benn C.R., Vigotti M., Carballo R., González-Serrano J.I., Sánchez S.F., 1998, MNRAS, 295, 451

Carballo R., Sánchez S.F., González-Serrano J.I., Vigotti M., Benn C., 1998, AJ, 115, 1234

Francis P.J., Hewett P.C., Foltz C.B., Chaffee F.H., Weymann R.J., Morris S.L., 1991, ApJ, 373, 465

Natali F., Giallongo E., Cristiani S., La Franca F., 1998, AJ, 115, 397

O'Brien P.T., Gondhalekar P.M., Wilson R., 1988, MNRAS, 233, 801

Serjeant S., Rawlings S., 1996, Nature, 379, 304

Vigotti M., Vettolani G.V., Merighi R., Lahulla J.F., Pedani M., 1997, A&AS, 123, 1

Wandel A., Petrosian V., 1988, ApJ, 329, L11

Webster, R.L., Francis, P.J., Peterson, B.A., Drinkwater, M.J., & Masci F.J. 1995, Nature, 375, 469

PKS2250–41 AND THE ROLE OF JET CLOUD INTERACTIONS IN POWERFUL RADIO GALAXIES.

M. VILLAR–MARTÍN AND C. TADHUNTER
Dept. of Physics, University of Sheffield
Sheffield S3 7RH, UK

R. MORGANTI
Istituto di Radioastronomia
Via Gobetti 101, 40129 Bologna, Italy

D. AXON
Dept. of Physics and Astronomy, University of Manchester
Oxford Road, Manchester, UK

AND

A. KOEKEMOER
Space Telescope Science Institute
3700 San Martin Drive, Baltimore, MD21218, USA

Abstract. We have studied the effects of the interaction between the radio jet and the ambient gas in the powerful radio galaxy PKS2250–41 ($z = 0.31$). Our results show that the gas has been accelerated, compressed, heated and collisionally ionized by the shock. This study helps us to understand the processes which determine the observed properties of many high redshift radio galaxies.

1. Introduction

The observed properties of radio galaxies at $z > 0.7$ are very different to those at low redshift. The extended optical structures (line emission and continuum) are closely aligned with the radio axis and often highly collimated. Although several explanations are popular to explain this alignment effect [1,2], the interaction between the radio jet and the ambient gas is certainly playing a role. In spite of the much observational evidence for such interactions, not only in distant radio galaxies, but also in some closer radio galaxies and Seyferts, the physics involved is not well understood. We

Astrophysics and Space Science is the original source of publication of his article. It is recommended that this article is cited as: *Astrophysics and Space Science* **263**: 67–70, 1999.

are carrying out a program to study a sample of low-intermediate redshift ($z < 0.4$) radio galaxies undergoing jet-cloud interactions, in order to study in detail how the interaction happens and its consequences for the observed properties of the objects.

We present here results of the study of PKS2250–41, a radio galaxy in which a very strong interaction is taking place between the radio jet and an arc-like structure to the west of the main galaxy [3,4].

2. Observations

Optical spectra were obtained with the RGO spectrograph at the AAT (AAO). The slit was located at different positions parallel and perpendicular to the radio axis so that the whole arc (the region where the interaction is strongest) was covered. By fitting the emission lines, we have isolated the different kinematic components and have studied their kinematic, flux and physical properties. Our goal is to discriminate the dominant physical mechanism for each gaseous component.

3. Results

3.1. KINEMATICS AND IONIZATION LEVEL

The kinematics are complex across the arc. The [OIII] $\lambda\lambda$5007,4959 and Hβ lines split into several kinematic components for which FWHM, flux and [OIII]5007/Hβ were studied. The two main components are:

A *broad component* characterized by:

- morphological association with the radio lobe;
- large line widths (FWHM $\sim$ 500–900 km s^{-1});
- line width and flux peak at the position of the radio hot spot;
- low ionization level ([OIII]/Hβ $\sim$1–3).

A *narrow component* characterized by:

- its presence across the whole surface of the arc and far beyond the edge of the radio lobe;
- small line width (FWHM $\sim$ 60–200 km s^{-1});
- relatively high ionization level ([OIII]/Hβ $\sim$ 4–9).

3.2. ELECTRONIC TEMPERATURE

We have used the measured [OIII](λ5007/λ4363) ratios to calculate the electronic temperatures for the narrow and broad components [5]. The equivalent width of the [OIII]λ4363 line is large in the regions of interest and systematic effects related to the subtraction of the underlying stellar continuum are not a significant source of error.

We measure [OIII](λ5007/λ4363) $= 50 \pm 20$ for the narrow component and 13 ± 4 for the broad component, which yield $T_{\rm eff} = 15000^{+1300}_{-1000}$ and 41000^{+10000}_{-8000} respectively. Therefore, *the gas emitting the broad component is hotter.*

3.3. DENSITY

We have used the [SII]$\lambda\lambda$6716,6731 doublet to calculate the electronic densities for the narrow and broad components [5]. We have assumed the $T_{\rm eff}$ values calculated in the previous section. We measure [SII](λ6716/λ6731) $= 1.47 \pm 0.19$ for the narrow component and 1.13 ± 0.08 for the broad component, which yield $n_e < 120$ and $n_e = 570^{+260}_{-190}$ respectively. Therefore, *the gas emitting the broad component is denser.*

3.4. IONIZATION MECHANISM

Two main mechanisms can ionize the gas: 1) AGN (active nucleus) photoionization, 2) Shocks via collisional ionization of the gas and further radiative cooling and via the generation of a strong ionizing UV continuum. Which mechanism is responsible for the emission of the narrow and the broad components resolved in PKS2250–41?

The diagnostic diagram [OIII]λ5007/λ4363 *versus* HeII/Hβ provides a strong method to discriminate between AGN and shock models [3]. We have compared the measured values for the line ratios of the broad and narrow components with the model predictions (see Fig. 1). Notice the clear separation between the two kinematic components due to the low [OIII]λ5007/λ4363 (high $T_{\rm eff}$) and faintness of the HeII emission of the broad component. The AGN photoionization models predict too high [OIII] 5007/4363 ratios (i.e. low electron temperatures) and do not appear in the diagram. Adding matter bounded models can make the position of the narrow component consistent with AGN models [6]. However these cannot explain the line ratios of the broad component, which on the other hand is consistent with shock (radiative cooling) models. Therefore, the line ratios of the broad component suggest that it *represents collisionally ionized gas which is cooling behind a shock front.* The line ratios of the *narrow component suggest that it has been ionized by a strong UV continuum* (AGN and/or shock generated).

4. Discussion

Hydrodynamical models for an advancing jet in a gaseous medium show that the shocked gas is compressed and accelerated and the temperature raised. The gas will be collisionally ionized and will then cool radiatively.

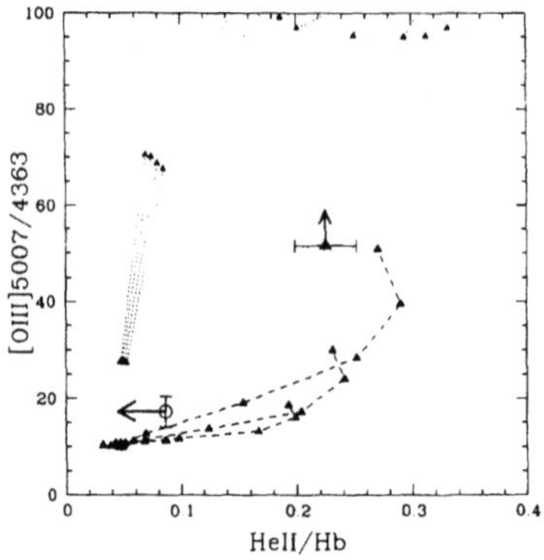

Figure 1. [OIII]λ5007/[OIII]λ4363 *versus* HeII/Hβ. The broad (open circle) and the narrow (solid triangle) components are clearly separated in this diagram. Dashed lines: shock models (shocked gas cooling radiatively). Dotted lines: Shock models (adding the contribution of ionization by the UV shock continuum).

All the properties of the broad component are consistent with it being shocked gas: Higher temperature and density imply heating and compression. Low ionization level implies compression and/or shock ionization. High velocities imply acceleration. Faint HeII and high temperature implies shock ionization. On the other hand, the narrow component is probably ambient gas whose properties can be explained in terms of pure gravitational motions and photoionization by a hard continuum source. This gas has not interacted with the radio jet.

5. Conclusions

We have studied the effects of the interactions between the radio jet and the ambient gas in PKS2250–41. The results provide clues to understanding the observed properties of many distant radio galaxies.

We have resolved kinematically the emission from the gas interacting with the radio jet and the emission from the non shocked ambient gas. The very different properties of these two gaseous components show the important effects of the interaction on the physical, morphological, ionization and kinematic properties of the gas.

References

1. Chambers et al., 1987, Nature, 329, 604
2. McCarthy et al., 1987, ApJ, 319, L39
3. Clark et al., 1997, MNRAS, 286, 558
4. Villar-Martín et al., 1999, MNRAS, submitted
5. Osterbrock, 1989, *The Astrophysics of Gaseous Nebulae and Active Galactic Nuclei*
6. Binette, L., et al., 1996, A&A, 312, 365

CHEMICAL EVOLUTION IN THE EARLY UNIVERSE: CAN DAMP LYα SYSTEMS BE SPIRAL GALAXIES?

ÁNGELES I. DÍAZ
Dpto de Física Teórica, C-XI, Universidad Autónoma de Madrid, Cantoblanco, 28049-Madrid, Spain

MERCEDES MOLLÁ
Département de Physique, Université Laval, Québec, Qc, G1K 7P4, Canada

AND

FEDERICO FERRINI
Dipartimento di Fisica, Università di Pisa, Piazza Torricelli 2, 56100 Pisa, Italia

Abstract. The distribution of element abundances with redshift in Damped Lyα (DLA) systems can be adequately reproduced by the same models reproducing the halo and disk components of the Milky Way Galaxy at different galactocentric distances. In this sense, DLA systems are well represented by normal spiral galaxies in their early evolutionary stages.

1. Introduction

There have been suggestions that the low ionization metal absorption lines in Damped Lyα Systems (DLAs) constitute direct evidence for disks (Wolfe 1995). In order to test this hypothesis – *i.e.* if DLAs show abundances consistent with those expected for high redshift spiral galaxies – we have compared the predictions of a specific model of the chemical evolution of the Galaxy at early times with recent observations of elemental abundances in DLAs.

We have used self-consistent chemical evolution models (Mollá, Ferrini & Díaz 1996, and references therein). These models reproduce the observational features obtained from the solar neighbourhood and the whole Galaxy structure studies.

Astrophysics and Space Science is the original source of publication of his article. It is recommended that this article is cited as: *Astrophysics and Space Science* **263**: 71–74, 1999.

The data have been taken from the recent works of Lu *et al.* (1997) and Pettini *et al.*(1997). We have assumed the efects of dust to be negligible. Dust seems to be present in some high redshift galaxies, but always in lower concentrations than in the Milky Way interstellar medium (Pettini *et al.* 1997). Galaxies at the highest redshifts (lower metallicities) are probably dust free, but abundances derived for high-metallicity (lower redshift) systems might have been underestimated.

To compare directly model results with observational data we have assumed values of q_o and H_o of 0.1 and 50 km s^{-1} Mps^{-1} respectively. The redshift z corresponds to the redshift of Damped Lyα Systems.

2. The Chemical Evolution Model

For our comparison we have used the evolutionary history of the Milky Way, taken to be a typical spiral galaxy. We assume that the primordial gas of the spheroidal protogalaxy collapses to form the halo. Star formation there proceeds providing the recycled gas out of which the disk forms at a rate inversely proportional to the collapse time, an input parameter of the model. The early galactic sphere is divided into concentric annular cylindrical regions 1 kpc wide, perpendicular to the disk. The numerical code calculates the time evolution of the halo and disk components of every cylindrical region.

Star formation is a two-step process: first the diffuse gas forms molecular clouds, and then stars form by cloud-cloud collisions or by the interaction of massive stars with clouds. The rate of occurrence of each ot these processes is a function of the galactocentric radius. They are proportional to the corresponding efficiencies representative of the probabilities of cloud formation, cloud-cloud collisions, and massive star-cloud interaction. The two first vary with the galaxy morphological type since gravitation through the spiral density wave changes the conditions of star formation in each subgalactic region. However, the interaction of molecular clouds with nearby massive stars seems to be a local process and its corresponding efficiency can be taken as constant for every zone and galaxy.

We have taken the multiphase model results from Ferrini *et al.* (1994) for the Milky Way Galaxy, where we computed the element distributions in the halo and in the disk by varying the radial galactocentric distance of every considered region. We have extracted the outputs for the regions centered on 8 kpc (*solar*) and on 5 kpc and 14 kpc (here-on called *inner* and *outer* zones) from the galactic center. Assuming that the absorbing gas in the line of sight to the Damped Lyα Systems may belong to both halo and disk components, we evaluate the abundance by averaring the abundances for the two populations with their corresponding weights.

3. Results and Discussion

The distribution of [Fe/H] as a function of redshift is shown in the left panel of Fig 1. It can be represented by two bands roughly centered at [Fe/H] $\simeq$ -2.0 and [Fe/H] $\simeq$ -1.0. The more metallic systems seem to be limited to $z < 3$ ([Fe/H] $\geq$ -2.0), while the less metallic ones extend to lower redshifts. This suggests the possibility that the gas in DLAs belongs to early galactic systems with two levels of enrichment and it could be explained if two different star formation episodes take place: an initial one providing a certain level of enrichement that would account for the low metallicities encountered at high redshifts, and a second one, starting from an already enriched gas, that would explain the higher metallicities observed in some of the systems.

In our models this sort of bimodal abundance distribution arises naturally since there is a delay of a few gigayears between the onset of star formation in the halo and the disk. Indeed, the beginning of the star formation in the disk is not inmediate: the system needs to form clouds as a step prior to star formation which produces a natural delay. It can be seen (Ferrini, Mollá & Díaz 1997) that the evolution of the halo population reproduces the lower metallicities observed while the evolution of disk regions well contain the higher metallicity data. Furthermore, in the two cases the evolution has been faster in the central regions than in the outer ones and this difference in the evolution of the different galactic regions can explain the spread in the observations.

In left panel of Fig. 1 we show the model results for the average abundance – halo and disk populations – as a function of redshift for the three defined regions: inner, solar and outer. The inner and solar region evolutions represent better the higher abundance data points, while the outer region evolution fits the lower observed abundances. Again, the history of the different regions can reproduce the spread in the observations.

We should also take into account that the observations possibly average, not only on the two populations – halo and disk – at a given radius, but over a large part of the galaxy structure, thus including regions at different radii: the dashed line in the figure shows our average for both disk and halo component at all galactic radii.

In the right panel we show the neutral hydrogen column density as a function of redshift which also has contributions of the gas present both in the spheroidal and disk components of the galaxy. The column density is sensitive to the inclination angle of the disk respect to the line of sight: the two solid lines represent the evolution for a face-on galaxy (lower line) and for an edge-on galaxy (upper line). The distribution of observed values is bound by these simple estimates.

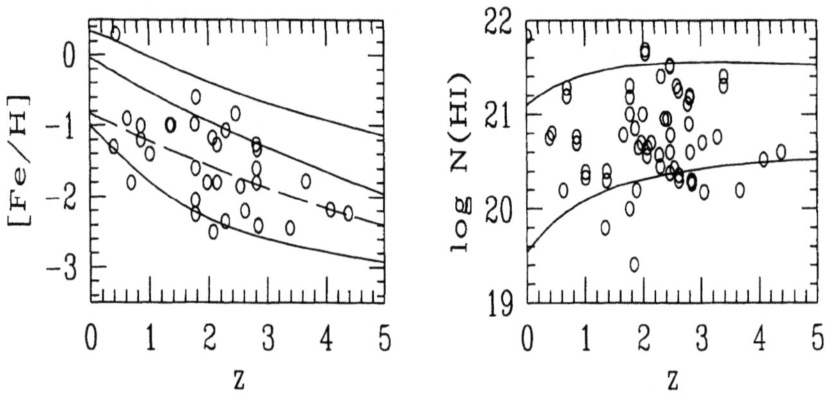

Figure 1. Left) [Fe/H] *versus* redshift. Solid lines are the results for the three defined regions. Right) Neutral gas column density for face–on and edge–on Milky Way twins. Circles are data from Lu *et al.* (1997) and Pettini *et al.* (1997)

4. Conclusions

The possibility that Damped Lyα Systems be indeed normal spiral galaxies (Wolfe 1995) must be considered: observations probably refer to gas belonging to different components and therefore the observed abundances are averaged over the halo and disk gas, if not over the whole galactic content. We confirm that the large metallicity spread observed in Damp Lyα systems can be explained by the heterogeneous mixture of galactic regions and galaxy types.

References

Ferrini, F., Mollá, M., Pardi, M.C., Díaz, A.I., 1994, ApJ 427, 745
Ferrini, F., Mollá, M. & Díaz, A.I. 1997, ApJ 487, L29
Lu, L., Sargent, W.L.W., Barlow, T.A., Churchill, C. W., Vogt, S. S., 1996, ApJS 107, 475
Mollá, M., Ferrini, F., Díaz, A.I., 1996, ApJ 466, 668
Pettini, M., King, D.L., Smith, L. J., Hunstead, R. W., 1997, ApJ 478, 536
Wolfe, A. M., 1995, in *QSO Absorption Lines*, p. 13, eds. G. Meylan (Berlin: Springer).

QSO LYMAN-ALPHA ABSORBERS AND GALAXY HALOS

X. BARCONS
Instituto de Física de Cantabria (CSIC-UC)
Facultad de Ciencias, 39005 Santander, Spain

K.M. LANZETTA, H.-W. CHEN AND N. YAHATA
Department of Physics and Astronomy
SUNY, Stony Brook, NY 11794-3800, USA

AND

J.K. WEBB, A. FERNÁNDEZ-SOTO AND A. ORTIZ-GIL
School of Physics, University of New South Wales,
Sydney, NSW-2052, Australia

Abstract. We review the current status of our survey of galaxies in fields of HST target QSOs, which has allowed us to identify the galaxies responsible for a number of Lyman-α absorption systems. We emphasize the use of QSO absorption lines to study the structure and kinematics of the large gaseous halos that virtually all galaxies appear to possess.

1. Introduction

A major goal in recent studies of Lyman-α absorption lines seen towards QSOs has been to establish a link between the absorbers and galaxies, particularly at low redshifts. While all surveys agree that Lyman-α absorbers follow the large-scale structure of the Universe as traced by galaxies (Morris et al 1993, Le Brun, Bergeron & Boissé 1996, Lanzetta et al 1995, Bowen, Blades & Pettini 1996), the issue on whether Lyman-α absorbers *arise* in galaxy halos has been extensively debated. These pioneering studies relied upon small numbers of galaxies identified and in many cases the absorber-galaxy associations were done in incomplete galaxy samples. The existence of an anti-correlation between Lyman-α absorption equivalent width (W) and projected distance from the galaxy centre to the QSO line of sight appears to be a crucial test for a physical link between absorbers and galaxies.

Astrophysics and Space Science is the original source of publication of his article. It is recommended that this article is cited as: *Astrophysics and Space Science* **263**: 75–78, 1999.

Different surveys found different results on this point. The compilation of a large number of absorber-galaxy associations that we are conducting in order to assess the reality of the physical link between galaxies and Lyman-α absorbers will also provide a tool to study the structure and kinematics of galaxy halos where the gaseous component is so tenuous that it can only be seen in absorption but not in emission.

2. Current status of the survey

We are surveying 24 fields centered on QSOs which have been spectroscopically observed in the UV by the Faint Object Spectrograph on the HST. Analysis of these spectra yield a total of 230 Lyman-α absorbers at a median redshift of $z \sim 0.35$.

The survey consists of deep imaging followed by multi-slit spectroscopy of galaxies in a field of variable size (from 5 to 10 arcmin on a side) around each QSO. A Principal Component Analysis of the spectra has been performed (Yahata et al 1998) to identify the galaxies and find their redshifts. The survey is by no means complete and it contains 352 galaxy redshifts.

Excluding Lyman-α absorbers falling within 3000 km s^{-1} from the QSO redshift (to avoid the proximity effect), we have found a total of 50 absorber-galaxy pairs. The definition of these pairs is based on a highly reliable method which finds those pairs in the Δv (velocity difference between absorber and galaxy)- ρ (galaxy impact parameter or projected distance to the line-of-sight to the QSO) plane which lie in the region where the absorber-galaxy cross-correlation function exceeds 1. This ensures a minimal number of spurious pairs and also ensures that a small number of real pairs will be missed. Most of these galaxies have been observed with the WFPC2 on HST.

In Chen et al (1998) the most recent compilation of the survey is presented together with a detailed analysis of the W vs ρ anticorrelation. This is seen to be highly significant thus confirming the physical link between galaxies and absorbers, rather than a loose relationship through the large-scale distribution of matter in the Universe. But furthermore, Chen et al (1998) show that the galaxy luminosity also matters, in the sense that the halo sizes appear to scale as $R \propto L^{0.4}$. For an $L_\star$ galaxy the halo size for Lyman-α absorption exceeds $100h^{-1}$ kpc. No other galaxy properties of the galaxies (spectral type, morphology, surface brightness, inclination etc.) appear to be correlated with absorption at the sensitivity of the sample. This supports a picture where Lyman-α absorption arises in extended gaseous halos of the galaxies, the halo size depending basically on galaxy luminosity (most probably on galaxy mass), but not obviously on other properties like star formation history.

3. Groups and clusters of absorbers

A particularly interesting result has emerged from this survey, namely the discovery of groups or clusters of Lyman-α absorption lines at low redshift associated with groups or clusters of galaxies at the same redshift. Lanzetta, Webb & Barcons (1996) first reported on a resolved Lyman-α absorption feature in the FOS spectrum of the QSO 1545+2101 at a redshift $z \approx 0.26$, coincidental with the redshift centroid of a group of 7 galaxies with distances to the line of sight from 7 to $300\,h^{-1}$kpc. Further GHRS studies of this object show that the feature splits in, at least, 8 distinct Lyman-α absorption lines spanning a velocity range $\sim 150\,\mathrm{km\,s^{-1}}$ similar to that spanned by the redshifts of the galaxies identified in the group or cluster (Ortiz-Gil et al 1998).

Shull et al (1998) also find a cluster of Lyman-α absorption lines at a median redshift of $z \sim 0.057$ towards the QSO PKS 2155-304. They identify a group of 5 HI galaxies close to the line of sight spanning a similar redshift range, but with large impact parameters ($500-1000\,h^{-1}$kpc). This probably implies a loose association between absorbers and galaxies in this particular case.

It is not clear whether the multiple Lyman-α absorptions seen in these groups or clusters are produced by gas directly associated to the galaxies themselves in such dense environments or by absorbing clouds distributed in the group or cluster. The fact that the most likely galaxy-absorber associations follow the general W vs ρ anticorrelation (Ortiz-Gil et al 1998), and that these absorbers are in no way different to 'field' Lyman-α absorbers suggest a common origin for both.

4. The structure of galaxy halos

One of the most interesting potential uses of the association between Lyman-α absorbers and galaxies is the use of absorption towards background QSOs to study the structure and kinematics of the gaseous envelopes of galaxies, where the gas is too tenuous to be seen in emission. Lyman-α absorption can be detected down to very low column densities ($\sim 10^{13}\,\mathrm{cm^{-2}}$) if a bright enough background QSO can be found close enough to the galaxy under study.

The structure of galaxy halos can be traced with Lyman-α absorbers if spatially resolved moderate-resolution spectroscopy of the absorbing galaxy is obtained. A handful of such objects have been published in the literature so far: 2 in Barcons, Lanzetta & Webb (1995), another one in Lanzetta et al (1997) and a further 2 from Van Gorkom et al (1996). The first 3 use the Hα emission line to trace the kinematics while Van Gorkom et al (1996) use the 21cm line. A further object studied by Van Gorkom et al (the source

F21569-330) is probably a misidentification since it falls $610\,h^{-1}$kpc apart from the line of sight to the corresponding QSO. In all the other 5 cases, the Lyman-α absorption redshift falls within the redshift range spanned by the rotation of the corresponding galaxy. This reinforces the arguments in favour of a physical link between absorbers and galaxies. Initial claims that the Lyman-α absorbing gas might be co-rotating with the galaxy disk (Barcons, Lanzetta & Webb 1995) are not to be considered general, since the extremely metal-poor damped Lyman-α absorber towards the QSO 0850+4400 found at $z = 0.163$ appears to counter rotate with respect to the motion of the corresponding galaxy at $\sim 16\,h^{-1}$kpc.

A further study is being carried out in galaxy absorber pairs where the moderate to high resolution spectroscopy of the galaxy cannot be spatially resolved. Precise galactocentric redshifts can then be compared with precise absorption redshifts sampling different impact parameters. The velocity difference between both Δv can then be sampled for separations from from $\sim 10\,h^{-1}$kpc to over $\sim 200\,h^{-1}$kpc. Preliminary results show that the mean square root velocity separation between galaxies and the corresponding absorbers is of the order of $\sim 100\,\mathrm{km\,s^{-1}}$, a factor of 2 smaller than for galaxy pairs. Although this is based on a small number of pairs it might be indicative of a closer link between galaxies and absorbers than what is derived from the large-scale distribution of galaxies. Scenarios in which the Lyman-α absorbing clouds are in hydrostatic equilibrium in the galaxy halos would be favoured if this result is confirmed.

Acknowledgments

XB acknowledges financial support by the DGES under project PB95-0122.

References

Barcons, X., Lanzetta, K.M., Webb, J.K., 1995, Nature, 376, 321
Bowen, D.V., Blades, J.C., Pettini, M., 1996, ApJ, 464, 141
Chen, H.-W., Lanzetta, K.M., Webb, J.K., Barcons, X., 1998, ApJ, 498, 77
Lanzetta, K.M., Bowen, D.V., Tytler, D., Webb, J.K., 1995, ApJ, 442, 538
Lanzetta, K.M., Webb, J.K., Barcons, X., 1996, ApJ, 456, L17
Lanzetta, K.M. et al, 1997, AJ, 114, 1337
Le Brun, V., Bergeron, J., Boissé, P., 1996, A& A, 306, 691
Morris, S.L. et al, 1993, ApJ,419, 524
Ortiz-Gil, A., Lanzetta, K.M., Webb, J.K., Barcons, X., Fernández-Soto, A., 1998, ApJ, submitted
Shull, J.M. et al, 1998, AJ, in the press (astro-ph/9807246)
Van Gorkom, J., Carilli, C.L., Stocke, J.T., Perlman, E.S., Shull, J.M., 1996, AJ, 112, 1397
Yahata, N. et al (1998), in preparation

ON THE NATURE OF DAMPED LYα SYSTEMS: CLUES FROM DETERMINATIONS OF ELEMENTAL ABUNDANCE RATIOS

M. CENTURIÓN
Instituto de Astrofísica de Canarias
38200 La Laguna, Tenerife, Spain

AND

P. BONIFACIO, P. MOLARO AND G. VLADILO
Osservatorio Astronomico di Trieste
Via G.B. Tiepolo 11, 34100 Trieste, Italy

Abstract. We present a study of the [α/Fe-peak] and [N/α] abundance ratios in Damped Lyα (DLA) systems. By using ratios of undepleted elements ([S/Zn] and [N/S]) or when abundances of refractory elements are corrected by dust depletion ([Si/Fe]$_{corr}$), the resulting ratios do not resemble those observed in the Galactic metal-poor stars, showing instead similarities with those ratios observed in dwarf galaxies. These results challenge the idea that these absorbers are the progenitors of the present-day spiral galaxies, and suggest an origin in low-mass galaxies for the systems up to now investigated.

1. Introduction

DLA systems are complexes of neutral gas detectable by a damped Lyα absorption in the spectra of QSOs. They have low metallicities (down to $\simeq$ 0.1% solar) and high redshift (up to z $\simeq$ 4) and therefore DLAs offer a unique tool for studying early stages of chemical evolution. The exact nature of the DLAs has yet to be clarified even though it has always seemed likely that DLAs are the progenitors of the present-day spiral galaxies. Imaging in the field of the background quasars indicate that bright spiral galaxies do not dominate the DLA population (Le Brun et al. 1997). Studies of abundances in DLA absorbers give important clues for understanding the origin of these QSO intervening systems. The precision obtained in DLA's

Astrophysics and Space Science is the original source of publication of his article. It is recommended that this article is cited as: *Astrophysics and Space Science* **263**: 79–82, 1999.

abundance determinations is remarkable and comparable to that attained in the galactic ISM. However, the interpretation of the observed abundance ratios has led to contradictory conclusions. From one side, claims have been made that DLAs have halo-like abundance pattern, associating this absorbers with intervening galactic disks (Lu et al. 1996, LSBCV; Prochaska & Wolfe 1997). From the other side, Molaro et al. (1996,1998; MCV98) have claimed that the abundance ratio of α-elements over iron-peak elements are about solar in DLAs, [α/Fe-peak] ~ 0, in contrast with what is observed in the Milky Way at comparable look-back time. The controversy on the nature of DLAs has in part arisen because depletions of refractory elements onto dust have been commonly considered negligible, hiding the intrinsic abundance pattern of these galaxies as it has been stressed by Vladilo (1998). Recently Pettini et al. (1999; PESB) reached the same conclusion – advanced by Molaro and collaborators – after correcting by dust the [Si/Fe] ratios observed in 3 DLAs. Evidences for dust in DLA galaxies come from: reddening studies of QSOs with and without foreground DLA absorbers (Pei et al. 1991), the relative abundances of Zn to Cr in DLAs (Pettini et al. 1997), and the detection of the 2175 Å extinction bump in high redshift MgII absorbers (Malhotra 1997). The DLA's abundance ratios presented here of undepleted elements and of refractory elements corrected by dust depletions, indicate that DLAs have undergone a different chemical enrichment to that experience by our Galaxy.

2. The abundance ratio of α over iron-peak elements in DLAs

In the course of the Milky Way evolution the abundance ratio of α over iron-peak elements have decreased from [α/Fe]$\simeq +0.6$ typical of metal-poor stars with [Fe/H] $\simeq -2$ (Israelian et al. 1998) down to the solar value, owing to the time delay between the explosions of α-rich type II supernovae and Fe-rich type Ia supernovae. If DLA galaxies are protospirals it is expected to observe the characteristic α-enhancement observed in halo stars formed at comparable look-back time. The α/Fe ratio most commonly studied in DLAs is [Si/Fe] and is found to be consistent with an halo-like pattern (LSBCV; Prochaska & Wolfe 1999, PW). However, in all 3 DLAs (see Table 1: Q0528–25, Meyer & Roth 1990; Q1331+17, Kulkarni et al. 1996; PHL957, MCV98) where a dust free diagnostic such as [S/Zn] is used, the [α/iron-peak] ratios are found solar while [Si/Fe] ratios emulate the halo-like abundance pattern. We interpret that this is caused by differential dust depletion, which also explains the enhanced [Si/Fe] ratios observed in the nearby ISM. If [Fe/Zn]$\simeq 0$ in DLAs – as it is observed in galactic stars of all metallicities (Sneden et al. 1991) – and being Zn almost undepleted in the galactic ISM (Roth & Blades 1995) also the differences between [Fe/H] and [Zn/H] shown in Table 1 indicate that the missing Fe is locked up in dust grains. Moreover, once the observed [Si/Fe] ratios (LSBCV, PW, MCV98, PESB) are corrected by depletion following Vladilo (1998), the [α/iron-peak] ratios obtained are solar (see Fig. 1a). These results suggest that DLA galaxies have a chemical enrichment history different from that of the Milky Way and do not conform the notion that DLA absorbers are the progenitors of the present-day spiral galaxies. [α/iron-peak] $\simeq$ 0 at low metallicities are observed or/and predicted to exist by chemical evolution models when the star formation proceeds in bursts separated by quiescent periods, as it happens in dwarfs galaxies (Matteucci et al. 1997) and when star formation is not as fast as in our Galaxy as it happens in LSB galaxies and in the outer regions of disks (Jiménez et al. 1999).

TABLE 1. [Si/Fe] and [S/Zn] ratios in DLA galaxies

QSO	z_{abs}	[Fe/H]	[Zn/H]	[Si/Fe]	[S/Zn]
PHL 957	2.31	−2.02	−1.55	+0.66	−0.05
0528-2505	2.78	−1.26	−0.76	+0.51	−0.12
1331+1700	1.78	−2.08	−1.22	+0.65	−0.08

3. The abundance ratio of nitrogen over α-elements

Nitrogen is also a key element to understand the early stages of chemical evolution, even though its production is not fully understood and deviates from the secondary behaviour expected from simple evolutionary models, and just observed at high metallicities. In our Galaxy the [N/O] ratio in metal-poor stars is about constant for [O/H] $\leq$ -1 (Tomkin & Lambert 1984; dashed line in Fig. 1b), indicating a primary nitrogen production at low metallicities. In DLA galaxies [N/α] ratios show a large spread not observed in metal-poor stars (see Fig. 1b: large empty symbols from Pettini et al. 1995; Lu et al. 1998 and filled symbols from Centurión et al. 1998). This scatter cannot be attributed to dust depletion since we have used [N/S] ratios for DLAs and these two elements are expected to be undepleted as it is observed in the Galactic ISM (Meyer et al. 1997, Savage & Sembach 1996). The [N/α] ratios in DLAs recall the behaviour in dwarf irregular galaxies (dots in Fig. 1b: from Kobulnicky & Skillman 1996; van Zee et al. 1996). Although a few [N/α] values are too high or to low to be explained by conventional models of chemical evolution (Centurión et al. 1998), in general they can be ascribed to the delayed release of nitrogen in the course of evolution when the star formation proceeds in bursts separated by quiescent periods (Garnett 1990), suggesting an origin in low-mass galaxies for the intervening DLA absorbers.

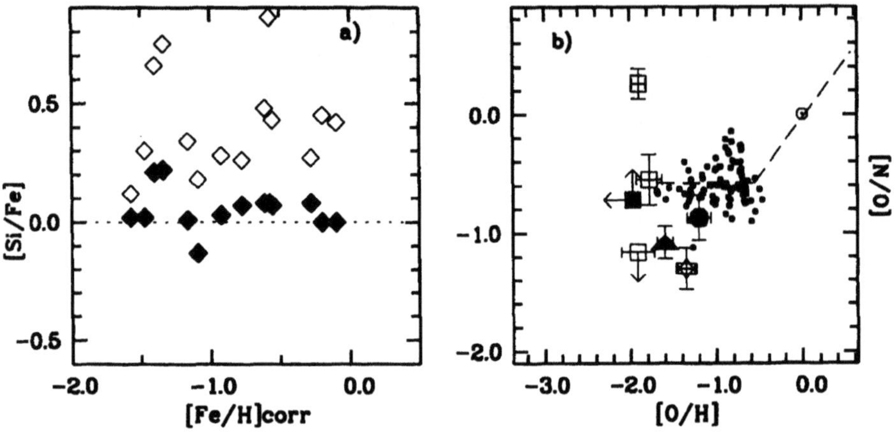

Figure 1. Abundance ratios in DLA galaxies: a) Observed [Si/Fe] in DLAs (empty symbols) and after correction for dust depletion (filled symbols) following Vladilo (1998). b) [N/α] in DLAs (large symbols) and other astrophysical sites of low metallicities: dwarf irregular galaxies (dots) and Galactic halo stars (horizontal dashed line).

References

Centurión, M., Bonifacio, P., Molaro, P., Vladilo, G., 1998, ApJ, 509, 620
Garnett, D.R., 1990, ApJ, 363, 142
Israelian, G., García, R., Rebolo, R., 1998, ApJ, 507, 805
Jiménez, R., Bowen, D.V., Matteucci, F., 1999, ApJ, in press (astro-ph/9812040)
Kobulnicky, H.A., Skillman, E.D., 1996, ApJ, 471, 211
Kulkarni, V., Huang, K., Green, R., Bechtold, J., Welty, D., York, D., 1996, MNRAS, 279, 197
Le Brun, V., Bergeron, J., Boisse, P., Deharveng, J.M., 1997, A&A, 321, 733
Lu, L., Sargent, W.L.W., Barlow, T.A., Churchill, C.W., Vogt S., 1996, ApJS, 107, 475
Lu, L., Sargent, W.L.W., Barlow, T.A., 1998, ApJ, 115, 55
Malhotra, S., 1997, ApJ, 488, L101
Matteucci, F., Molaro, P., Vladilo, G., 1997, A&A, 321, 45
Meyer, D.M., Cardelli, J.A., Sofia, U.J., 1997, ApJ, 490, L103
Meyer, D.M., Roth, K.C., 1990, ApJ, 363, 57
Molaro, P., Centurión, M., Vladilo, G., 1998, MNRAS, 293, L37
Molaro, P., D'Odorico, S., Fontana, A., Savaglio, S., Vladilo, G., 1996, A&A, 308, 1
Pei, Y.C., Fall, M., Bechtold, J., 1991, ApJ, 348, 6
Pettini, M., Lipman, K., Hunstead, R.W., 1995, ApJ, 451, 100
Pettini, M., Ellison, S.L, Steidel, C.C., Bowen, D.V., ApJ, 1999, in press (astro-ph/9808017)
Pettini, M., King, D.L., Smith, L.J., Hunstead, R.W., 1997, ApJ, 478, 536
Prochaska, J.X., Wolfe, A.M., 1999, ApJ, in press, (astro-ph9810381)
Prochaska, J.X., Wolfe, A.M., 1997, ApJ, 474,140
Roth, K.C., Blades, J.C., 1995, ApJ, 445, L95
Savage, B.D., Sembach, K.R., 1996, ARA&A, 34, 279
Sneden, C., Gratton, R., Crocker, D.A., 1991, A&A, 246, 354
Tomkin, J., Lambert, D.L., 1984, ApJ, 279, 220
van Zee, L., Haynes, P.M., Salzer, J.J., Broeils, A., 1996, AJ, 112, 129
Vladilo, G., 1998, ApJ, 493, 583

THE ROLE OF COOLING FLOWS IN THE STAR FORMATION HISTORY OF CENTRAL CLUSTER GALAXIES

N. CARDIEL AND J. GORGAS
Departamento de Astrofísica
Universidad Complutense de Madrid, 28040-Madrid, Spain

AND

A. ARAGÓN-SALAMANCA
Institute of Astronomy
Madingley Road, Cambridge CB3 0HA, UK

Abstract. Using a recent determination of the X-ray luminosity function of galaxy clusters, accurate mass deposition rates and L_X in cooling flow clusters from the literature, and reliable constraints on the star fomation episodes in cooling flows obtained by our group, we quantify the relevance of such star formation in comparison with the SFR in the local universe. We conclude that the SFR density in CF is $\lesssim 0.1\%$ the total local reference[1].

1. Cooling Flows and Central Cluster Galaxies

A large fraction ($\sim$ 70–90%) of the X-ray detected galaxy clusters exhibit centrally concentrated X-ray emission from hot gas (Edge et al. 1992; Peres et al. 1998). This emission, which constitutes an important energy loss, forces the gas to increase its density, accelerating the emission rate. At the center of these clusters, the characteristic cooling time is shorter than the age of the cluster, and the gas, pushed inwards by the hotter intracluster medium, may settle towards the central cluster galaxy (CCG), giving rise to what is usually known as a cooling flow (CF). A more realistic scenario involves a multiphase atmosphere, which cools over a wide range of radii, depositing cooled material throughout the inner $\sim$ 100–200 kpc of the cluster, with typical total mass deposition rates $\dot{M} \sim$ 10–1000 $M_{\odot}$ yr^{-1} (see review in Fabian 1994).

[1]We assume throughout that $H_0 = 50$ km s^{-1} Mpc^{-1}.

Astrophysics and Space Science is the original source of publication of his article. It is recommended that this article is cited as: *Astrophysics and Space Science* **263**: 83–86, 1999.

2. Star Formation in Cooling Flows

Although accurate X-ray data show that CF's are highly inhomogeneous (in fact, mass deposition profiles are well fitted by $\dot{M}(< r) \propto r$; Thomas et al. 1987; Peres et al. 1998), it is likely that at least a fraction of the cooled gas could accumulate as molecular clouds at the center of CCG's, forming new stars. The search for this star formation has led many authors to seek spectral evidences in the central regions of CCG's (see short review in McNamara 1997). The pionering works of Johnstone et al. (1987) and Gorgas et al. (1990) established the suitability of the analysis of absorption line-strength features to infer the stellar population properties of CCG's inmersed in CF's. In this sense, we have conducted a long-term project devoted to obtain high-quality line-strength gradients in a relatively large sample of CCG's, inmmersed in clusters with and without CF (Cardiel et al. 1995, 1998). We are not repeating here a full description of the results derived from this study (we refer the interested readers to the original papers), although we want to highlight that the analysis of the radial variation of the λ4000Å break and the Mg_2 index reveals that a large fraction, *probably most*, of the gas accreted in the central $\sim$ 20 kpc of CCG's is converted into stars. This fraction typically represents $\sim$ 10% of the total X-ray derived mass deposition rate for the CF clusters.

3. The relevance of Star Formation in Cooling Flows

Using the Hα luminosity as an estimator of the star formation rate (SFR), Gallego et al. (1995) derived the SFR density of the local universe to be 0.013 $M_\odot$ yr^{-1} Mpc^{-3}. An important question is whether the SFR associated to CF's can be relevant compared to the mentioned value.

Three results allow us to obtain such estimation. First, there is a correlation between the X-ray derived mass deposition rate $\dot{M}$ and the X-ray luminosity, for clusters with $\dot{M} > 0$ $M_\odot$ yr^{-1} (e.g. White et al. 1997; Peres et al. 1998). In Figure 1 we have represented the $\dot{M}$ values derived by Peres et al. (1998) in a X-ray flux-limited sample containing the brightest clusters over the sky in the 2–10 keV band, versus the X-ray luminosity of the galaxy clusters in the same band published by David et al. (1993). We have modelled this relationship by fitting a straight line to the clusters with $\dot{M} > 0$ $M_\odot$ yr^{-1} (see figure caption). Thus, the derived expression is an overestimation of the mass deposition rate, since we have left out clusters without CF. The result of the fit is $\log \dot{M}(L_X) = 1.78 + 0.69 \log L_X$, with $\dot{M}$ and L_X in units of $M_\odot$ yr^{-1} and 10^{44} erg s^{-1}, respectively.

Additionally, Ebeling et al. (1997) have recently published the X-ray luminosity function (XLF) of the *ROSAT* Brightest Cluster sample. In the 2–10 keV band, the XLF is well represented by a Schechter function

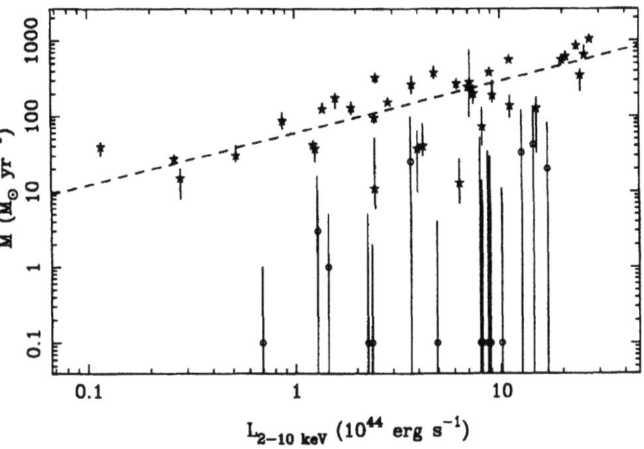

Figure 1. Relation between the mass deposition rate and the X-ray luminosity of galaxy clusters in the 2–10 keV band (this figure is similar to Fig. 2 in Peres et al. 1998). Open circles correspond to clusters with $\dot{M}$ consistent with zero (considering the error bars). Clusters with $\dot{M} = 0$ M$_\odot$ yr^{-1} are plotted with a value of 0.1 only for display purposes. The dashed line is the error weighted least-squares fit of a straight line to the clusters with $\dot{M}$ significantly larger than zero.

(Schechter 1976) of the form $\phi(L_{\rm X})\ {\rm d}L_{\rm X} = A\ \exp[-L_{\rm X}/L_{\rm X}^{\star}]\ L_{\rm X}^{-\alpha}\ {\rm d}L_{\rm X}$, where $\phi(L_{\rm X})\ {\rm d}L_{\rm X}$ is the number density of galaxy clusters with luminosities in the range $L_{\rm X}$ and $L_{\rm X}+{\rm d}L_{\rm X}$, $A = 2.35 \times 10^{-7}$ Mpc^{-3} (10^{44} erg s^{-1})$^{\alpha-1}$, $L_{\rm X}^{\star} = 12.6 \times 10^{44}$ erg s^{-1}, and $\alpha = 1.54$.

The mass deposition rate per unit volume due to cooling flows, $\dot{\mu}$ (in M$_\odot$ yr^{-1} Mpc^{-3}), can be obtained through the integration of the XLF, multiplied by the relation fitted in Fig. 1, i.e.

$$\dot{\mu} = \int_{L_{\rm Xmin}}^{L_{\rm Xmax}} \phi(L_{\rm X})\ \dot{M}(L_{\rm X})\ {\rm d}L_{\rm X} = 10^{1.78}\ A\ [L_{\rm X}^{\star}]^{1.69-\alpha} \int_{t_{\rm min}}^{t_{\rm max}} {\rm e}^{-t}\ t^{0.69-\alpha}\ {\rm d}t,$$

with $t \equiv L_{\rm X}/L_{\rm X}^{\star}$. For $L_{\rm Xmin} = 0$ and $L_{\rm Xmax} = \infty$, the last integral is $\Gamma(1.69-\alpha)$ and $\dot{\mu} \approx 1.3\ \times 10^{-4}$ M$_\odot$ yr^{-1} Mpc^{-3} (note that using for $L_{\rm Xmin}$ and $L_{\rm Xmax}$ the limits exhibit by the data in Fig. 1 reduces the integral by a factor ~ 2).

If we define $f_{\dot{M}}$ as the fraction of the total $\dot{M}$ which is transformed into stars, and ϵ as the efficiency of the conversion of cooled gas into stars, we obtain that, for the local universe,

$$\frac{\rho_{\rm SFR(cooling\ flow)}}{\rho_{\rm SFR(H\alpha)}} = \frac{\dot{\mu}\ f_{\dot{M}}\ \epsilon}{\rho_{\rm SFR(H\alpha)}} \approx \frac{1.3 \times 10^{-4}\ f_{\dot{M}}\ \epsilon}{1.3 \times 10^{-2}} = 0.01\ f_{\dot{M}}\ \epsilon.$$

Thus, even assuming $\epsilon = 1$, and considering that we have neglected the contribution of clusters with $\dot{M} = 0$ M$_\odot$ yr^{-1} and extrapolated the fit in

Fig. 1 from 0 to ∞, the rate of stellar formation due to CF's is $\lesssim 1\%$ of the SFR in the local universe. In addition, taking into account the result mentioned in Section 2, a reliable $f_{\dot{M}}$ value is ~ 0.1, which lowers the previous estimate by one order of magnitude.

At the light of this result, the relevance of CF's in the star formation history of the universe at present time is small ($\lesssim 0.1\%$). However, it is important to note that, using the [O II] λ3727 emission line luminosity in cluster and field galaxies, Balogh et al. (1998) conclude that the mean SFR of cluster galaxies brighter than $M_r = -19.0$ is found to vary from 0.68 ± 0.08 $M_\odot$ yr^{-1} (at 3–4 Mpc from the cluster center) to 0.00 ± 0.04 $M_\odot$ yr^{-1} (at the cluster center, excluding CCG's), whereas this value is 1.56 ± 0.04 $M_\odot$ yr^{-1} for galaxies in the field. These numbers reveal that, although the impact of CF's in the star formation history of the local universe is small (due to scarcity of CCG's inmersed in CF's in comparison with normal galaxies), CCG's are hosting very important episodes of star formation ($\sim$ 1–100 $M_\odot$ yr^{-1}) in comparison with normal galaxies.

Acknowledgments

We are indebted to J. Gallego for valuable discussions. This work was supported by the Spanish 'Programa Sectorial de Promoción del Conocimiento' under grant No. PB96-610. AAS acknowledges generous financial support from the Royal Society.

References

Balogh, M.L., Schade, D., Morris, S.L., Yee, H.K.C., Carlberg, R.G., Ellingson, E., 1998, ApJ, 504, L75
Cardiel, N., Gorgas, J., Aragón-Salamanca, A., 1995, MNRAS, 277, 502
Cardiel, N., Gorgas, J., Aragón-Salamanca, A., 1998, MNRAS, 298, 977
David, L.P., Slyz, A., Jones, C., Forman, W., Vrtilek, S.D., Arnaud, K.A., 1993, ApJ, 412, 479
Ebeling, H., Edge, A.C., Fabian, A.C., Allen, S.W., Crawford, C.S., Böhringer, H., ApJ, 1997, 479, L101
Edge, A.C., Stewart, G.C., Fabian, A.C., 1992, MNRAS, 258, 177
Fabian, A.C., ARA&A, 32, 277
Gallego, J., Zamorano, J., Aragón-Salamanca, A., Rego, M., 1995, ApJ, 455, L1
Gorgas, J., Efstathiou, G., Aragón-Salamanca, A., 1990, MNRAS, 245, 217
Johnstone, R.M,. Fabian, A.C., Nulsen, P.E.J., 1987, MNRAS, 224, 75
McNamara, B.R., 1997, in *Galactic and Cluster Cooling Flows*, ed. N. Soker, ASP Conf. Ser., 115, 109
Peres, C.B., Fabian, A.C., Edge, A.C., Allen, S.W., Johnstone, R.M., White, D.A., 1998, MNRAS, 298, 416
Schechter, P., 1976, ApJ, 203, 297
Thomas, P.A., Fabian, A.C., Nulsen, P.E.J., 1987, MNRAS, 228, 973
White, D.A., Jones, C., Forman, W., 1997, MNRAS, 292, 419

TURBULENCE AND MAGNETIC FIELDS IN CLUSTERS OF GALAXIES

F.J. SÁNCHEZ-SALCEDO, A. BRANDENBURG AND A. SHUKUROV
Department of Mathematics, University of Newcastle
Newcastle upon Tyne, NE1 7RU, England

Abstract. We consider turbulence generated by galaxies moving transonically through the intracluster gas. We show that neither the gravitational drag nor the gas stripping from the galaxies are able, by themselves, to generate turbulence at a level required to feed the dynamo in the intracluster gas. Some implications for cluster radio halos are discussed.

1. Introduction

Many clusters of galaxies have been searched for extended radio halos. Magnetic fields in the hot (10^8 K), ionized intracluster medium (ICM) are revealed by diffuse radio emission, by excess of Faraday rotation and by asymmetric depolarization of radio sources embedded in the cluster (Kronberg 1994). There are two separate problems with no current consensus: the acceleration of relativistic electrons and the amplification of the magnetic field in the ICM (e.g. Sarazin 1986). Here we will focus on the second one.

Polarization measurements give information on the mean line-of-sight component of the field weighted by the gas density. It is inferred that magnetic fields in the ICM are random, tangled at scales 10–20 kpc, and are detected out to 500 kpc in radius, considerably larger than the cluster core. In the core, the observed local magnetic field strengths are 0.2–8 μG for different clusters (e.g. Feretti et al. 1995; Henriksen 1998).

It seems plausible that the magnetic field in the ICM should be maintained by some turbulent small-scale dynamo capable to convert kinetic energy of intracluster turbulence into magnetic energy. Two different sources have been suggested to produce the required turbulence: disturbances from the cluster galaxies, and ongoing merger of other cluster substructures. The energy released in a major merger event is certainly sufficient to maintain

Astrophysics and Space Science is the original source of publication of his article. It is recommended that this article is cited as: *Astrophysics and Space Science* **263**: 87–90, 1999.

the observed magnetic fields. Still, it is quite natural to ask if the cluster galaxies are able to magnetise the cluster medium up to a few μG, or otherwise, a recent merger event is a necessary condition for the existence of radio halos (Tribble 1993).

The r.m.s. turbulent velocity v_0 required to explain the observed magnetic field H can be roughly estimated from equipartition between kinetic and magnetic energies as $v_0 \simeq H/\sqrt{4\pi\rho}$, where ρ is the gas density. For $H = 1\,\mu$G and ρ corresponding to 10^{-3} particles/cm^3, we obtain $v_0 \simeq 250$ km/s.

2. Stirring of the ICM by cluster galaxies

The origin of the sources of turbulence in the ICM has received considerable attention to explain not only the magnetic fields but also the reacceleration of relativistic electrons, the properties of radio sources in clusters or the heating of the central cooling core.

Galaxies may produce turbulent motions in the ICM mainly by two different interactions. Firstly, every galaxy produces a gravitational wake behind it, with the subsequent drag heating of the cluster gas and, secondly, the gas in galactic halos can be stripped by ram pressure and thereby contributes to turbulence. The high metallicities found in the ICM provide clear indication that significant amounts of interstellar gas have been removed to the intergalactic space.

For the dynamo action proposed by Jaffe (1980) and Ruzmaikin et al. (1989), the turbulent motions in the cluster gas generated by the merged turbulent wakes control the magnetic field strength. Hence it is necessary to estimate the typical scale l_0 and the velocity v_0 of the turbulence in a steady state.

Independently of the model of turbulence and regardless of the nature of the interaction, we require that at least one galaxy must have come across any volume of size l_0^3 in the eddy turnover time, i.e.

$$n_{\rm G} l_0^2 v_{\rm G} \approx v_0/l_0, \tag{1}$$

where $n_{\rm G}$ is the number density of galaxies, and $v_{\rm G}$ is the typical velocity of a galaxy with respect to the ICM (Goldman & Rephaeli 1991). Further, the rate of energy transfer into turbulence must be equal to that injected by the galaxies, $\dot{E}_{\rm G}$, thus

$$\dot{E}_{\rm G} \approx \frac{1}{2}\frac{\rho v_0^3}{l_0}, \tag{2}$$

where ρ is the gas density of the ICM. Our aim is to estimate $\dot{E}_{\rm G}$ for both the gravitational disturbances and the gas pressure stripping, and then to obtain l_0 and v_0 from Eqs. (1) and (2).

2.1. GRAVITATIONAL DISTURBANCES

For a supersonic galaxy moving through gas of mass density ρ, the energy loss rate is given by Chandrasekhar's formula with the Coulomb logarithm, $\ln\Lambda$, depending on the Mach number:

$$\dot{E} = n_{\rm G} \frac{4\pi G^2 M_{\rm G}^2 \rho \ln\Lambda}{v_{\rm G}} \tag{3}$$

(Rephaeli & Salpeter 1980). If all the energy is transferred into turbulent motions then $v_0 \approx 40$ km/s, where we have used Eqs. (1) and (2), and assuming a typical galactic mass of 10^{11} $M_\odot$, $v_{\rm G} \simeq 1500$ km/s and $n_{\rm G} \simeq 1.4 \times 10^{-6}$ kpc^{-3}. However, the driving force for these motions is potential, so the motions should be irrotational, presumably in the form of sound waves (shocks do not develop if the accretion radius is smaller than the physical size of the galaxy). Irrotational motions are thought to be inefficient in generating magnetic fields at low Mach numbers even if they are random (Kazantsev et al. 1985). So we conclude that gravitational wakes can only contribute negligibly into amplification of intracluster magnetic fields.

2.2. MASS-OUTFLOW DISTURBANCES

Consider galaxies in the cluster as moving solid bodies of effective radius $a_{\rm eff}$. The turbulent wake dissipates a power of $\frac{1}{2}\lambda_3 \pi \rho v_{\rm G}^3 a_{\rm eff}^2$, with λ_3 a parameter of efficiency, which results in $v_0 \approx 150$ km/s for comfortable values of $a_{\rm eff} \approx 2$ kpc and $\lambda_3 \approx 0.25$. Therefore, it is tempting to identify the galactic gas as the natural source of turbulence.

In fact, it is known that in a short time, a steady state is achieved in which the rate of stripping, for a typical galaxy in cluster, is equal to the rate of gas replenishment by stellar winds within the galaxy (Portnoy et al. 1993). Therefore, the initial gas halo associated with the galaxy does not play any role and only the replenishment of gas at a rate $\dot{\rho}$ is important. In that case, an estimate of the turbulent velocity in the head of the wake, v, is simply given by imposing the conservation of momentum in one dimension as

$$\frac{4}{3}\pi R_{\rm c}^3 \dot{\rho}\, v_{\rm G} \approx \rho \pi R_{\rm c}^2 v_{\rm G} v, \tag{4}$$

where $R_{\rm c}$ is the core radius of the stellar mass distribution. In the equation above the galactic wind has been neglected. Note that in the inner parts of the galaxy where the gravitational force and cooling are important, Eq. (4) is inapplicable. We may write $v \approx R_{\rm c}\dot{\rho}/\rho$ and, consequently,

$$\dot{E}_{\rm G} = \frac{1}{2} n_{\rm G} \rho \pi R_{\rm c}^2 v_{\rm G} v^2 \approx \frac{2}{3} n_{\rm G} \dot{M} \frac{\dot{\rho}}{\rho} R_{\rm c} v_{\rm G}, \tag{5}$$

where $\dot{M}$ is the rate of mass loss by a single galaxy. For typical mean values of $\dot{M} \simeq 0.2$ $M_{\odot}/yr^{-1}$ and $R_c \simeq 16$ kpc, and using Eqs. (1)–(2), we obtain a mean turbulent velocity of $v_0 \simeq 10$–15 km/s. This velocity corresponds to an equipartition magnetic field $B_{eq} \equiv \sqrt{4\pi\rho v_0^2} \leq 0.15$ μG. Thus, the mass outflow from galaxies cannot account for the required amplitude of turbulence.

These estimates only apply outside the accretion radius of the galaxy. While gas is stripped from the outer parts of the galaxy, a cooling flow may be formed in the inner parts if there is enough gas replenishment (Balsara, Livio & O'Dea 1994). In that case galaxies can support a bow shock at some radius that we can identify with a_{eff}. Due to the strong dependence of the results on the adopted a_{eff}, the combination of the gravitational potential, gas replenishment and cooling may result in higher levels of turbulence.

3. Conclusions

From the above discussion it is apparent that efficient transfer of energy from galactic motions to intracluster turbulence is required to explain the magnetic fields observed in radio halos. The efficiency depends dramatically on the effective galactic radius adopted. A definite test requires the inclusion of the gravitational and mass outflow disturbances simultaneously into a model allowing for a range of transonic velocities of galactic motion. Turbulence driven by merger events remains an attractive possibility.

Acknowledgements

FJSS acknowledges financial support from a TMR Marie Curie grant from the European Comission.

References

Balsara, D., Livio, M., O'Dea, C.P., 1994, ApJ, 437, 83
Feretti, L., Dallacasa, D., Giovannini, G., Tagliani, A., 1995, A&A, 302, 680
Goldman, I.,Rephaeli, Y., 1991, ApJ, 380, 344
Henriksen, M., 1998, astro-ph/9807049
Jaffe, W., 1980, ApJ, 241, 925
Kazantsev, A.P., Ruzmaikin, A.A., Sokoloff, D.D., 1985, Exper. Teor. Fiz., 88, 487; JETP, 61, 285
Kronberg, P.P., 1994, Rep. Prog. Phys., 57, 325
Portnoy, D., Pistinner, S., Shaviv, G., 1993, ApJ, 86, 95
Rephaeli, Y., Salpeter, E.E., 1980, ApJ, 240, 20
Ruzmaikin, A., Sokoloff, D., Shukurov, A., 1989, MNRAS, 241, 1
Sarazin, C. L., 1986, Rev. Mod. Phys., 58, 1
Tribble, P.C., 1993, MNRAS, 263, 31

THE SLOAN DIGITAL SKY SURVEY

F. J. CASTANDER FOR THE SDSS COLLABORATION
University of Chicago
Astronomy & Astrophysics Department, 5640 S Ellis Ave., Chicago, IL 60637, USA

Abstract.
The Sloan Digital Sky Survey (SDSS) is going to carry out a uniform survey of π steradians of the sky in the Northern Galactic cap and ~ 225 deg^2 in the Southern Hemisphere. The survey consists of a photometric and a spectroscopic survey. The SDSS will generate accurate photometry in five bands of approximately tens of millions of galaxies, tens of millions of stars and roughly a million quasars. It will also take spectra and measure redshifts of approximately a million galaxies and ten thousand quasars.

The main characteristics and components of the survey are a dedicated 2.5m telescope, wide field correctors for photometry and spectroscopy that will provide a field of view of ~ 3 degrees, a photometeric camera with 30 photometric and 22 astrometric CCDs, and two fibre-fed spectrographs of 320 fibres each. The survey will produce a publicly available science database of Terabytes dimensions.

1. Survey Overview

General information about the survey can be found at http://www.sdss.org/. Technical details are available at http://www.astro.princeton.edu/BBOOK/. Here, we briefly outline some of the main characteristics and components of the survey.

1.1. THE SDSS TELESCOPE

The SDSS telescope has a 2.5m primary mirror and a 1m secondary, giving a focal ratio of f/5; and two corrector lenses to provide an undistorted focal plane. It is an altitude-azimuth telescope with a modified Ritchey-Crétien design that works at the Cassegrain focus (Waddell *et al.*, 1998). It

Astrophysics and Space Science is the original source of publication of his article. It is recommended that this article is cited as: *Astrophysics and Space Science* **263**: 91–94, 1999.

is situated at Apache Point Observatory, New Mexico, USA. The telescope is housed in an enclosure which rolls off for observing, and is encased in co-rotating wind and light baffles. This design does not introduce dome-induced seeing and protects the telescope from wind and stray light.

A dedicated 0.5m Photometric Telescope, situated beside the SDSS telescope, will operate in conjunction with the 2.5m telescope to calibrate the survey data.

1.2. THE IMAGING CAMERA

The imaging camera (Gunn *et al.*, 1998) consists of 54 CCDs that cover 2.4° on the sky. Thirty of these are 2048×2048 SITe CCDs ($24\mu m$ pixel size) that will be used to carry out the imaging survey. They are arranged in 6 dewars of 5 CCDs each. The camera operates in drift-scan mode with the sky passing in succession through each of the 5 CCDs (each one with a different filter). The survey defines a new photometric system of five broad passbands: u′, g′, r′, i′ and z′ (Fukugita *et al.*, 1996).

In addition to the imaging CCDs, there are 22 2048×400 CCDs that will reach brighter magnitudes and will allow us to tie the imaging data to existing astrometric catalogues (in particular, Hipparcos) to an expected accuracy of 50-100 mas. There are two additional CCDs that will be used to monitor the telescope focus.

1.3. THE SPECTROGRAPHS

Spectroscopy will be taken with two identical fibre-fed spectrographs of 320 fibres each. The spectrographs give a resolution of $\lambda/\Delta\lambda \sim 2000$ or ~ 150 km/s. Each spectrograph is divided into two channels (blue and red) each having a 2048×2048 CCD ($24\mu m$ pixel size) and will cover an approximate spectral range of 3900-9000 Å (blue: 3900-6000 Å, red: 6000-9000 Å).

The fibres are $180\mu m$ diameter ($3''$ on the sky) and will be held in the focal plane by plug-plates. The minimum separation allowed between fibres is $55''$.

1.4. DATA PROCESSING

The bulk of the data reduction will be at Fermilab. The images will be recorded to DLT tapes at the mountain and sent by overnight courier to Fermilab, where they will be processed. There will be $\sim$ 200 GBy generated per imaging night and $\sim$ 2 GBy per spectroscopy night.

The data processing has been divided into several pipelines which perform a variety of tasks. An operational database regulates the data flow. First, the imaging data are analysed. Objects are automatically identified, measured, classified and calibrated with the photometric and astrometric calculated solutions. When a significant area of the sky has been imaged, objects are carefully selected for spectroscopic follow-up. After spectroscopy is obtained, the data is automatically analysed. We expect to obtain redshits for 99% of the objects without human intervention. Once analyses are finished, the data are loaded and organised into an object-oriented science database, which will be accessible by the astronomical community once the scientific quality of the data is assured. The pipelines are mostly written in C and TCL and integrated into a specially-written environment, called Dervish.

2. Survey Strategy

The SDSS consist of two surveys, one photometric and one spectroscopic. Both surveys will cover the same area of the sky: 10000 deg^2 centred roughly on the North Galactic Pole and 225 deg^2 in the South, that will be observed in Autumn when the North Galactic cap is not accessible.

The imaging survey will only take approximately 20% of the total survey time and will be done in the best conditions. The photometic telescope, that will continuosly observe photometric standards, will provide the photometric calibration for the imaging camera data. The absolute photometric accuracy is expected to be 2%. The astrometric CCDs of the camera will provide the astrometric calibration.

The imaging camera will operate at sidereal rate resulting in an integration time of 55 s and expected limiting magnitudes of 22.3 in u$'$, 23.3 in g$'$, 23.1 in r$'$, 22.5 in i$'$ and 20.8 in z$'$ (5 σ for a point source in 1$''$ seeing). In the Southern survey, several scans will be obtained of the same area, resulting in approximately 2 magnitudes fainter limits. These multi-scans will also allow us to study variable objects.

Targets for spectroscopy will be carefully selected from the photometric data. Targets are divided into four broad categories: galaxies, quasars, bright red galaxies and other objects. Galaxies will be fully sampled (except for a small percentage due to fiber separation constraints) down to the limiting magnitude. The fibres will be manually plugged into the plates.

The typical spectroscopic exposure time will be of 45 minutes to reach an approximate Petrossian magnitude limit of $r' \sim 18$ for galaxies. Quasars, however, will be targeted at fainter magnitudes ($i' \sim 19$) and will be selected according to their distinctive colours. In addition, spectra will be taken of bright red galaxies approximately one magnitude fainter than the main galaxy survey. In total, we expect to target complete samples of $\sim 9 \times 10^5$ galaxies, $\sim 10^5$ QSOs and $\sim 10^5$ brightest red galaxies. In addition, many thosands of stars, radio and X-ray counterparts and "serendipitous" objects will be targeted.

3. Science

The plethora of data which will result from the SDSS will allow the scientific community to carry out numerous studies ranging from stars to large scale structure. The main scientific goals of the project are the study of the local large scale structure of the universe (power spectrum, correlation functions, topology,...), galaxy clusters (clustering, evolution, morphology-density relation,...), quasars (identification of high redshift QSOs, clustering, absorption systems,...), galaxies (clustering, luminosity function, morphologies,...), galactic structure, stars, extinction, minor planets and comets.

4. Current Status

The 2.5m telescope and imaging camera had first light on 9 May 1998, and are now in an intensive commissioning phase. Data have also been obtained in the June and September dark periods (the October dark run was clouded out) and observations are currently planned for the November dark period. The telescope primary mirror will be re-aluminised in December.

The spectrographs are at the site. One has been completely assembled and some test spectra have been taken with it. A few test plug plates have been drilled and intensive commissioning on the sky will start in January.

All data reduction pipelines are written, with ongoing work on minor bug-fixes, speed-ups and integration of the entire data processing system.

The intent of the project is to make the survey data available to the astronomical community in a timely fashion. An early partial release of the commisioning data will be done at the American Astronomical Society Meeting next January.

Acknowledgements. The Sloan Digital Sky Survey is a joint project of the University of Chicago, Fermilab, the Institute for Advanced Study, the Japan Participation Group, the Johns Hopkins University, Princeton University, the United States Naval Observatory, and the University of Washington. Apache Point Observatory, site of the SDSS, is operated by the Astrophysical Research Consortium. Funding for the project has been provided by the Alfred P. Sloan Foundation, the SDSS member institutions, the National Science Foundation and the U.S. Department of Energy.

References

Gunn, J.E., Carr, M.A., Rockosi, C.M., Sekiguchi, M. et al., 1998, AJ, in press.

Fukugita, M., Ichicawa, T., Gunn, J.E., Doi, M., Shimasaku, K., Schneider, D., 1996, AJ, 111, 1748.

Waddell, P., Mannery, E. J., Gunn, J., Kent, S. 1998, Proc SPIE, Vol. 3352.

OPTIMIZING RESULTS FROM AUTOMATIC OBJECTIVE-PRISM SURVEYS

C. E. GARCíA DABÓ, J. GALLEGO & J. ZAMORANO
Dpto. Astrofísica.
Universidad Complutense de Madrid E-28040. Spain

Abstract.

One of the most important results in observational cosmology is the galaxy star formation rate distribution, or its equivalent observable, the $H\alpha$ luminosity function. The UCM-CIDA project has grown to address the problem of the poorly-known faint star forming galaxy population using the objective-prism technique along with a CCD mosaic in a Schmidt telescope.

However, one tough problem when computing statistical properties of a selected sample are the selection effects. Here we present tools developed by us to undertake the analysis of objective-prism (OP) selected samples in an unbiased way and quantify the sensitivity of the survey to every object. The main tools for this purpose are simulations.

1. Introduction

The complete description of the star-forming (SF) population of galaxies and specifically the global properties of faintest objects ($M_B > -16$) requires to carry out a large survey with a statistically significative number of galaxies. This is the scope of the UCM-CIDA survey using an OP, a 4×4 CCD mosaic and discovering 2000 new SF galaxies up to $m_r = 20$ ([2]).

In addition to the use of the OPERA automatic method ([3]) to process and analyze the data , we need methods to extract the required results (eg. luminosity function) as well as predicting tools. The calibration of CCDs claims for a precise analysis and rejection of systematic effects. We show that simulations are the way to go. Next sections face up these problems.

2. Simulating the survey results

The best way to check and monitoring the performance of the survey is by means of simulations. They are a powerful tool for three main purposes:

Astrophysics and Space Science is the original source of publication of his article. It is recommended that this article is cited as: *Astrophysics and Space Science* **263**: 95–98, 1999.

- Expectation of the scientific profit. With the help of the simulations we have designed the best filter in the midway between low sky background and useful wavelength range (i. e. redshift limit).
- Control the automatic method reliability. Several features like object overlapping or emission-line detection have been added and checked in benefit of the good performance of the automatic selection.
- Taking into account selection effects: section §3.

2.1. SCHEMATIC DESCRIPTION

Simulations are fed by parameters divided into two main groups: instrumental and galaxy parameters. The former being the telescope and detector characteristics. Galaxy parameters must be well chosen along with a good physical modeling in order to have a reliable simulation. The output image is built to resemble that of the final system output.

We call $I_0(x, y, \lambda)$ the initial distribution of light, where x, y are appropriate coordinates on the sky. The spatial and wavelength information of the galaxy is contained in this function. The effect of seeing is represented by a 2d-gaussian filter in $I_0(x, y, \lambda)$ yielding to $I_s(x, y, \lambda)$. Next, light is deviated with different angles, depending on the wavelength, by the objective-prism according to the dispersion law of the prism $\Delta\delta_x(x, y, \lambda)$, and is focused on the focal plane. D : telescope diameter, f : focal length, px : pixel size:

$$I_{CCD}(X,Y,\lambda) = \pi\,(D/2)^2 \left(\frac{206265\,px}{f}\right)^2 I_s(x - \Delta\delta_x, y - \Delta\delta_y, \lambda) \qquad (1)$$

Then, energy input must be converted to electrons depending on the instrumental response and detector quantum efficiency for each wavelength.

$$N_{CCD}^{e^-}(X,Y) = \int_{-\infty}^{\infty} I_{CCD}(X,Y,\lambda)\frac{\lambda\, t_{exp}}{hc}T(\lambda)QE(\lambda)d\lambda + DARK\, t_{exp} \qquad (2)$$

ADUs in every pixel of the CCD, i.e. final output image, are calculated using a Poisson event distribution for electrons and the parameters of the CCD (Read out noise (RON), gain, bias and previously dark current) ([4]):

$$N_{ADU}(X,Y) = \frac{Poisson(N_{CCD}^{e^-}(X,Y) + Gauss(RON))}{GAIN} + BIAS \qquad (3)$$

Figure 1 illustrates the result of a simulation with the CIDA instrumentation where galaxies with diverse apparent magnitudes and equivalent widths (EW) have been generated.

3. Selection function

Most of the optical surveys have been characterized by a unique parameter: the limiting magnitude. But samples selected by emission lines in objective-prism data are not so simply selected and a more complex scheme is needed.

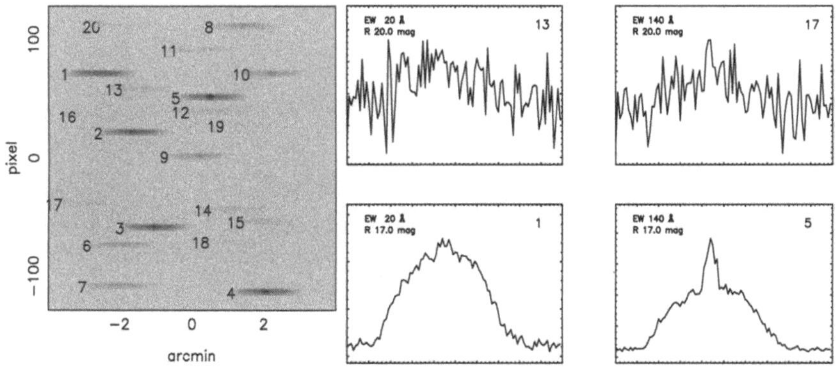

Figure 1. Simulated objective-prism field for galaxies with different magnitudes and equivalent widths and their respective extracted spectra (right).

A synthetic magnitude combination of continuum flux (m_r) and EW has been proposed and used ([5]). But if we want to be precise it does not reflect all the factors involved in the selection of an object.

A complete description of the selection effects is contained in the so called selection probability function p_D. It just gives the probability of a certain galaxy to be detected. It obviously depends on the specific parameters of the galaxy which have an influence in the detection process. This means that a *good modeling* of the galaxies is needed if we want to quantify the selection effects.

We call $p_D(m_r, ..)$ the probability for a galaxy with parameters m_r, EW, z, *size* to be detected. These set of parameters have been chosen because simulations have proved they are the most influencing in the detection.

The important point is that this selection function p can be calculated with simulations. If the whole instrumental system would be fed by M objects getting N objects out of the process, then $p = N/M$ and $\Delta p = \sqrt{p(1-p)/M}$. The same process is done for every bin in the n-dimensional parameters space to construct $p_D(m_r, ..)$. In figure 2 we have plotted a projection of p_D in m_r and EW planes.

4. $H\alpha$ luminosity function calculation

Standard methods like $1/V_{max}$ are designed for magnitude-limited samples. Since this is not our case, we have developed a generalization of the maximum likelihood STY ([6]) and SWML ([1]) methods, which have the best performance among the standard methods ([7]).

Let's call $\Phi(L_{H\alpha}, z, EW, ...)$ the multiparametric distribution function, that is, the luminosity function being $\phi(L_{H\alpha}, z) = \int\int ..\Phi dEWdz....$ The

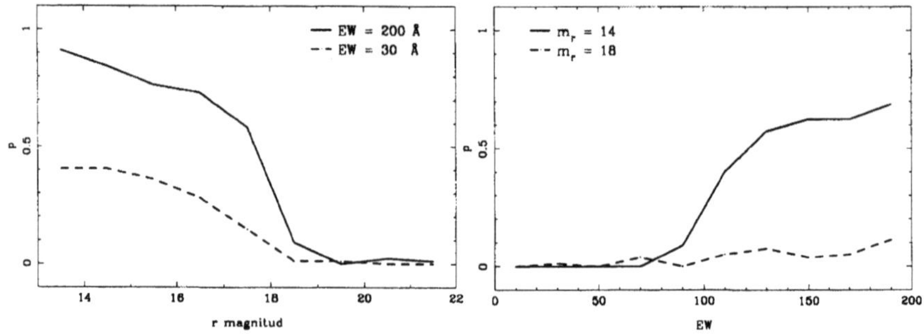

Figure 2. Probability function projected along apparent magnitude (left) and equivalent width (right) ($\Delta p \approx 0.2$). There is not a sharp cut-off at the limiting magnitude. 1 minute exposure, 1.5m Schmidt telescope and 3.5 deg objective-prism have been used.

probability of *existing* a galaxy with parameters $M^i, z^i, EW^i, ...$ will be $p_E \propto \Phi(L_{H\alpha}, z, EW, ...)$. Now the selection probability function $p_D(L_{H\alpha}, ..)$ gives the probability of *detecting* a galaxy with given parameters $L_{H\alpha}, z, ..,$ thus it may be best written as the conditional probability $p_D(detect|param = L_{H\alpha}, ..)$. We want to maximize the probability of finding in the survey a galaxy with those parameters, i.e. $p_S^i(param = L_{H\alpha}, ...|detect)$:

$$p_S^i \propto p_E p_D = \frac{\Phi(L_{H\alpha}^i, z^i, EW^i, ...)p_D(L_{H\alpha}^i, z^i, EW^i, ...)}{\int .. \int \Phi(L_{H\alpha}, z, EW, ..)p_D(L_{H\alpha}, z, EW, ..)..\frac{dV}{dz}dL_{H\alpha}dz} \tag{4}$$

p_S^i is maximized for a parameterization of Φ and $\phi(L_{H\alpha})$. With this formulation the method is suited to calculate multiparametric luminosity or distribution functions using a generalization of the SWML method.

5. Conclusions

Simulations of objective-prism images are presented as a predicting tool also needed to check calibrations and quantify selection effects.

A new method to calculate luminosity functions extracts all the available information out of the survey in an unbiased way. The method is valid in general for every kind of survey where the wanted statistical results depend on parameters which also impose observational restrictions.

References

1. Efstathiou G., Ellis R. S., Peterson B. A., 1988, MNRAS 232, 431
2. Gallego J., et al, 1998, in preparation
3. García Dabó C. E., Gallego J., 1998, this proceedings
4. McLean I. A., 1997, *Electronic imaging in astronomy.* Ed. John-Wiley
5. Salzer J. J., 1989 ApJ 347,152
6. Sandage A., Tammann G. A., Yahil A., 1979, ApJ 232, 352
7. Willmer C. N. A. 1997 AJ 114,898.

OPERA: OBJECTIVE-PRISM REDUCTION PACKAGE

C. E. GARCíA DABÓ AND J. GALLEGO
Dpto. Astrofísica.
Universidad Complutense de Madrid E-28040. Spain

Abstract.

Surveys of star-forming galaxies are of vital importance to constrain galactic evolution theories. One of the most successful searching methods is the objective-prism (OP) technique, which can register thousands of low resolution spectra in large areas. The UCM-CIDA (Universidad Complutense de Madrid-Centro de Investigación F. J. Duarte) project expects to cover 150 square degrees and detect $\sim$ 2000 Hα emission line galaxies.

OPERA (Objective Prism Enhanced Reduction Algorithms) is an automatic method to handle and analyze such amount of data. It goes trough the detection of the objects in the image, extraction of one-dimensional spectra and identification of the emission line feature.

The main advantages of a method of this kind are: 1) to avoid subjectivity inherent to visual inspection used in past studies; and 2) the ability to obtain physical parameters without follow-up spectroscopy.

1. Introduction

The main goal of the UCM-CIDA project is the identification of the actual star-forming galaxies population. The observational setup used combines an OP, the 1-m Schmidt telescope at Llano del Hato (Venezuela) and a mosaic of 4×4 CCDs We present here the analysis tools we have developed for this survey. Many parameters can be adjusted so it is a high configurable package. Figure 1 shows schematically the main steps in the data processing.

Best performance of this technique is get using a direct image (without OP) of the same field We shall assume this is the case from now on.

2. First step: Object identification

After common preprocessing (bias, flat, etc) the first step is the identification of sources on both direct and OP images. This involves the previous determination of the local sky background and associated noise. Images are

Astrophysics and Space Science is the original source of publication of his article. It is recommended that this article is cited as: *Astrophysics and Space Science* **263**: 99–102, 1999.

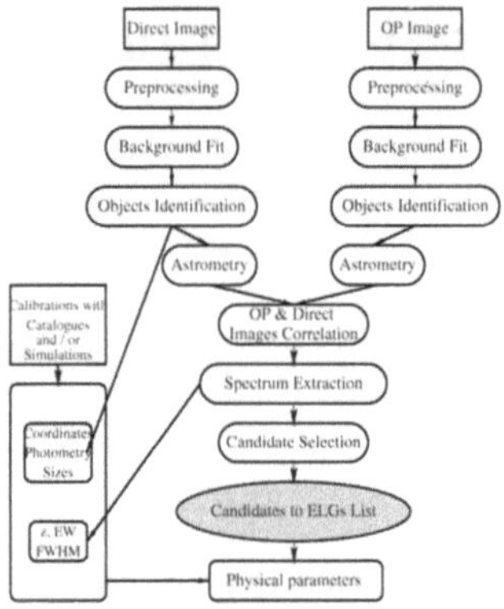

Figure 1. Basic scheme of the reduction process. Direct and OP images follow the same steps until spectra extraction is performed. Different physical parameters are obtained out of both images using external calibrations and/or simulations.

divided in user-defined small boxes to allow for sky background changes along the image. Inside these subsets the mode of the pixel values is assigned to the local background, while noise is assigned to standard deviation.

Objects are found in the image following these steps: (1) the algorithm begins when it finds a pixel $N\sigma$ above the local background ($N \approx 1$ usually). (2) neighboring pixels are added if the global S/N ratio is improved after adding them. This test is made for every surrounding pixel. (3) neighboring pixels cannot differ significatively from a gaussian seeing profile among each other. (4) Minimum number of pixels for an object can be prefixed.

Star-galaxy separation is performed using the algorithm described in [5]. However line search is performed on the whole catalog to allow for star-like objects like quasi-stellar objects or ultracompact galaxies ([2]).

Once the objects on the image have been identified and astrometric solutions have been computed, a cross-correlation may optionally be performed between direct and OP catalogs to eliminate scratches and other stuff in the images. Moreover, the final end of this correlation is to allow us to use an optimal spectra extraction. We also attempt to search emission lines in unpaired objects from the OP image for the case of very strong lines and weak continuum objects in the detection limit.

3. Second step: Spectra extraction

One approach is to add the central scans of the OP image but little information about the spectrum can be extracted in this way. Several methods have been applied to make wavelength calibration, for instance, the common cut-off of filter response (see [1]), two OP with perpendicular dispersion (with additional time costing ([6]), etc.

We will use another method using astrometric calibrations, therefore, translating positions from OP image to direct one. We only need a reference wavelength for the OP image which is easy to calibrate.

Shall we call $D(x, y)$ and $OP(x, y)$ respectively to the direct and OP images. The position coordinates x, y have been put to a common system using the astrometric calibration. Let us call $\Delta(\lambda)$ the known dispersion curve of the prism. The relationship between these quantities is :

$$OP(x, y) \propto \int D(x + \Delta(\lambda), y) \quad F(\lambda) \quad T(\lambda) \quad d\lambda \tag{1}$$

where $F(\lambda)$ is the true spectrum and $T(\lambda)$ the transmission of the whole system. This equation can be solved for $F(\lambda)$ using the deconvolution method described in [4] under the assumption that the object emits the same spectrum over the x direction. If this is not the case, the method recovers the mean spectrum. Due to a degeneracy between position and wavelength impossible to solve, this is the best we can do. Using the direct image is possible to correct for overlapping spectra as well. It is important to note that this method allows us to get spectra limited only by spectral resolution, getting rid of the spatial profile systematic effect.

4. Third step: Searching emission lines

Emission lines are detected using a maximum likelihood method (MLM). Let us call $f(\lambda)$ the observed spectrum. We shall call $g(\lambda)$ a synthetic spectrum created upon some basis on the spectrum features. In our case we can suppose a flat spectrum due to the narrow coverage in wavelength (less than 500 Å) plus an Hα emission line but the method remains for more complicated parameterizations. Thus we have a $g(\lambda)$ of the form:

$$g(\lambda) = 1 + \frac{EW}{\sqrt{2\pi}\,\sigma} \exp\left\{ -\frac{(\lambda - \lambda_{\mathrm{H}\alpha})^2}{2\,\sigma^2} \right\} \quad ; \quad FWHM = 2.35\sigma \tag{2}$$

The idea is to compare pixel by pixel the two spectrum and calculate the joined probability p for the whole wavelength range that the two spectra are just the same. Supposing a gaussian error distribution this yields to:

$$\log p = \sum \log \left[N \left(f(\lambda_i); \overline{x} = g(\lambda_i), \sigma(f(\lambda_i)) \right) \right] \tag{3}$$

Those values of EW (equivalent width), FWHM (full width half maximum) for which p is maximum are the most likely to represent the spectrum. We have made some simulations to check the output. Input spectra with variable S/N, EW, FWHM have been simulated. Figure 2 shows how the EW is recovered for a given $S/N = 3$ which is fairly out of the human recognition level.

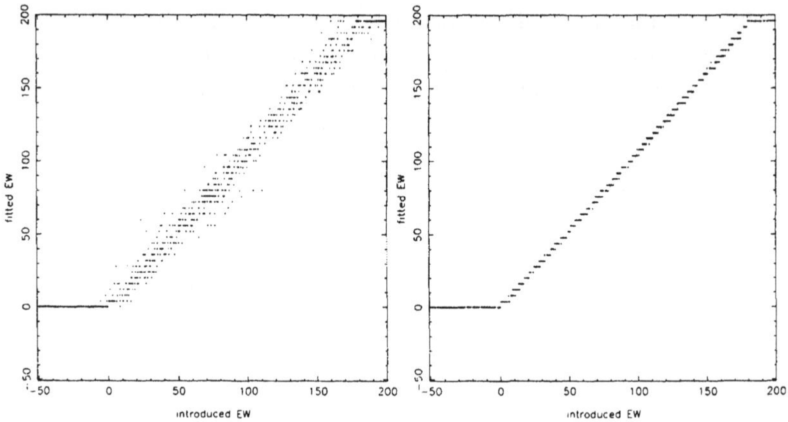

Figure 2. Simulations showing the searching algorithm behavior. Different EW are simulated for $S/N = 3$ (left) and $S/N = 20$ (right) being well identified by the algorithm. Discretization in EW grid is visible. If we restrict to $EW >= 0$, it is found that MLM assigns $EW = 0$ to absorption lines.

5. Summary

We have developed an automatic method for the identification of emission line objects in objective-prism images. The method is a general-purpose software for similar data (grism ...) suited to work with large CCD images.

Different features have been added like variable sky background, optimal extraction, wavelength calibration, spectra overlapping, maximum likelihood identification of emission lines, etc. All of them have been widely tested by means of simulations ([3]) to provide the best performance for the UCM-CIDA survey project.

After the method has been applied to a set of images, it can readily provide apparent magnitudes, sizes and emission line fluxes for every candidate along with luminosity functions and other statistical information. Errors are propagated throughout the whole process so that we have a probability for the reliability of the candidate and an error for every extracted parameter.

This work has been supported by the Spanish DGES grant PB96-0645. CEGD would like to acknowledge a UCM fellowship grant.

References

1. Alonso, O., Zamorano, J., Rego, M., Gallego, J., 1995, A&AS, 113, 399
2. Drinkwater, M.J., Philipps, S., Jones, J.B., Gregg, M.D., Parker, Q.A., Smith, R.M., Davies, J.I., Sadler, E.M., astro-ph/9810294
3. García Dabó, C.E., 1998, these volume
4. Lucy, L.B., 1974, AJ, 79,745
5. Sebok, W.L., 1979, AJ, 84,1526
6. Stock, J., Osborn, W., 1980, AJ, 85,1366

SEYFERT GALAXIES: A PERSPECTIVE WITH THE INFRARED SPACE OBSERVATORY (ISO)

A.M. PÉREZ GARCÍA AND J.M. RODRÍGUEZ ESPINOSA
Instituto de Astrofísica de Canarias
C/ Vía Láctea s/n, La Laguna, Tenerife, Spain

Abstract. We present data from the Infrared Space Observatory (ISO) of a complete set of Seyfert galaxies (Seyfert 1 and Seyfert 2) in a wavelength range seldom explored so far. These data allow a detailed study of the far IR Spectral Energy Distribution of these sources.

A Bayesian inversion method has been used to invert the Spectral energy Distribution (SED) of these sources yielding two fundamental results, namely that the far IR SED of Seyfert Galaxies can be explained solely through thermal reradiation of high energy photons by dust. And, that this thermal emission is made up of two or three independent components.

Comparisons between the parameters obtained from the analysis of the IR SEDs and the emission in other wavelengths ranges (radio and X-rays) show that the host galaxies are key elements in the understanding of the IR emission of Seyferts. Besides, the nuclear properties seem to be adequately explained with the Unified Models of AGN, that state that the differences seen between the broad (Seyfert 1) and narrow (Seyfert 2) line objects are due to geometrical orientation effects.

1. Observations and results

The study of how the mid and far IR emission is produced in Seyfert galaxies is of great interest for the understanding of the emission mechanism that give rise to the large IR luminosities observed in Seyfert galaxies. We have observed a complete sample of Seyfert galaxies with ISO: the CfA Seyfert sample (Huchra and Burg 1992). Data consists of broad band observations through ISO filters at 16, 25, 60, 90, 120, 135 and 200 μm. (Pérez García and Rodríguez Espinosa 1998).

Astrophysics and Space Science is the original source of publication of his article. It is recommended that this article is cited as: *Astrophysics and Space Science* **263**: 103–106, 1999.

We start from the assumption that the mid and far IR emission of Seyfert galaxies is thermal. This is supported by two previous results (Rodríguez Espinosa et al. 1996; Rodríguez Espinosa and Pérez García 1997), i.e., that a preliminary fit of the IR emission to two blackbodies produces satisfactory results, and that the long λ IR emission correlates with the optical emission from the galaxy discs. The method employed to analyse the Spectral Energy Distributions (SED) of the CfA sample is the Inverse Planckian Transform, an inversion method that is based on the iterative approach developed by Richardson (1972) and Lucy (1974). Details of the method are explained in Pérez García et al. (1998). Results are shown in Figure 1 for two objects of the sample. The results are similar for all the objects: we find three thermal components, namely, (1) a warm component, corresponding to dust at a T$\sim$120-170 K, heated by the active nucleus and/or massive starforming regions near the nucleus; (2) a cold component at T$\sim$40-70 K, heated by starforming regions in the disc of the host galaxies and (3) a very cold dust component at T$\sim$20-25 K heated by the general interstellar radiation field.

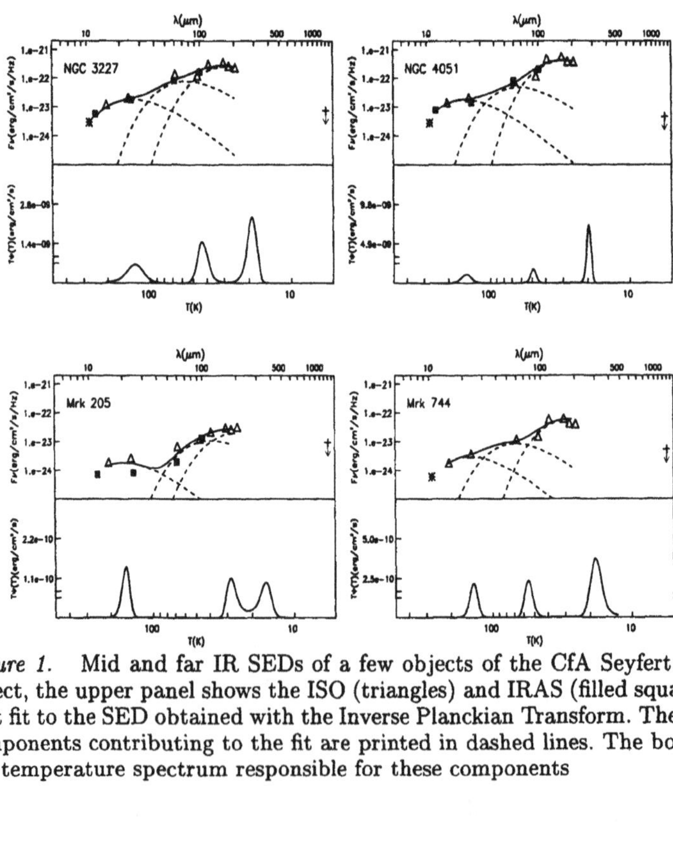

Figure 1. Mid and far IR SEDs of a few objects of the CfA Seyfert sample. For each object, the upper panel shows the ISO (triangles) and IRAS (filled squares) data and the best fit to the SED obtained with the Inverse Planckian Transform. The different thermal components contributing to the fit are printed in dashed lines. The bottom panel shows the temperature spectrum responsible for these components

2. Discussion

2.1. INFRARED-RADIO CORRELATION

A correlation between the IR and radio emission has been observed in most galaxy types. In normal galaxies, the models explain the radio emission as synchrotron radiation produced by accelerated cosmic rays in the galaxy magnetic field, and the FIR emission as due to absorption and re-emission by dust of the interstellar radiation field (Mas-Hesse 1992). These models do not explain the correlation found in active galactic nuclei, since their radio emission is dominated by the nucleus and circumnuclear starforming regions (Pérez Olea and Colina 1995), while in normal galaxies the nuclear emission represents only a 10% of the total radio emission. We have tested the starburst model of Pérez-Olea and Colina (1995). The model explains the IR-radio correlation in active galactic nuclei assuming that both the radio and mid and far IR emission are produced mainly by the nucleus. We have found a tight linear correlation between the warm IR emission and the radio emission at 20 cm using data from Edelson (1987). Figure 2 shows this correlation. Therefore, the warm IR and the radio emission have the same spatial origin and are probably produced by similar physical processes.

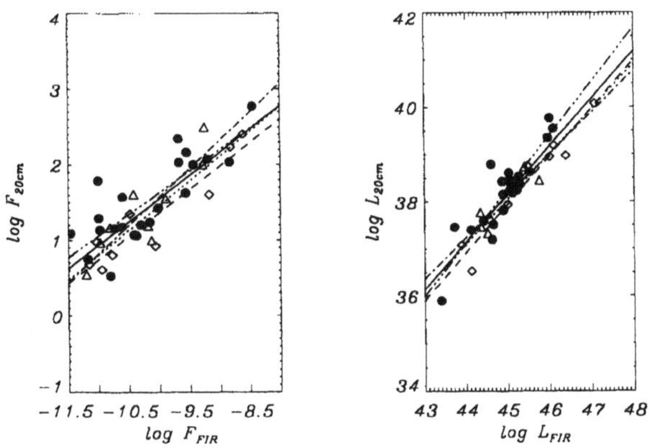

Figure 2. Flux and luminosity at 20 cm *versus* nuclear IR fluxes and luminosities. The lines represents the best fits to data.

The ratio obtained between the warm infrared and the radio emission ($\log(L_{\rm IRwarm}/\ L_{\rm 20cm}) = 6.3$) is well explained by the model, i.e., the warm IR emission and the radio emissions are originated in the nucleus and circumnuclear starforming regions. However, if instead we calculate the ratio between the total IR luminosity and the radio luminosity we obtain $\log(L_{\rm IRtotal}/L_{\rm 20cm}) = 6.8$, well above the range predicted by the model. Hence, the IR emission of the host galaxy has to be added in the model, as the host galaxies have an important contribution to the far IR emission of Seyfert galaxies.

3. X ray-IR correlation

We have also compared the X ray and the far IR emission of these objects. We choose the warm IR component, because it is the component connected with the nuclear emission. ROSAT data between 0.2 and 4 keV have been used (Rush et al. 1996). We have used censured statistics to account for the presence of data with upper limits. We obtain that the correlation between the X ray and the warm IR luminosity is linear. Therefore it is not necessary to invoke emission mechanisms with different efficiencies in the X-ray and the IR band.

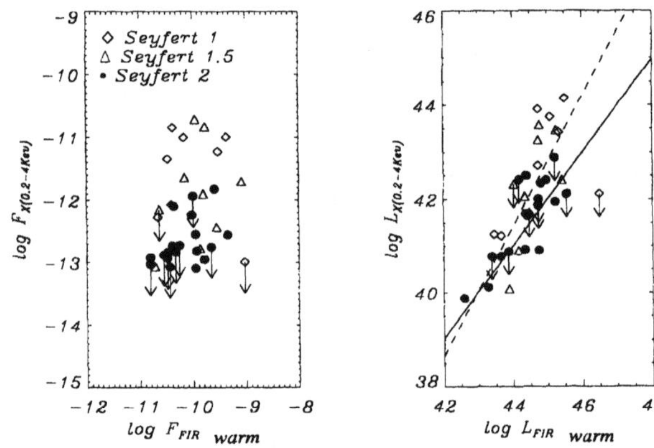

Figure 3. X-ray flux and luminosity *versus* nuclear IR fluxes and luminosities. The lines represents the best fits to data.

References

Edelson, R.A., 1987, ApJ, 313, 651
Huchra, J., Burg, R., 1992, ApJ, 393, 90
Lucy, L.B., 1974, AJ, 79, 745
Mas-Hesse, J.M., 1992, A&A, 253, 49
Pérez García, A.M., Rodríguez Espinosa, J.M., Santolaya Rey, A.E., 1998, ApJ, 500, 685
Pérez Olea, D.E., Colina, L., 1995, MNRAS, 277, 857
Richardson, W.H., 1972, JOSA, 62, 55
Rodríguez Espinosa, J.M., Pérez García, A.M., 1997, ApJ, 487, L33
Rodríguez Espinosa, J.M., Pérez García, A.M., Lemke, D., Meisenheimer, K., 1996, A&A, 315, L129
Rodríguez Espinosa, J.M., Pérez García, A.M., 1998, (in preparation)
Rush, B. Malkan, M.A., Fink, H.H., Voges, W., 1996, ApJ, 471, 190

CCD SURFACE PHOTOMETRY OF NEARBY ACTIVE GALAXIES

M. SÁNCHEZ PORTAL, A.I. DIAZ AND M. ÁLVAREZ ÁLVAREZ
Dpto. de Física Teórica, C-XI, Universidad Autónoma de Madrid, Cantoblanco, 28049-Madrid
ROBERTO TERLEVICH
Institute of Astronomy, Madingley Road, Cambridge CB30HA, United Kingdom
AND
ELENA TERLEVICH AND ITZIAR ARETXAGA
INAOE, Tonantzintla, Puebla, México

1. Introduction

Up to date, only a reduced number of studies have been devoted to compare the structural properties of the host galaxy with those of normal galaxies of similar properties. Nevertheless, bars and interactions have been proposed as mechanisms to carry fuelling material to the central engine, a need shared by all AGN models. Therefore, it seems clear that a detailed knowledge of the structural parameters of active galaxies would contribute to our understanding of the nature and mechanisms of the various degrees of nuclear activity.

2. Observations

The CCD observations were performed in two runs (1988 and 1990) using the Jacobus Kapteyn Telescope (JKT) at La Palma Observatory. Broadband *VRI* and narrow-band H_α interference filters were used. Our sample comprises 27 galaxies and was selected in such a way that it included a wide range of morphological types (spirals, lenticulars, ellipticals, irregulars) and all degree of nuclear activity (7 Seyfert 1, 8 Seyfert 2, 6 LINERs, 3 starbursts and 3 normal galaxies). Nearby objects were chosen in order to have a good spatial resolution (typically, that of the NLR scale size) with our instrumental configuration and expected seeing conditions. A detailed

Astrophysics and Space Science is the original source of publication of his article. It is recommended that this article is cited as: *Astrophysics and Space Science* **263**: 107–110, 1999.

description of the sample, data reduction techniques used and the analysis performed can be found in Sánchez-Portal et al. (1998,1999).

3. Morphological characteristics

We have revised the morphological types of the sample objects given by de Vaucouleurs et al. (1991) (RC3). The highest abundance corresponds to early-type spirals. A fraction of 20% of the objects is composed by lenticular galaxies. A slight tendence for Seyfert 2 nuclei to inhabit galaxies of later morphological type than Seyfert 1 ones can be guessed from our sample data. We have searched for bars in the sample galaxies using several criteria (inspection of the images, ellipticity and position angle profiles analysis), and found that bars are present in a high fraction[1] of the objects with any kind of activity (i.e. excluding normal galaxies), about 71%. This proportion is even higher, about 78%, if only Seyfert galaxies are considered. These fractions are higher than those found in other samples (e.g. Xanthopoulos, 1996). These results could reinforce the clues about the connection between bars and nuclear activity. Nevertheless, there is not universal agreement on this subject (see for instance, Mulchaey & Regan, 1997).

We have studied interactions in the sample galaxies using two criteria: presence of near neighbours and morphological peculiarities. We define an Interaction Activity Class (IAC), following Dahari (1985) as an integer ranging from 1 (isolated symmetrical galaxies) to 6 (strongly disturbed or overlapped systems). We find that Seyfert 1 galaxies have a mean IAC = 2.57, higher than Seyfert 2 (IAC = 1.87) and LINERs; the latter have the same IAC than normal galaxies (IAC = 1.67). No Seyfert, LINER or normal galaxy is observed in the highest interaction class. On the other hand, our sample starburst galaxies seem to be very disturbed objects, with a mean IAC = 4.67

4. Structural parameters

We have computed bulge parameters (r_e, μ_e) and disk parameters (r_0, μ_0) by means of an iterative fitting method (Kormendy, 1977). The bulge to disk (B/D) relationship[2] has been also computed. Figure 1 shows the relationship between I band bulge (left panel) and disk (right panel) parameters for our sample objets. We have also included results from Mediavilla et al. (1989) and Xanthopoulos for comparison. We have observed a linear correlation between bulge parameters $\log r_e$ and μ_e that is also observed between

[1]For instance, in the Shapley-Ames catalog (Sandage and Tammann, 1981) 26% of all spiral galaxies and 36% of Sa0-Sbc galaxies are barred

[2]Defined as the ratio of integrated luminosity of bulge to disk

disk parameters $\log r_0$ and μ_0 that can be explained as a luminosity selection effect. On the other hand, we did not find any correlation between disk and bulge parameters and nuclear activity. Nevertheless, we observed that the mean effective surface luminosities of Seyferts are higher than in other nuclear types, being the highest that of Seyferts 1. No correlation between the B/D relationship value distribution and the nuclear type is observed.

We have found that three out of the fifteen Seyfert galaxies seem to present type II disks (according to Freeman, 1970) that could be due to the presence of an additional component, like a lens, or to a real deficience of disk material towards the center (maybe a result of the fuelling mechanism). We did not find type II disks in other nuclear types.

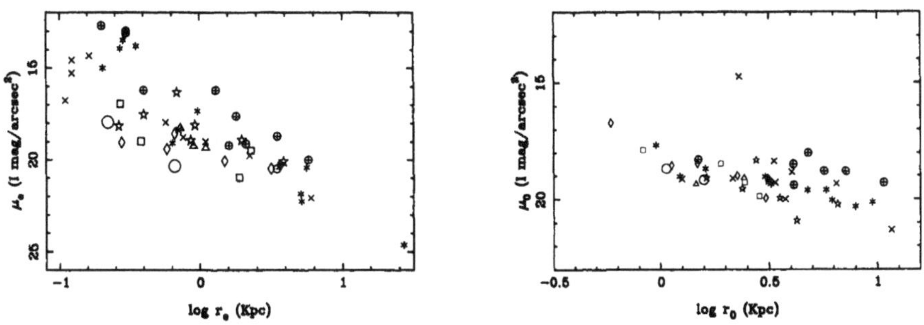

Figure 1. Relationship between I band bulge (left) and disk (right) parameteres for the sample galaxies: Seyfert 1 ($\star$), Seyfert 2 ($\square$), LINER ($\diamond$), Starburst ($\circ$) and normal galaxies ($\triangle$). The Seyfert 1 ($*$) and Seyfert 2 ($\times$) galaxies from Xanthopoulos (1996) and the Seyfert 1 ($\oplus$) and Seyfert 2 ($\odot$) from Mediavilla *et al.* (1989) are also included

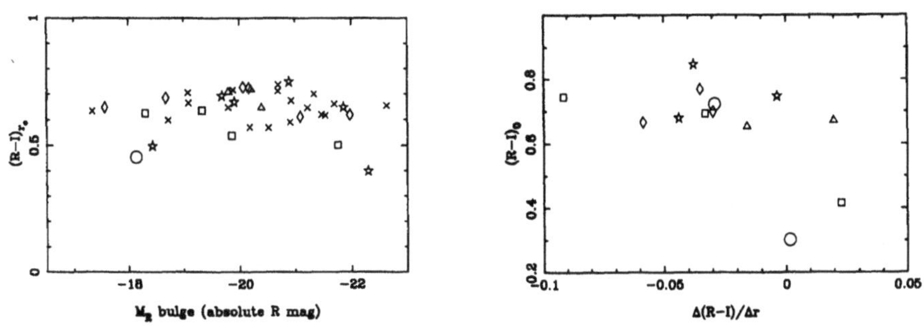

Figure 2. Left: Relationship between characteristic bulge colours and absolute R band magnitude: Seyfert 1 ($\star$), Seyfert 2 ($\square$), LINER ($\diamond$), Starburst ($\circ$) and normal galaxies ($\triangle$). We have also included normal galaxies ($\times$) from Balcells & Peletier (1994) for comparison. Right: $R-I$ colour gradients. Symbols as above.

We have tried to characterize the bulge and disk colours of the sample galaxies avoiding the line emission contamination. We define the "characteristic" bulge colour as that measured at r_e, and the disk colour gradient computed as the slope of a linear fit of colour vs. radial distance. Figure 2 (left panel) shows the $R - I$ characteristic bulge colour for our sample galaxies, along with comparison data from Balcells & Peletier. It can be seen that the characteristic colours are uniformly distributed in all nuclear types and are similar to those of normal galaxies[3]. This information suggests that the ages and/or metallicites of our sample bulges are similar to those of normal galaxies. On the other hand, the reduced amount of reliable disk colour gradient data (figure 2, right panel) precludes the possibility of a detailed analysis. Only a trend of Seyfert 1 central colours to be redder than those of other nuclear types can be guessed from our sample data.

5. Conclusions

The results from our morphological and structural analysis suggest that there is not correlation between nuclear activity and main structural properties (bulge and disk parameters), but with the existence of gravitational perturbations like bars and interactions. On the other hand, the characteristic colours of the host galaxies are similar in all nuclear types, and comparable with those of normal galaxies of similar morphological type.

Acknowledgements

The JKT is operated at the island of la Palma by the RGO at the Observatorio del Roque de los Muchachos of the Instituto de Astrofísica de Canarias. We would like to thank CAT for awarding observing time.

References

Balcells, M., Peletier, R. F., 1994, AJ, 107, 135
Dahari, O., 1985, ApJS, 57, 643
de Vaucouleurs, G., et al., 1991, Third Reference Catalog of Bright Galaxies (RC3)
Freeman, K.C., 1970, ApJ, 160, 811
Kormendy, J., 1977, ApJ, 217, 406
Mediavilla, E., Pastoriza, M.G., Battaner, E., 1989, Ap&SS, 157, 145
Mulchaey, J.S., Regan M.W., 1997, ApJ, 482, L135
Sánchez-Portal, M., Díaz, A. I., Terlevich, R. J., Terlevich, E., Álvarez-Álvarez, M., Aretxaga, I., 1998, MNRAS, submitted
Sánchez-Portal, M., Díaz, A.I., Terlevich, R.J., Terlevich, E., Álvarez-Álvarez, M., Aretxaga, I., 1999, MNRAS, in preparation
Sandage, A., Tammann, G.A., 1981, *A revised Shapley-Ames Catalog of Bright Galaxies*
Xanthopoulos, E., 1996, MNRAS, 280, 6

[3]only one Seyfert 1 galaxy with a very compact bulge shows a bluer characteristic colour, but it is contaminated by H_α emission from the NLR

KINEMATICS AND STELLAR POPULATIONS OF SEYFERT GALAXIES

LUIS JIMÉNEZ-BENITO AND ANGELES DÍAZ
Dpto. de Física Teórica, C-XI, Universidad Autónoma de Madrid, Cantoblanco, 28049-Madrid, Spain
ROBERTO TERLEVICH
Institute of Astronomy, Madingley Road, Cambridge CB30HA, United Kingdom
AND
ELENA TERLEVICH
Instituo Nacional de Astronomía, Óptica y Electrónica, Tonantzintla, Puebla, Mexico

Abstract. We have analyzed high signal-to-noise spectra of 10 active galaxies in order to investigate the stellar populations of these objects and the kinematics of the gas and stars in their nuclei.

1. Introduction

In the last years, several groups have found evidence for the presence of young stars in type 2 Seyfert nuclei (e.g. González-Delgado *et al.* 1998). Spectral features of young stars have also been found in LINERs (Maoz *et al.*, 1998). If unification models are correct, young stars should also be found in Seyfert 1 galaxies. We have looked for signs of young red supergiants in the spectra of Seyfert 1 galaxies by observing the near-IR CaII triplet (CaT) (Díaz *et al.*, 1989). On the other hand, kinematical studies can also be used to test if Seyferts of types 1 and 2 are essentially the same kind of galaxy. The kinematics of Seyfert galaxies has been studied by Nelson & Whittle (1995, 1996) who concluded that the stellar kinematics of Seyfert galaxies is normal, and the gas motions in their nuclei are dominated, in the majority of the objects, by the gravitational field of the bulge. Our aim here is to try to confirm these results.

Astrophysics and Space Science is the original source of publication of his article. It is recommended that this article is cited as: *Astrophysics and Space Science* **263**: 111–114, 1999.

2. Sample, Observations and Methods of Measurement

Our sample includes 6 type 1 Seyfert galaxies (NGC 4235, NGC 5940, NGC 6104, Mk 766, Mk 885 and Mk 1098), two Seyfert 2 (Mk 423, Mk 270), a compact narrow line radio galaxy (3C305) and a starburst galaxy (Mk 759). To increase the number of galaxies in our analysis, we have added the Seyfert galaxies from the sample of Terlevich *et al.* (1990).

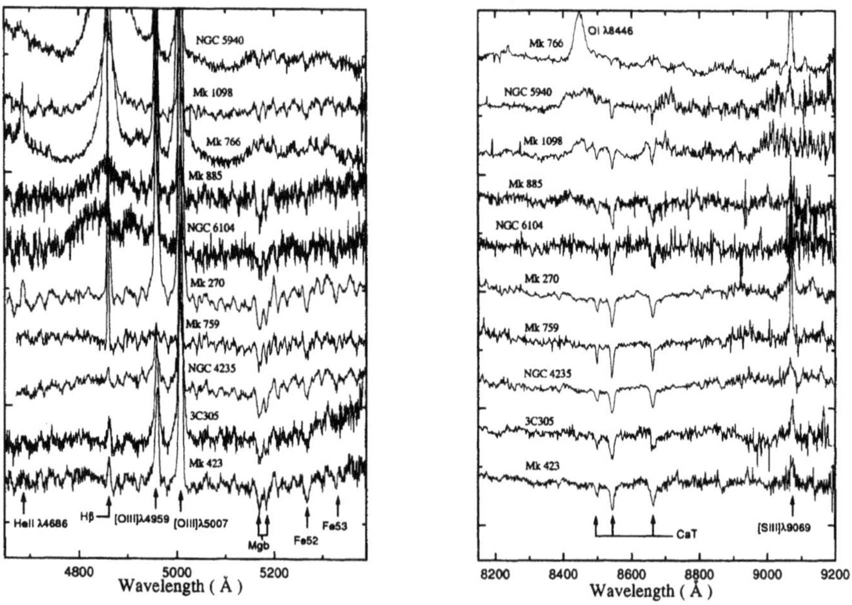

Figure 1. Blue (left) and near-IR (right) spectra of the observed galaxies showing the stellar absorption features.

High resolution spectra of the sample galaxies were obtained in 1991 May with the WHT using the ISIS spectrograph. Spectra were taken in two spectral ranges, one centered on the CaT ($\sim$ 8600Å), and the other centered on the Mgb feature at 5180 Å. The corresponding spectra are presented in Fig. 1.

We have measured the emission line widths by fitting a gaussian to their cores. The nuclear stellar velocity dispersions were measured using the cross-correlation method (Tonry & Davis, 1979). The equivalent widths (EWs) of the Mgb, Fe52, Fe53 and CaT(2+3) indices were also measured. A complete description of the measurement process and the definitions of the continuum and central band-passes of the spectral indices can be found in Palacios *et al.* (1997).

3. Discussion and Results

3.1. CAT DILUTION IN SEYFERT 1

We have compared the CaT dilution vs. the dilution of the blue indices in Fig. 2. The reference values of the indices have been extracted from the results of Terlevich *et al.* (1990) and Idiart *et al.* (1996) on normal early type spiral galaxies. In this figure we can see two groups of type 1 Seyfert galaxies. In the first one the indices are not diluted. In this group we can find Seyferts like NGC 4235 or NGC 6104 that have spatially resolved nuclei (Malkan *et al.*, 1998). The galaxies of the second group, on the contrary, show unresolved central point sources and their nuclear spectral indices are diluted by an extra continuum. The EW(CaT) of the galaxies in this last group seems to be smaller than we expected from the dilution of their blue indices. This can be explained if part of the extra continuum (referred to in the literature as the *blue featureless continuum*, BFC) of these galaxies is originated in a young starburst (see Terlevich *et al.*, 1990); alternatively, the blue indices could be diluted not by the BFC but by the strong FeII permited emission lines present in the spectra of these objects.

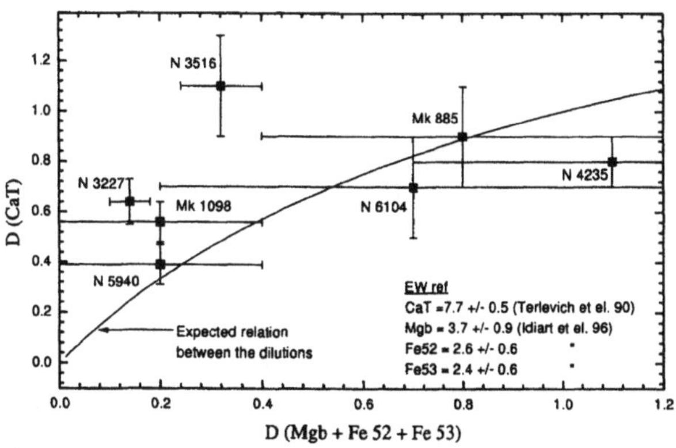

Figure 2. CaT and Mgb dilution in Seyfert1

3.2. EMISSION LINE WIDTHS VS. STELLAR VELOCITY DISPERSIONS

We have analyzed the kinematical data on our sample galaxies together with those of Nelson & Whittle (1995). In Fig. 3 we plot the difference between the width of the [OIII]λ5007Å emission line and the nuclear stellar velocity dispersion of the galaxies of both samples. In the majority of the objects,

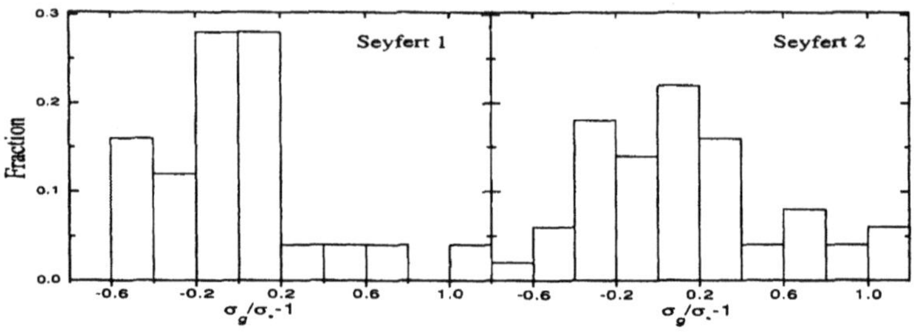

Figure 3. Relation between the emission lines width and the stellar velocity dispersion in Seyfert galaxies

gas and stellar velocity dispersions are similar. This confirm the previous result of Nelson & Whittle (1996). Nevertheless, we have found evidence for a group of Seyfert 2 galaxies which have emission line widths much broader than their stellar velocity dispersion. This subgroup of type 2 Seyferts had previously been found (Terlevich *et al.*, 1990), and their behaviour has been related to the presence of strong radio-jets, or to merging processes (Nelson & Whittle, 1996). We have also found some type 1 Seyfert galaxies which show the opposite behaviour. The kinematics of these galaxies can be explained by projection effects or by pure morphological reasons, since Seyfert 1 galaxies are, in average, of earlier morphological type than type 2 Seyferts (Malkan *et al.*, 1998). In early type galaxies, the stellar velocity dispersions are greater than those derived from the emission lines, whereas in late type spirals the widths of both, absorption and emission lines, are similar (Vega *et al.*, 1998).

References

Díaz, A.I., Terlevich, E., Terlevich, R., 1989, MNRAS, 239, 325
González Delgado, R.M., Heckman, T.M., Leitherer, C., Meurer, G., Krolik, J., Wilson, A.S., Kinney, A., Koratkar, A., 1998, ApJ, 505, 174
Idiart, T.P., de Freitas Pacheco, J.A., Costa, R.D.D., 1996, AJ, 112, 2541
Malkan, M.A., Gorjian, V., Tam, R., 1998, ApJS, 117, 25
Maoz, D., Koratkar, A., Shields, J.C., Ho, L.C., Filippenko, A.V. & Sternberg, A., 1998, AJ, 116, 55
Nelson, C.H., Whittle, M., 1995, AApJS, 99, 67
Nelson, C.H., Whittle, M., 1996, ApJ, 465, 96
Palacios, J., García-Vargas, M.L., Díaz, A.I., Terlevich, R. & Terlevich, E., 1997, A&A, 323, 749
Terlevich, E., Díaz, A.I., Terlevich, R., 1990, MNRAS, 242, 271
Tonry, J., Davis, M., 1979, AJ, 84, 1511
Vega Beltrán, J.C., Pignatelli, E., Zeilinger, W., Bertola, F, Beckman, J.E., Pizzella, A., Corsini, E., 1998, in D.R. Merritt, M. Valluri & J.A. Sellwod (eds.), *Galaxy Dynamics*, ASP Conference Series, in preparation.

X-RAY OBSERVATIONS OF ACTIVE GALACTIC NUCLEI IN THE COSMOLOGICAL CONTEXT

X. BARCONS, R. CARBALLO, F.J. CARRERA, M.T. CEBALLOS AND J.I. GONZÁLEZ-SERRANO
Instituto de Física de Cantabria (CSIC-UC)
Facultad de Ciencias, 39005 Santander, Spain

A.C. FABIAN
Institute of Astronomy
Madingley Road, Cambridge CB3 0HA, UK

AND

R.S. WARWICK
Department of Physics and Astronomy
University of Leicester, Leicester LE1 7RH, UK

Abstract. We review recent discoveries of obscured Active Galactic Nuclei at significant redshifts with X-ray observations. Most, if not all, the X-ray luminous galaxies found at faint X-ray fluxes could in fact be obscured AGN, where the obscuring material is also forming stars copiously. Given that AGN of various types appear to dominate the source populations at all X-ray fluxes, we also discuss their clustering properties. Current results imply that the population of X-ray selected AGN is not strongly biased, and therefore X-ray observations can be used to trace the distribution of matter out to redshifts ~ 2.

1. Introduction

Active Galactic Nuclei (AGN) are known to dominate the high-galactic latitude soft X-ray source counts down to fairly faint fluxes. Deep surveys carried out with ROSAT and subsequent follow-up at other wavebands reveal that 60-70% of the sources down to a 0.5-2 keV flux of $\sim 5\times 10^{-15}\mathrm{erg\,cm^{-2}\,s^{-1}}$ are AGN, mostly of type I (Hasinger et al 1997, Schmidt et al 1998, McHardy et al 1998, Boyle et al 1994).

Astrophysics and Space Science is the original source of publication of his article. It is recommended that this article is cited as: *Astrophysics and Space Science* **263**: 115–118, 1999.

These and other ROSAT surveys also find a significant fraction of X-ray luminous galaxies at the faintest flux levels (Carballo et al 1995). These objects are in principle a mixed bag of Seyfert 2 galaxies, star-forming galaxies and other galaxies with not very obvious signs of activity in their optical spectra. The narrow-line X-ray emitting galaxies (NLXGs) are the largest sub-class of these objects and their main distinctive property with respect to type 1 AGN is their flatter X-ray spectrum (Almaini et al 1996, Romero-Colmenero et al 1996). This immediately suggests that photoelectric absorption might operate in the NLXGs.

In this paper we review our current understanding of the NLXGs as obscured AGNs and also the X-ray selected AGN clustering properties which indicate that the distribution of these objects is not strongly biased with respect to the underlying matter distribution.

2. Obscured Active Galactic Nuclei

There is now a handful of obscured AGN at significant redshifts discovered in ROSAT and ASCA X-ray observations. The highest redshift one ($z = 2.35$) was discovered by ROSAT (Almaini et al 1995) and shows narrow Lyα and CIV emission lines. Infrared spectroscopy of this object shows broad Hα and no detectable Hβ which argues in favour of a moderate absorption column for 'normal' dust to gas ratios (Georgantopoulos et al 1998). ASCA observations of this object show a rather large photoelectric absorbing column. The fact that X-ray photoelectric absorption appears to infer larger extinction than what is observed, has also been found in Seyfert 2s (see Simpson 1998 for NGC 3281). Granato, Danese & Franceschini (1997) attribute this to the fact that most of the gas responsible for photoelectric absorption lies very close to the central engine, where the ambient temperature is above the dust sublimation point.

ASCA has also discovered a few obscured AGN: among them one at $z = 0.9$ (Ohta et al 1996) and another one at $z = 0.672$ (Boyle et al 1998). In Barcons et al (1998) we report on the discovery and multiwavelength observations of a radio-loud obscured X-ray selected AGN at $z = 1.246$. This source (RX J1011.2+5545) was discovered in the ROSAT UK Medium Sensitivity Survey (Carballo et al 1995), as being the hardest one in the survey and relatively bright (0.5-2 keV flux $\sim 10^{-13} \mathrm{erg\,cm^{-2}\,s^{-1}}$). Early optical imaging revealed a unique faint optical counterpart within the PSPC error circle with $R = 21.0$ which also points towards an obscured source. Optical spectroscopy reveals a narrow-line AGN with strong high ionisation lines ([NeV]3426) and no broad components in the CIII]1909 line. A very weak and broad MgII 2800 feature is barely detected, confirming that the nucleus is heavily obscured. The strong [OII]3727 feature is also suggestive of copious star formation. An ASCA observation of this source, together with the ROSAT data, can be fitted with an absorbed power law with photon spectral index $\Gamma \sim 1.45$ (typical of radio-loud objects) and moderate absorption $N_{HI} \sim 2 \times 10^{21}\mathrm{cm}^{-2}$. Archival radio observations at various frequencies show that this object is radio-loud and with a lobe-dominated structure. The overall energy spectral distribution confirms the absorbed character of this AGN (with a 2-10 keV luminosity $\sim 10^{45}\mathrm{erg\,s}^{-1}$) which is the first radio-loud version of this type of source.

3. Are the NLXGs obscured AGN?

Iwasawa et al (1997) have revealed the link betweek obscured AGN and NLXGs in the source AXJ1749+684 at $z = 0.09$. Diagnostic diagrams with optical emission lines place this object in the borderline between starburst and active galaxies, but its hard X-ray luminosity clearly indicates the presence of an AGN. Similar arguments apply to NGC 7582 (Schachter et al 1998).

Fabian et al (1998) suggest that NLXGs might just be the low luminosity version of obscured AGNs. A cloud of gas and dust around the central engine would have the desired property of both absorbing the X-rays and the broad line region and forming stars. This surrounding extended nuclear starburst can survive if its mass exceeds that of the central engine. In this case, a small virialised phase of the gas could feed the accretion disk through Bondi accretion, at a sub-Eddington rate. When the mass of the black hole is much larger, the obscuring star-forming cloud can no longer survive and we have a standard AGN.

According to this, Fabian & Iwasawa (1998) estimate that $\sim 95\%$ of the accretion in the Universe occurs in this absorbed mode, and the energy is then re-radiated somewhere in the infrared. On-going and future surveys with ISO and FIRST in the infrared and SCUBA and LSA/MMA in the submillimeter should be able to test this model.

4. Clustering of X-ray selected AGN

Since it appears that a relevant fraction of the sources that make the 30-50% yet unresolved component of the X-ray background are also AGN, it is clear that AGN of various types dominate the extragalactic X-ray population. Attempts to trace the large-scale structure of the Universe with X-ray observations (Carrera, Fabian & Barcons 1997; Barcons, Fabian & Carrera 1998) require therefore the knowledge of how biased is the AGN population with respect to the underlying matter distribution.

Carrera et al (1998) have carried out the first successful approach to measure the clustering properties of X-ray selected AGN, using a sample of 235 objects from both the RIXOS survey (Mason et al 1998) and the ROSAT deep survey (Boyle et al 1994). Clustering is detected below $40-80\,h^{-1}$Mpc ($H_0 = 100\,h\,\mathrm{km\,s^{-1}\,Mpc^{-1}}$) at low redshifts, but not at high redshift. This constrains evolution to be approximately faster than comoving and the correlation length r_0 to lie between 1.5 and $5.5h^{-1}$ Mpc. This is in agreement with the clustering of IRAS sources (which are thought to be unbiased with respect to the dark matter distribution). We then conclude that X-ray selected AGN trace mass with a bias parameter $b_X \leq 0.7 - 1.2$. This is very encouraging in terms of using X-ray observations to measure the structure of the Universe at redshifts $z \sim 1 - 2$ where the X-ray volume emissivity peaks.

Acknowledgments

Partial financial support for this work was provided by the DGESID under project PB95-0122. ACF thanks the Royal Society for support.

References

Almaini, O. et al, 1995, MNRAS, 277, L31
Almaini, O. et al, 1996, MNRAS, 282, 295
Barcons, X., Fabian, A.C., Carrera, F.J., 1998, MNRAS, 293, 60
Barcons, X., Carballo, R., Ceballos, M.T., Warwick, R.S., González-Serrano, J.I., 1998, MNRAS, in the press
Boyle, B.J. et al, 1994, MNRAS, 271, 639
Boyle, B.J. et al, 1998, MNRAS, 297, L53
Carballo, R. et al, 1995, MNRAS, 277, 1312
Carrera, F.J., Fabian, A.C., Barcons, X., 1997, MNRAS, 285, 820
Carrera, F.J. et al, 1998, MNRAS, 299, 229
Fabian, A.C., Barcons, X., Almaini, O., Iwasawa, K., 1998, MNRAS, 297, L11
Fabian, A.C., Iwasawa, K., 1998, MNRAS, submitted
Georgantopoulos, I. et al, 1998, MNRAS, in the press
Granato, G.L., Danese, L., Franceschini, A., 1997, ApJ, 486, 147
Hasinger, G. et al, 1998, A&A, 329, 482
Iwasawa, K., Fabian, A.C., Brandt, W.N., Crawford, C.S., Almaini, O., 1997, MNRAS, 291, L17
Mason, K.O. et al, 1998, MNRAS, submitted
McHardy, I.M. et al, 1998, MNRAS, 295, 641
Ohta, K. et al, 1996, Nature, 382, 426
Romero-Colmenero, E., et al, 1996, MNRAS, 282, 94
Schachter, J.F. et al, 1998, ApJ, 503, 123
Schmidt, M. et al, 1998, A&A, 329, 495
Simpson, C., 1998, Preprint astro-ph/9807191

A METHOD FOR OBTAINING THE NUCLEAR COMPONENT IN ACTIVE GALAXIES

MIGUEL SANCHEZ PORTAL
Dpto. de Física Teórica, Universidad Autónoma de Madrid, Cantoblanco, 28049-Madrid, Spain

LAEFF/INTA, P.O. Box 50727, 28080-Madrid, Spain

AND

ANGELES I DIAZ BELTRAN
Dpto. de Física Teórica, Universidad Autónoma de Madrid, Cantoblanco, 28049-Madrid, Spain

Abstract. We present a new method for the obtention of the nuclear component of active galatic nuclei, that relies on the convolution of a two-dimensional model of the host galaxy, based on seeing-free morphological parameters, with the seeing PSF. The procedure overcomes some limitations of the methods currently in use and does not assume that the nucleus is unresolved.

1. Introduction

Almost all the methods currently used for nuclear-component deconvolution in active galaxies are based upon the assumption that the nucleus of the galaxy is not resolved (standard AGN model). Therefore, can be analytically represented as $I(r) = F_T\delta(r) \otimes PSF(r)$. The procedure for derivation of the main structural parameters then relies on a χ^2 minimization of the five free parameters $(F_T, r_e, \mu_e, r_0, \mu_0)$ that represent the nuclear component, bulge and disk of the galaxy along with a convolution with the seeing PSF. A problem of this approach is that the parameters are tightly coupled and therefore the final result has a strong dependence on the initial guess. Attempting to overcome this limitation, we have tried a different approach that avoids the simultaneous derivation of all five parameters and does not assume any analytical expression for the nuclear profile.

Astrophysics and Space Science is the origirlal source of publication of his article. It is recommended that this article is cited as: *Astrophysics and Space Science* **263**: 119–122, 1999.

2. The deconvolution method

Our deconvolution method consists of five steps that can be summarized as follows: (a) Derivation of the bulge and disk parameters by means of an iterative procedure (Kormendy, 1977). The details of the method are described elsewhere (Sánchez-Portal, 1996; Sánchez-Portal et al., 1998c) We assume that beyond certain radius $r_{ini}(r_e, \sigma_{PSF})$ from the center the seeing efect is negligible and therefore the bulge (r_e, μ_e) and disk parameters (r_0, μ_0) will be essentially seeing-free. r_{ini} is estimated using an inital guess of r_e and σ_{PSF} and the curves from Capaccioli & de Vaucouleurs (1983). (b) Construction of a two-dimensional model of the galaxy, assuming that the galaxy is accurately representd by the sum of the bulge and disk components. Furthermore, we assume that the galaxy is thin and the surface brightness distribution is symmetric with respect to an axis perpendicular to the galactic plane. Therefore, the isophotes are concentric ellipses with $1 - \epsilon \equiv b/a = \cos i$, where i is the inclination angle. This last assumption forces us to select objects whose isophotes keep a roughly constant position angle or have a low inclination. (c) Calculation of the PSF fitting a sum of gaussians ($I(r) = I_0 \left[\exp\left(-r^2/2\sigma_1^2\right) + k_1 \exp\left(-r^2/2\sigma_2^2\right)\right]$) to an azimuthally averaged profile of a mean field star image. (d) Convolution of the galaxy model with the the PSF model in the sky plane. (e) Extraction of a radial surface brightness profile from the convolved model and subtraction from the observed profile. The residual should give the nuclear component.

3. Results

We have tested this method with VRI CCD images of nearby low-inclination galaxies taken with the Jacobus Kapteyn Telescope (JKT). The pixel size is 0.3 arcsec. All the observations where done in good seeing condition, FWHM $\sim$ 1 - 1.5 arcsec.

3.1. NGC 6340

We first applied the method to an I band image of anormal S0/a galaxy with regular morphology like NGC 6340 in order to have a check of validity. The maximum deviation between the observed profile and the seeing-convolved model is $\sim$ 7 % and can be attributed to a seeing overestimation. Figure 1 (left panel) shows the fitting results.

3.2. NGC 4151

NGC 4151 is a nearby early type spiral Seyfert 1.5 galaxy with strong nuclear emission. The results of applying our method are shown in figure 2. It

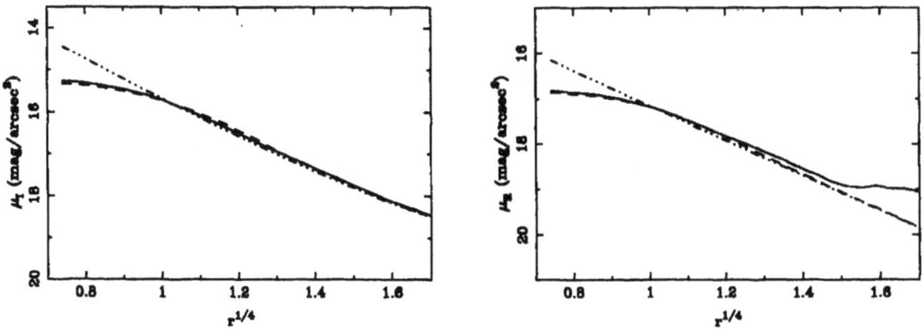

Figure 1. Left: observed profile (solid line), model (dashed-doted line) and convolved model (dashed line) for NGC 6340 (I band); right: the same for NGC 3982 (R band)

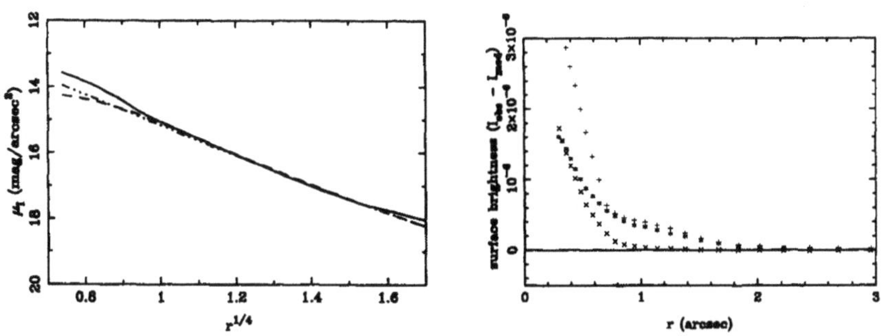

Figure 2. Left: model for NGC 4151 (I band); legends as in fig. 1. Right: surface brigthness profiles for the nuclear component in V (*), R (+) and I (×) bands

can be seen that the procedure is sensitive to the extension of the emission region. In fact, our narrow-band H_α photometry reveals a NLR of about 2 arcsec, in good agreement with the resulting nuclear profile. A simple gaussian fit to the I nuclear profile gives FWHM = 0.751" ± 0.003"; the same fit to field stars gives FWHM = 0.714". This means that the profile is compatible (within the uncertainties involved in the calculations) with a point-like (unresolved) source, but also with the predicted size of the nuclear cluster postulated by the starburst model (Terlevich & Melnick, 1985; Terlevich et al., 1987; Terlevich, 1989). The predicted observed (i.e., seeing-convolved) size for the nuclear starburst (assuming for simplicity that can be represented by a gaussian profile) is FWHM $\simeq$ 0.738" (Terlevich *et al.*, 1991; Sánchez-Portal, 1998), in good agreement with the nuclear size value obtained from our calculations. We have also derived the luminosity contributions due to nucleus and host galaxy (contained in a 3 arcsec radius): 40.3% for the V band, 35% for the R band and 7.1% for the I band.

3.3. NGC 3982

The right panel of figure 1 shows the results of applying the method to the Seyfert 2 galaxy NGC 3982, an early-type spiral. We have used only the bulge to model the central part of the galaxy, since it seems to present a Freeman type II disk (Freeman, 1970; Sánchez-Portal et al., 1998c). As can be seen, no detectable nuclear excess is observed; this result is compatible with our surface photometry, where only weak H_α emission is found and no [OIII] signature is detected in colours (Sánchez-Portal et al., 1998b). A possible explation to this lack of nuclear excess in our analysis is that the nuclear emission and continuum are too faint to be detected within the uncertainties involved in the PSF and morphological parameters calculations.

4. Conclusions

We have presented a new method for nuclear-component deconvolution. It does not assume that the nucleus is unresolved. When applied to strong active galaxies, it reveals to be sensitive to the emission extension. Nevertheless, its validity is limited when working with faint nuclear sources. The results presented here refer to low-inclination galaxies. Currently we are applying the same method to high inclination ones.

Acknoledgements

The JKT is operated at the island of la Palma by the RGO at the Observatorio del Roque de los Muchachos of the Instituto de Astrofísica de Canarias. We would like to thank CAT for awarding observing time.

References

Capaccioli, M., de Vaucouleurs, G., 1983, ApJS, 52, 465
Kormendy, J., 1977, ApJ, 217, 406
Freeman, K. C., 1970, ApJ, 160, 811
Sánchez-Portal, M., 1996, Ph. D. Thesis, Univ. Autónoma de Madrid
Sánchez-Portal, M., 1998, MNRAS, submitted
Sánchez-Portal, M., Díaz, A. I., Terlevich, R. J., Terlevich, E., Álvarez-Álvarez, M., Aretxaga, I., 1998, MNRAS, submitted
Sánchez-Portal, M., Díaz, A. I., Terlevich, R. J., Terlevich, E., Álvarez-Álvarez, M., Aretxaga, I., 1998, MNRAS, in preparation
Terlevich, R. J., Melnick, J., 1985, MNRAS, 195, 839
Terlevich R. J., Melnick J., Moles M., 1987, In Khachikyan E., Fricke K. J., Melnick J. (eds.) Observational Evidence of Activity in Galaxies p. 149. Reidel, Dordretch
Terlevich R. J., 1989, In Tenorio-Tagle G., Moles M., Melnick J. (eds.) Structure and Dynamics of the Interstellar Medium p. 343. Springer-Verlag, Berlin
Terlevich, R. J., Sánchez-Portal, M., Díaz, A. I., Terlevich, E., 1991, MNRAS, 249, 36

SPECTROSCOPY OF THE EXTENDED EMISSION LINE REGION IN NGC 4388

L. TOMAS-CALDERON AND J.A. ACOSTA-PULIDO
ISO Data Centre
Astrophysics Division of ESA, Madrid

B. VILA-VILARO
Nobeyama Radio Observatory, Nagano, Japan

AND

I. PEREZ-FOURNON
Instituto de Astrofisica de Canarias, La Laguna, Tenerife

Abstract. We present preliminary results of long-slit spectroscopic observations of the extended emission line region (hereafter, EELR) in NGC 4388. We detect broad emission lines in the line complexes Hα+[NII] and [SII] which can be modelled by the superposition of several narrow components. Multicomponent line profiles appear at positions very close to the ionization cone axis. There is also some correspondence with the radio emission, which suggests that there is interaction between the radio jets and the ionized gas. We do not confirm the claim of a broad Hα component (FWZI=4000 km/s) by Shields and Filippenko (1996).

1. Introduction

NGC 4388 is a high-inclination Sab galaxy, containing a Seyfert 2 nucleus. High excitation gas emission (hereafter EELR) has been detected outside the nuclear region, defining a bi–conical morphology (Corbin *et al.* , 1988; Pogge 1988). The axis of the cones is almost perpendicular to the galaxy disk and aligns well with the axis of the radio emission. The excitation of the gas is mostly due to photoionization by the nuclear radiation field, which escapes preferentially along the cone axis.

Astrophysics and Space Science is the original source of publication of his article. It is recommended that this article is cited as: *Astrophysics and Space Science* **263**: 123–126, 1999.

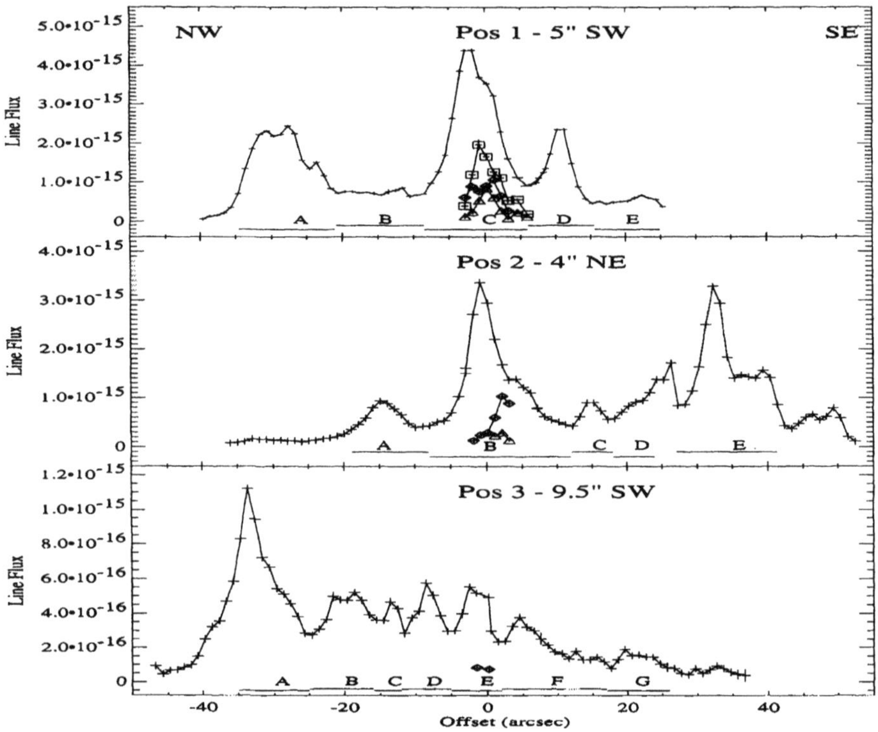

Figure 1. Spatial distribution of Hα line intensity along the slit for three observed positions.

2. Observations and Data Reduction

The EELR of NGC 4388 was observed using the ISIS spectrograph at the 4.2 m William Herschel Telescope of the Observatorio del Roque de los Muchachos, La Palma. Several spectra were taken crossing the nucleus and at several offset positions (a detailed study will be presented elsewhere). In this paper we present preliminary results corresponding to three off–nuclear slit positions: Pos #1 – 5" SW, Pos #2 – 4" NE and Pos #3 – 9.5" SW from the optical nucleus. In all cases the slit was oriented at $P.A. = 102.5^o$.

The observations were reduced using standard procedures within the IRAF software package. The line flux and radial velocity were measured by fitting Gaussian profiles, using the task FITLINES (developed by J.A. Acosta–Pulido) within IRAF and DIPSO (Howart, Murray, Mills & Berry, 1997) when multicomponent fitting was necessary.

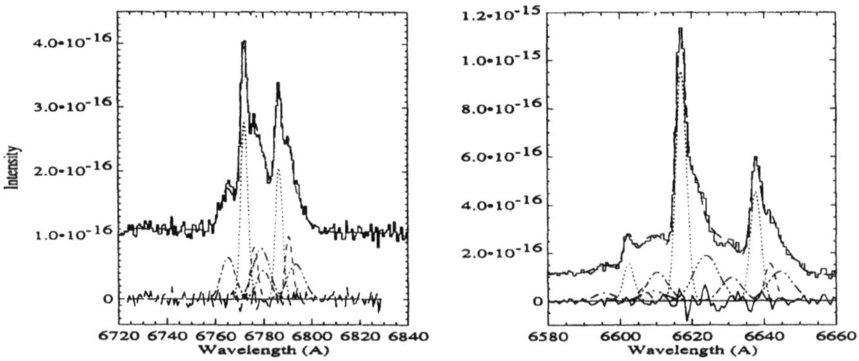

Figure 2. Decomposition of the complex [SII]λ6716, 6731 in the spectrum corresponding to Pos #1. The lines are descomposed into four components. We have plotted the observed line profile (histogram-like), the model, the individual components and the residuals. Each kinematical component is marked by a different line style.

3. Results

We have measured the line flux and radial velocity corresponding to Hα, [NII]λ6584 and [SII]$\lambda\lambda$6716, 6731 along the slit for the different positions. The Hα line intensity spatial distribution is presented in Fig. 1.

In most cases the emission lines are well modelled by a simple narrow component (unresolved). However, at certain slit positions the lines broaden and the line profiles are well reproduced by a multi-component model, in agreement with our previous results (Perez–Fournon *et al.* 1994). At those positions we have first modelled the complex [SII]$\lambda\lambda$6716, 6731 using several components and provided a minimum number of constrains (fixed wavelength separation, and same width for each pair of components). Once a satisfactory fit is found for those lines we extend the same kinematical model to the Hα + [NII] complex (the theoretical line ratio for the two [NII] lines was also used). We have found that in all analyzed regions the line profiles can be modelled by several kinematical components, consistently for both the [SII] and Hα + [NII] complex (see Fig 2). Therefore, we do not need a broad component for Hα as claimed by Shields and Filippenko (1996).

The regions where several components appear are limited to 5" around the cone axis (see Fig. 1) and coincide with the extension of the radio emission. The dominant component follows the rotation curve of the galaxy and appears unresolved. Other components appear as radial flows with velocity differences up to 250 km/s, blueshift dominates in the NE and redshift in the SW. At Pos #1 and #2 broader components (FWHM 150–360 km/s) appear.

We have looked at emission line ratios integrating upon selected regions, as marked in Fig. 1. Despite the limited number of line ratios presented here, the regions closer to the cone axis seems to show larger values of the line ratios [NII]/Hα and [SII]/Hα, as well as an slight increase of the density for some kinematical components.

4. Discussion

We have found that most of the line emission in the EELR of NGC 4388 can be attributed to relatively quiet gas following the rotation of the galaxy. The broad wings appearing in the Hα+[NII] and [SII] complexes can be explained as the superposition of several kinematical components which correspond to the same kinematical model. This result was already discussed by Pérez–Fournon *et al.* (1994) and seems to contradict the claim of extranuclear broad Hα emission (Shields & Filippenko, 1996). We cannot completely rule out their claim because of the limited spatial coverage of our spectra. The location of the multiple components is limited to regions very close (within 5") to the ionization cone axis, and distances $\leq$ 9" from the nucleus. Moreover we have found some correspondence between those regions and features detected in the radio emission, using the recently published VLA 3.5 cm map (Falcke, Wilson & Simpson, 1998). However, the comparison is not straighforward due to the fact that the radio nucleus is displaced $\sim$2" N with respect to the apparent optical one.

Several scenarios can be proposed as the origin of the different kinematical components. They could represent scattered radiation from a partially hidden NLR. The non–detection of polarized broad components in permitted lines either in the optical (Goodrich *et al.* 1994) nor in the near-IR (Veilleux, Goodrich & Hill, 1997) does not support the scattering scenario. Alternatively, the different components can be produced in situ, *e.g.* as a result of the interaction between the radio jets and the ISM. The good correspondence between HST emission–line images and the radio emission (Falcke, Wilson & Simpson, 1998) is strongly suggestive of such an interaction.

References

Corbin, R., Baldwin, J.A., Wilson A.S., 1988, ApJ, 334, 584
Falcke, H., Wilson, A.S., Simpson, C., 1998, ApJ, 502, 199
Goodrich, R.W., Veilleux, S., Hill, G.J., 1994, ApJ, 422, 521
Howart, D.I., Murray, J., Mills, D., Berry D.S., 1997, Starlink User Note 50,20
Perez-Fournon, I., Vila-Vilaro, B., Acosta-Pulido, J.A. et al., 1994. in 'The Nature of Compact Objects in AGN', Cambridge University Press, p.40
Pogge, R., 1988, ApJ, 311, 393
Shields, J.C., Filippenko, A.V., 1996, A&A, 311, 393
Veilleux, S., Goodrich, R.W., Hill, G.J., 1997, ApJ, 477, 631

BLUE COMPACT GALAXIES AT REDSHIFTS Z=0.2-1.3

R. GUZMÁN
Dept. of Astronomy, Yale University
260 Whitney Ave., New Haven, CT 06511

Abstract.
We study the nature of faint blue compact galaxies (BCGs) at redshifts $z \sim 0.2 - 1.3$ using Keck and HST. Despite being very luminous ($L_B \sim L^*$), most distant BCGs have masses $M \sim 10^{10} M_\odot$, i.e., they are *dwarf* stellar systems. The majority of these galaxies have colors, sizes, surface brightnesses, luminosities, velocity widths, excitations, star formation rates (SFR), and mass-to-light ratios characteristic of the most luminous nearby HII galaxies. The more massive BCGs form a more heterogeneous class of evolved starbursts, similar to local disk starburst galaxies. Without additional star formation, HII-like BCGs will most likely fade to resemble today's spheroidal galaxies such as NGC 205.

1. The Galaxy Sample and Data Set

Our galaxy sample consists of 45 compact emission-line galaxies selected from I_{814} HST images of the flanking fields (FF) around the Hubble Deep Field (HDF; [13]). These objects are faint ($I_{814} > 23.74$) and compact, in the sense that they have small apparent half-light radii ($r_{1/2} \leq 0.5$ arcsec) and high surface brightnesses ($\mu_{I814} \leq 22.2$ mag arcsec^{-2}). Spectra for these objects were obtained using LRIS at the Keck telescope with a slitwidth of 1.1 arcsec and a 600 l/mm grating. The effective resolution is ~3.1 Å FWHM. Typical exposure times were 3000s. The total spectral range is ~4000-9000 Å. The spectra of these distant galaxies exhibit prominent oxygen and Balmer emission lines and blue continua characteristic of vigorous star-forming systems at redshifts $z = 0.2 - 1.3$. We also obtained two 300s direct V-band exposures with LRIS in order to derive colors. Most of these objects are bluer than a typical Sbc galaxy. Thus, hereafter we refer to this sample as blue compact galaxies (BCGs).

Astrophysics and Space Science is the original source of publication of his article. It is recommended that this article is cited as: *Astrophysics and Space Science* **263**: 127–130, 1999.

The final data set includes: redshifts, colors, absolute blue magnitudes (M_B), half-light radii (R_e), average surface brightnesses within R_e (SB_e), velocity widths (σ), masses (M), mass-to-light ratios (M/L), O[III]/Hβ line ratios (for objects at $z < 0.7$), and SFRs. This data set is used to: i) identify the local counterparts of distant BCGs; ii) to assess their evolution with look-back time; and iii) to study their role on the star formation history of the universe. A full description of the results summarized here can be found in [5],[6],[7],[8],[9], and [10]. We adopt $H_0 = 50$ km s^{-1} Mpc^{-1} and $q_0 = 0.05$.

2. The Nature of Distant Blue Compact Galaxies

Global structural galaxy properties can be adequately described using the $SB_e - M_B$ and $R_e - \sigma$ diagrams ([4]). In these diagrams, various galaxy types define distinct correlations, albeit with large scatter (Figures 1a and 1b). Most of our BCGs follow the sequences defined by young starbursts, such as nearby HII galaxies. Despite having luminosities around L^*, most BCGs are *dwarf* stellar systems with masses $M \sim 10^{10} M_\odot$.

To parametrize the starburst properties of BCGs we use the M_B – [OIII]/Hβ and the M – SFR/M diagrams (Figures 2a and 2b). These diagrams discriminate among various types of starburst and active galaxies. Most BCGs with [OIII]/Hβ measurements lie in the moderate to high excitation regime populated by nearby HII galaxies. These objects also have SFR/M characteristic of HII galaxies. A second group of distant BCGs have [OIII]/Hβ and SFR/M similar to those of more evolved disk starbursts such as local DANS and SBNs. We have thus classified the BCG sample into HII-like and disk starburst-like galaxies, depending on their SFR/M (see Figure 2b). With this simple criterion, we find that ~60% of distant BCGs have global properties entirely consistent with those of the most luminous nearby HII galaxies, while the remaining 40% are similar to more evolved disk starbursts.

3. Implications for Theoretical Scenarios of Galaxy Evolution

Various theoretical scenarios have suggested that the faint BCGs are low-mass stellar systems experiencing their initial starburst at redshifts $z \leq 1$, some of which turn into the present population of spheroidal galaxies (Sph), such as NGC 205 ([1],[2]). To assess whether these distant young BCGs relate to nearby evolved stellar systems such as Sphs, we use the $R_e - \sigma$ diagram, since neither R_e nor σ depend strongly on the luminosity evolution of the stellar population. After studying the effect of various physical processes that may modify R_e and σ of BCGs as they evolve (such as dissipation, mergers, or galactic winds), we find no evidence against the idea

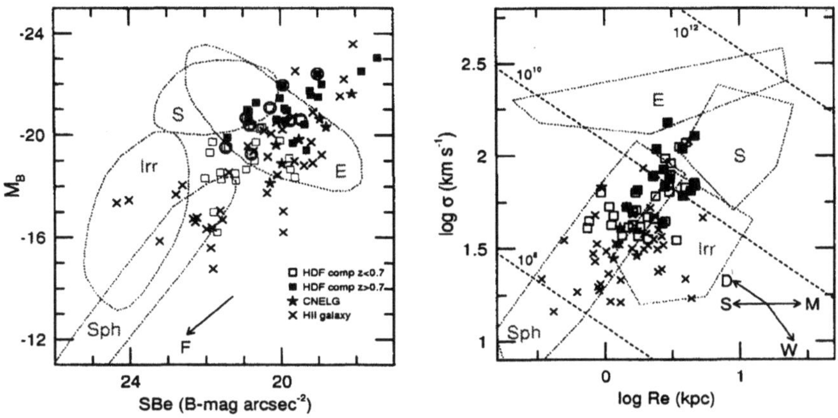

Figure 1. (a): SB_e vs. M_B. Dotted lines indicate the regions typically occupied by different classes of local galaxies (**E**llipticals, **S**pirals, **Irr**egulars, and **Sph**eroidals). Nearby HII galaxy sample (crosses) is from [12]. The arrow (**F**) represents the fading of the stellar population as BCGs evolve. (b): R_e vs. σ. Dashed lines represent constant-mass lines in $M_{\odot}$, estimated using the Virial Theorem. The arrows represent the effect of dissipation (**D**), mergers (**M**), stripping (**S**) and winds (**W**).

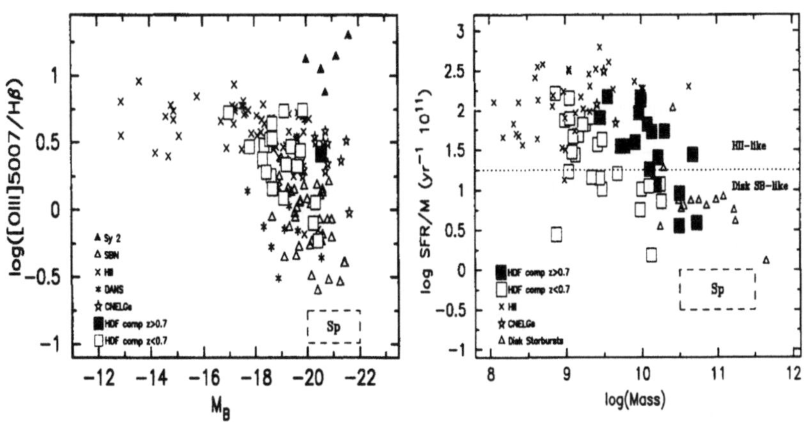

Figure 2. (a): M_B vs. [OIII]/Hβ. Local galaxy sample is from [3]; DANS: Dwarf Amorphous Nuclear Starbursts; SBN: Starburst Nuclei; Sy2: Seyfert 2 galaxies; HII: HII galaxies. Dashed lines represent the approximate location of spiral galaxies. (b): M vs. SFR/M. The dotted line represents the division between HII-like and disk starburst-like galaxies adopted in our classification.

that HII-like BCGs –mostly those with $M < 10^{10} M_{\odot}$– are related structurally and kinematically to the nearby dwarf galaxy population of Sph and Irr galaxies, rather than to giant systems such as E and S galaxies.

The evolution of HII-like BCGs into Sph or Irr galaxies may depend critically on their ability to retain part of the interstellar medium. Their extremely low M/L ratios suggest that the kinetic energy supplied by the current starburst is likely to be large enough, compared to the binding energy, to blow out most of the gas in galactic winds, halting thus the star-forming process. Without additional star formation they will fade within a few billion years.

The total amount of fading in BCGs will depend on the age/strength of the current burst *and* the extent of any underlying older population. Although it seems reasonable to expect that the current burst may be the last *major* star-forming episode, it is likely that it was not the first one. Metallicity estimates from emission-line ratio diagrams suggest previous chemical enrichment in BCGs (typical metallicities are ~0.7 solar). Indeed, BCGs show evidence for an older underlying population characterized by a lower surface brightness exponential profile and slightly redder colors than the starburst region. Using galaxy evolution models that best reproduce their observed M/L ratios and (B-V) colors ([11]), we estimate that the current starburst in BCGs may be, typically, ~ 0.5 Gyrs old, and involve only ~10% of the total galaxy mass. Assuming these burst parameters, the models predict that HII-like BCGs will fade by ~4 mag from their lookback times till the present (see fading vector in Figure 1a). After fading, they will have the same low luminosities and surface brightnesses characteristic of the brightest Sph galaxies. We thus conclude that HII-like BCGs are most likely the progenitors of today's bright Sph galaxies such as NGC 205.

Acknowledgements. This project is a collaborative effort of the DEEP team at UC Santa Cruz. Funding for this project is credited to NASA grants AR-06337.08-94A, AR-06337.21-94A, GO-05994.01-94A, AR-5801.01-94A, and AR-6402.01-95A from the STScI, and NSF grants AST 91-20005 and AST 95-29098. R. Guzmán also acknowledges support from the Hubble Fellowship grant HF-01092.01-97A.

References

1. Babul, A., Rees, M., 1992, MNRAS, 255, 346
2. Babul, A., Ferguson, H.C., 1996, ApJ, 458, 100
3. Gallego, J., Zamorano, J., Rego, M., Vitores A.G., 1997, ApJ, 475, 502
4. Guzmán, R., Lucey, J.R., Bower, R.G., 1993, MNRAS, 265, 731
5. Guzmán, R., Koo, D.C., Faber, S.M. et al. , 1996, ApJ, 460, L5
6. Guzmán, R., Gallego, J., Koo, D.C. et al. , 1997, ApJ, 489, 559
7. Guzmán, R., Jangren, A., Koo, D.C. et al. , 1998, ApJ, 495, L13
8. Koo, D.C., Bershady, M.A., Wirth, G.D. et al. , 1996, ApJ, 460, L5
9. Koo, D.C., Guzmán, R., Faber, S.M. et al. , 1995, ApJ, 440, L49
10. Phillips, A.C., Guzmán, R., Gallego, J. et al. , 1997, ApJ, 489, 54
11. Schmidt, A.A, Alloin, D., Bica, E., 1995, MNRAS, 273, 945
12. Telles, E., Terlevich, R., 1997, MNRAS, 286, 183
13. Williams, R.E. et al. , 1996, AJ, 112, 1335

THE STARBURST PROPERTIES OF THE INTERACTING SYSTEM ARP 299 (IC 694 + NGC 3690)

A. ALONSO-HERRERO, M. J. RIEKE AND G. H. RIEKE
Steward Observatory, Tucson, AZ 85721, USA

Abstract. We present a detailed study of the starburst properties of the components of the interacting system Arp 299. The burst ages range from 1 to 8 million years, with effective SFR from 60 to 300 $M_\odot$ yr^{-1}.

1. Introduction

Numerical simulations of collisions between gas-rich galaxies (Barnes & Hernquist 1996 and references therein) show that these processes are very efficient in transporting large quantities of molecular gas into the center of galaxies. As a result of the interaction a strong burst (or various bursts) of star-formation (SF) may occur, and an AGN may be activated, as observed in many interacting systems. Sanders et al. (1988) suggested an evolutionary connection between luminous infrared (LIR) galaxies, ultraluminous IR galaxies (ULIR) and optical QSOs. The interacting system Arp 299 is relatively nearby (distance $D = 42$ Mpc for $H_0 = 75$ km s^{-1} Mpc^{-1}) and has an IR luminosity ($L_{IR} = 5 \times 10^{11}$ $L_\odot$, which places this system in the LIR galaxy category) close to the limiting value for ULIR galaxies, and therefore it offers a unique opportunity to study in detail the effects of the interaction.

The starburst properties of Arp 299 have long been studied from the ground (e.g., Gehrz et al. 1983; Wynn-Williams et al. 1991). However these works lacked the necessary spatial resolution to isolate individual star-forming regions. The nucleus of IC 694 is component A; the components in NGC 3690 are B1 (the brightest IR source), B2 (the brightest optical source) and C (at the interface of the two galaxies). For this study we use recently obtained near-IR HST/NICMOS images combined with optical and near-IR spectroscopy (see Alonso-Herrero, Rieke & Rieke 1998).

Astrophysics and Space Science is the original source of publication of his article. It is recommended that this article is cited as: *Astrophysics and Space Science* **263**: 131–134, 1999.

2. The Models and the Properties of Arp 299

We make use of Rieke et al. (1993, R93) evolutionary synthesis models. Our choice for the initial mass function (IMF) and star formation rate (SFR) are the solar-neighborhood IMF3 (see table 5 in R93) and a single burst of SF with a Gaussian FHWM of 5×10^6 yr respectively. The peak of SF occurs at 5×10^6 yr after the beginning of the burst. The observables that we will try to reproduce with the evolutionary synthesis models are: bolometric luminosity $L_{\rm bol}$, the K-band luminosity L_K, the number of ionizing photons $N_{\rm Ly}$, the CO index (at 2.29 μm), the supernova rate (SNr) and and the equivalent width (EW) of Paα. As in R93, we normalize the output of the models to the observed values for each component, then we plot the evolution in time of the six quantities chosen in a diagram so that our target quantity will be unity. Mass-to-light ratios for both L_K and $L_{\rm bol}$ are used to calculate the contributions of the old underlying population to the total model K-band and bolometric luminosities to be compared with the observed values.

Since the output of the models depends on the input mass, we need to obtain an estimate of the available mass for each source. We use the dynamical mass of component A obtained from the CO index (Shier et al. 1996). This value is in reasonable agreement (but is larger than) the determination from radio emission lines (Zhao et al. 1997). We also assume a constant mass-to-IR luminosity ratio in the system to derive the masses of B1 and C. The $N_{\rm Ly}$ can be obtained from the NICMOS Paα line fluxes assuming case B for the hydrogen recombination lines. The (SNr) in starburst galaxies is often estimated from radio observations (assuming that the non-thermal radio flux is synchrotron emission produced in shocks associated with supernovae), and more recently from [Fe II] emission lines. We used both radio measurements and the NICMOS [Fe II]1.644 μm fluxes. The K-band luminosities and EW of Paα were measured from our NICMOS images, whereas the CO indices are from our IR spectroscopy for B1 and C, and from previous observations for A. Finally since Arp 299 is an IR luminous galaxy, that is, it emits most of its energy in the IR, we approximate the $L_{\rm bol}$ with the IR luminosities. We determined the extinction by carefully modeling the near-IR colors (see Alonso-Herrero et al. 1998 for more details). The quantities L_K and $N_{\rm Ly}$ were corrected for the observed extinction.

3. Results and Conclusions

In this section we describe the model fits for sources A, B1 and C in Arp 299. Component B2 shows very little or no line emission, and therefore if any SF is present it must be at a very low level.

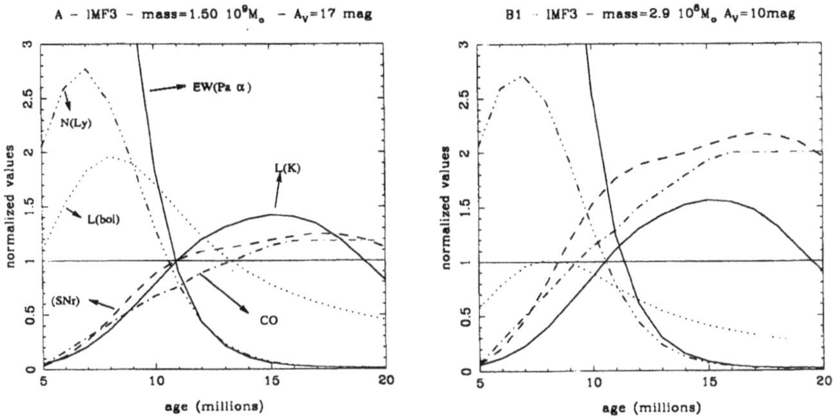

Figure 1. Evolutionary synthesis model fits for source A with IMF3 and extinction $A_V = 17\,\text{mag}$ (*left panel*), and for component B1, IMF3 and $A_V = 10\,\text{mag}$ (*right panel*). The solid line (declining) is the EW of Paα, solid line L_K, dotted line L_{bol}, the dashed line the (SNr), the dot-dash line the CO index and dot-dot-dot-dash line N_{Ly}.

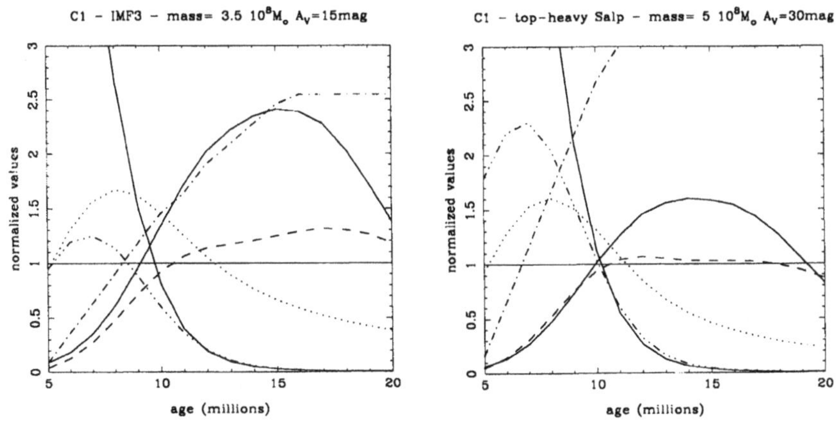

Figure 2. Evolutionary synthesis model fits for source C with IMF3 and $A_V = 15\,\text{mag}$ (*left panel*) and with a top-heavy Salpeter IMF and $A_V = 30\,\text{mag}$ (*right panel*).

Source A. — The fit for the observables of A (Figure 1, left panel) uses a mass in newly formed stars of $1.5 \times 10^9\,\text{M}_\odot$ ($\simeq 30\%$ of the total mass) and gives an age for the SF episode of $6-8 \times 10^6$ yr (after the peak of SF). The total mass would be reduced with little change in other parameters with a more top-heavy IMF, such as IMF8 derived for the M82 starburst (R93).
Source B1. — The best fit for B1 (Figure 1, right panel) is obtained with an age between 3 and 6.5 Myr after the peak of SF, with only 10% of the total mass in B1 being used in the burst.
Source C. — If we consider a conservative value of the extinction ($A_V =$

15 mag) we are able to fit the observables (Figure 2, left panel) using approximately 30% of the available mass, with an age between 3.5 and 7 Myr after the peak of SF. If we correct the observables for the highest possible extinction obtained from the near-IR spectroscopy ($A_V = 30$ mag), then the best fit (Figure 2, right panel) is only obtained with a top-heavy IMF. We obtained a fit with a Salpeter high mass slope and a lower mass cutoff of $m_{\rm L} = 5\,{\rm M}_{\odot}$. Similar results would be achievable with IMF8 (R93). Note that in order to fit $L_{\rm bol}$ we have to assume that 85% of the total $L_{\rm bol}$ is unaccounted for in the observed IR luminosity. The age derived from this fit is younger than for the previous fit (between 1 and 5 Myr after the peak of SF) and uses $\simeq$ 42% of the total mass available for this component. A reasonable fit with $A_V = 30$ mag is not possible with the IMF3 because it would require a starburst mass higher than the available mass. Another possibility to explain the poor fit of C would be the presence of an AGN.

The difficulties found in fitting the observables of C are not surprising, since C, at the interface of the two galaxies, is undergoing the youngest and strongest burst of SF involving approximately 42% of the total mass available. The effective SFR (mass of the newly formed stars over the FWHM of the burst) is high for A and C, 300 $M_{\odot}$ yr^{-1} and 100 $M_{\odot}$ yr^{-1} respectively, and moderate for B1, 60 $M_{\odot}$ yr^{-1}. We can conclude that the burst ages derived from the EW of hydrogen recombination lines are not very sensitive to the mass in newly formed stars. On the other hand, L_K, $L_{\rm bol}$ and $N_{\rm Ly}$ are good indicators for the burst for the ratio of the mass in newly formed stars to the total available mass. Finally, we would like to stress the importance of using the right extinction since the mass involved in the starburst and the SFRs are extremely sensitive to the estimated value of the extinction.

AAH acknowledges support from the National Aeronautics and Space Administration on grant NAG 5-3042 through the University of Arizona.

References

Alonso-Herrero, A., Rieke, M.J., Rieke, G.H., 1998, ApJ, in preparation
Barnes, J. E., Hernquist, L., 1996, ApJ, 471, 115
Gehrz, R.D., Sramek, R.A., Weedman, D.W., 1983, ApJ, 267, 551
Rieke, G.H., Loken, K., Rieke, M.J., Tamblyn, P., 1993, ApJ, 412,99 (R93)
Sanders, D.B. et al., 1988, ApJ, 325, 74
Shier, L.M., Rieke, M.J., Rieke, G.H., 1996, ApJ, 470, 222
Wynn-Williams, C.G. et al., 1991, ApJ, 377, 426
Zhao, J.-H. et al., 1997, ApJ, 482, 186

SPECTROPHOTOMETRIC ANALYSIS OF HII REGIONS IN NGC 4258

MARCELO CASTELLANOS AND ÁNGELES I. DíAZ
Universidad Autónoma de Madrid, Dpto. Física Teórica, Cantoblanco, 28049 Madrid, Spain.

ELENA TERLEVICH
INAOE, Tonantzintla, Apdo. Postal 51, 72000 Puebla, Mexico.

AND

MARíA LUISA GARCíA-VARGAS
Grantecan, S.A., 38200 La Laguna, Tenerife, Spain.

1. Introduction

Theoretical evolutionary models for ionizing star clusters predict the existence of high-excitation and high-metallicity HII regions as a result of the appearance of Wolf-Rayet stars at an age of about 3.5–4 Myr (García-Vargas & Díaz 1994; García-Vargas, Bressan & Díaz 1995; Stasińska & Leitherer 1996). However, no convincing observational evidence for the existence of these regions has been presented so far. We are carrying out a search for this HII region population in order to test current stellar evolution models.

In selecting our objects for observation, two main criteria have been followed: a high excitation, as deduced from the intensity of the [OIII] λ 5007 Å line ($\log([OIII]/H\beta) \geq 0.0$) and a solar or oversolar abundance as deduced from empirical calibrations. Here, we report the first of these observations corresponding to 8 HII regions in the Sb(s)II galaxy NGC 4258 (Sandage & Tammann 1981). Some of them were originally selected from the work of Oey & Kennicutt (1993). Four of the observed regions have been identified as 74C, 69C, 5N(A) and 5N(B) in Courtes et al. (1993). The observations were obtained at the Roque de los Muchachos Observatory in La Palma on 1994 March 12 and 13 using the WHT and the ISIS double spectrograph. The near infrared spectra of two regions of different excitation are shown in Figure 1. In both cases the [SIII] lines are easily observed.

Astrophysics and Space Science is the original source of publication of his article. It is recommended that this article is cited as: *Astrophysics and Space Science* **263**: 135–138, 1999.

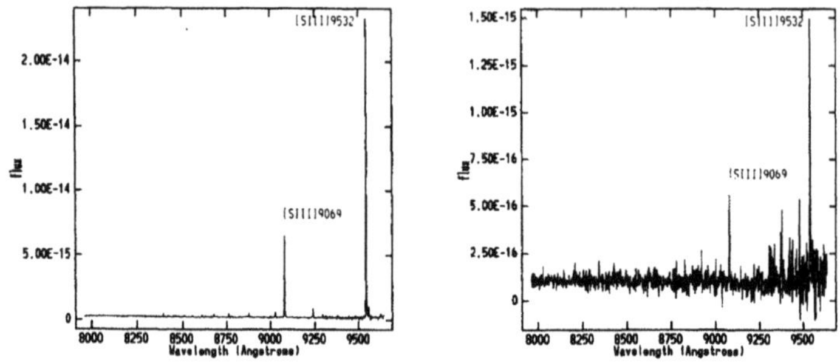

Figure 1. Near infrared spectra of high excitation region 74C (left) and region #2A (right). Flux units are: erg cm^{-2} s^{-1} Å^{-1}

2. Physical Conditions of the Gas in the Observed HII Regions.

Electron densities for each observed region were derived from the ratio of the [SII] $\lambda\lambda$6717,6731 lines, which implies values of $n_e \leq 100$ cm^{-3} for all the regions except #2A (160 cm^{-3}). For two of the regions, 74C and 69C, it has been possible to measure the weak auroral [SIII] λ6312 Å line which, together with the observed intensities of the nebular lines at $\lambda\lambda$9069,9532 Å, can be used to obtain a value for the electron temperature. Values of 8600 $\pm$ 400 K and 8800 $\pm$ 700 K are found for regions 74C and 69C respectively.

A prominent feature corresponding to WR stars is observed in region 74C (see below). According to recent theoretical models, the presence of these stars could lead to large temperature fluctuations (Pérez 1997). We have tried to evaluate these possible fluctuations by measuring the Paschen discontinuity at 8200 Å. Our measurement implies an electron temperature of 8000 $\pm$ 700 K which agrees with that derived from the [SIII] lines within the errors. Therefore, we can conclude that, despite the presence of WR stars, no important temperature fluctuations exist in this region.

Using our derived values for the electron density and temperature, we have calculated the elemental abundances for regions 74C and 69C following standard methods (*e.g.* Pagel et al. 1992). The oxygen abundances derived for these two regions are 0.3 and 0.6 times solar, significantly lower than those derived by Oey & Kennicutt (1993) by using empirical calibrations based on the optical oxygen lines.

For the rest of the regions no direct determination of the temperature was possible and the oxygen abundances have been derived using an empirical calibration based on the sulphur emission lines (Díaz & Pérez-Montero 1999). The derived oxygen abundances for all the observed regions, together with their estimated errors, are given in Table 1.

The ionization parameter for each region has been derived from different emission line ratios: [OII]$\lambda\lambda$3727,3729/Hβ, [SII]$\lambda\lambda$6716,6731/Hα and

TABLE 1. Derived parameters for HII regions in NGC4258

Region	n_e (cm^{-3})	t_e (10^4 K)	$12 + \log(O/H)$	$< \log U >$
74C	60	0.86 ± 0.04	8.17 ± 0.09	−2.70
69C	<50	0.88 ± 0.07	8.10 ± 0.10	−2.90
#2A	160	–	8.65 ± 0.15	−2.90
#2B	<50	–	8.47 ± 0.15	−2.96
#2C	<50	–	8.49 ± 0.15	−2.98
#2D	<50	–	8.48 ± 0.15	−2.82
5N(A)	<50	–	8.40 ± 0.15	−2.81
5N(B)	<50	–	8.39 ± 0.06	−2.87

[SII]$\lambda\lambda$6716,6731/[SIII]$\lambda\lambda$9069,9532 according to the calibrations given in Díaz (1999) and Díaz et al. (1991). The ionization parameters found are in all cases lower than -2.70.

Finally, we have used single star (Mihalas NLTE stellar atmospheres) photoionization models (CLOUDY, Ferland 1996) to constrain the mean effective temperature of the ionizing stars in each region. For all regions, we have assumed a plane parallel geometry and a constant electron density. We have used the previously derived electron density, ionization parameter and elemental abundances as input to the models. For most lines of interest, consistency is found for effective temperatures of the ionizing stars of about 35,000 K (Table 2).

3. Wolf-Rayet features

We have observed a prominent Wolf-Rayet feature in region 74C (Figure 2). Assuming a distance to NGC4258 of 5.5 Mpc (Martin et al., 1989) and a constant extinction value through this region, $c(H\beta) = 0.48$ as derived from the Balmer and Paschen recombination lines, the total luminosity of the broad HeII feature, without the contribution of the [FeIII] λ4658 Å line, is $(1.1 \pm 0.1) \times 10^{38}$ erg s^{-1}. This comprises the broad features of both NIII $\lambda\lambda 4634, 4640$ and HeII $\lambda 4686$Å. Assuming the contribution of the NIII lines to the WR bump to represent 0.4 times the total emission (Terlevich et al. 1996), we obtain for the HeII line luminosity: $L(\text{HeII}\lambda 4686) = (6.6 \pm 0.6) \times 10^{37}$ erg s^{-1}. The non-detection of NV $\lambda\lambda$4604,4620Å suggests that the WR stars are of type late N or intermediate. Using the calibration of Vacca & Conti (1992), we find 38 ± 5 WN stars in this region.

However, the estimated effective temperature from the photoionization

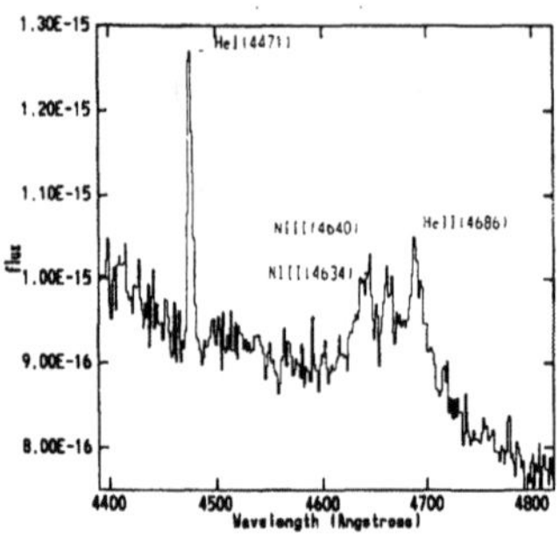

Figure 2. Wolf-Rayet feature in region 74C

TABLE 2. Derived parameters from photoionization models

Regions	T_{eff}(K)	log U	[OII](3727, 29)		[OIII](4959 + 5007)		[SII](6717 + 6731)		[SIII](9069 + 9532	
			Model	Observed	Model	Observed	Model	Observed	Model	Observed
74C	35000	−2.50	2.10	2.10 ± 0.04	1.55	1.52 ± 0.02	0.44	0.45 ± 0.02	1.42	0.93 ± 0.0[illegible]
69C	35400	−2.80	2.22	2.28 ± 0.09	1.25	1.22 ± 0.04	0.63	0.61 ± 0.02	1.30	0.67 ± 0.0[illegible]
5N(A)	35000	−2.83	3.24	3.30 ± 0.10	1.11	1.13 ± 0.04	0.60	0.53 ± 0.03	0.93	0.80 ± 0.0[illegible]
5N(B)	35000	−2.85	3.40	3.40 ± 0.20	1.11	1.12 ± 0.06	0.61	0.59 ± 0.04	0.88	0.70 ± 0.0[illegible]
#2A	35000	−2.65	2.32	2.34 ± 0.08	1.06	1.06 ± 0.03	0.51	0.99 ± 0.02	1.11	1.05 ± 0.0[illegible]
#2B	35000	−2.80	3.07	3.01 ± 0.09	1.14	1.12 ± 0.03	0.54	0.86 ± 0.02	0.87	0.63 ± 0.0[illegible]
#2D	35000	−2.95	2.96	2.80 ± 0.20	0.70	0.68 ± 0.04	0.86	0.77 ± 0.03	0.98	0.77 ± 0.0[illegible]

models (35000 K) does not seem to imply a hardening of the ionizing radiation due to the presence of these stars. A quantitative analysis of the ratio of O to WR stars is needed in order to test different stellar evolution models.

References

Courtes, G., Petit, H., Hua, C.T., Martin, P., Blecha, A., Huguenin, D., Golay, M., 1993, A&A, 268, 419

Díaz, A.I., 1999, in preparation

Díaz, A.I., Pérez-Montero, E., 1999, in preparation

Díaz, A.I., Terlevich, E., Vílchez, J.M., Pagel, B.E.J., Edmunds, M.G., 1991, MNRAS, 253, 245

Ferland, G.J., 1996, *HAZY: A Brief Introduction To CLOUDY*, Univ. Kentucky internal report

García-Vargas, M.L., Díaz, A.I., 1994, ApJS, 91, 553

García-Vargas, M.L., Bressan, A., Díaz, A.I., 1995, A&AS, 112, 13

Martin, P., Roy, J.R., Noreau, L., Lo, K.Y., 1989, ApJ, 345, 707

Oey, M.S., Kennicutt, R.C., 1993, ApJ, 411, 137

Pagel, B.E.J., Simonson, E.A., Terlevich, R.J., Edmunds, M.G., 1992, MNRAS, 255, 325

Pérez, E., 1997, MNRAS, 290, 465

Sandage, A., Tammann, G., 1981, *A Revised Shapley-Ames Catalog of Bright Galaxies*, p. 42

Stasińska, G., Leitherer, C., 1996, ApJS, 107, 661

Terlevich, E., Díaz, A.I., Terlevich, R., González-Delgado, R.M., Pérez, E., García-Vargas, M.L., 1996, MNRAS, 279, 1219

Vacca, W.D., Conti, P.S., 1992, ApJ, 401, 533

STAR FORMATION IN CIRCUMNUCLEAR RINGS

ÁNGELES I. DÍAZ AND MAR ÁLVAREZ-ÁLVAREZ
Dpto de Física Teórica, C-XI, Universidad Autónoma de Madrid, Cantoblanco, 28049-Madrid, Spain
MIGUEL SÁNCHEZ PORTAL
Dpto de Física Teórica, C-XI, Universidad Autónoma de Madrid, Cantoblanco, 28049-Madrid, Spain, and LAEFF/INTA, P.O. Box 50727, 28080-Madrid, Spain
ELENA TERLEVICH AND ITZIAR ARETXAGA
Instituto Nacional de Astronomía, Óptica y Electónica, Tonantzintla, Puebla, México
AND
ROBERTO TERLEVICH
Institute of Astronomy, Madingley Road, Cambridge CB30HA, United Kingdom

1. Introduction

Nuclear activity and the presence of circumnuclear structures with ongoing star formation, are simultaneous phenomena in a large number of galaxies (Arsenault 1989). Their mutual influence and the role played by each of them in the dynamics, and/or the energy balance of the galaxy, are still open questions. Three possibilities have been put forward: that the nuclear activity has been originated in the supernova remnants product of the intense star formation rate experienced by the circumnuclear regions of the galaxy; that the star formation in the circumnuclear regions has been induced by gas and radiation ejected by the active nucleus, or that the two phenomena are not related to each other.

A possible way to elucidate this matter may be through the study of the general properties and the evolutionary state of circumnuclear star forming regions in galaxies with different levels of nuclear activity. Specifically, their age and metallicity can provide important clues about the sequence of appearance of both phenomena: nuclear and circumnuclear activity.

Astrophysics and Space Science is the original source of publication of his article. It is recommended that this article is cited as: *Astrophysics and Space Science* **263**: 139–142, 1999.

For a first study we have selected four nearby galaxies with confirmed circumnuclear star forming rings and different nuclear type: Seyfert 1 (NGC 7469), Seyfert 2 (NGC 1068), LINER (NGC 7177) and Starburst (NGC 3310) to perform a comparative study. The characteristics of these galaxies are given in Table 1.

TABLE 1. Characteristics of the sample

Galaxy	Morph. Type	Pc/"	Nuc. Type	B_T	D (Mpc)
NGC 1068	(R)SA(rs)b	87.8	Sy2	9.61	18.1[1]
NGC 3310	SAB(r)bc pec	60.6	Starburst	11.15	12.5[2]
NGC 7177	SAB(r)b	74.2	LINER	12.01	15.3[3]
NGC 7469	(R')SAB(rs)a	315.1	Sy1	13.00	65.0[3]

Notes to Table 1:
[1] Sandage & Tammann (1975)
[2] Bottinelli et al. (1984)
[3] de Vaucouleurs et al. (1991) ($H_0 = 75$ km s^{-1} Mpc^{-1})

2. Observations and Measurement Techniques

Images in the broad bands V, R, and I and narrow Hα and continuum filters for every galaxy were obtained at the JKT (Observatorio del Roque de los Muchachos, La Palma). The data reduction was carried out using standard procedures. No internal absorption correction was attempted. The Hα frames were continuum subtracted in order to obtain a net line image. This continuum corrected frame was used afterwards to subtract the line contribution (Hα+[NII]) from the R image.

Star forming regions were identified on the Hα frames and were later located in the continuum and broad band frames. Subsequently, the equivalent widths of the Hα line and the broad-band V, R and I magnitudes were computed. The magnitudes were then subtracted to get the colors. Further details can be found in Díaz et al. (1999). A two-color diagram, as well as the Hα equivalent width *versus* continuum color (R-I), for the observed regions can be seen in Figs 1 and 2.

3. Results and Discussion

Our results can be compared with those published by Mayya (1994) on HII regions in galactic disks and by Telles & Terlevich (1997) on HII galaxies. HII galaxies, essentially a low metallicity population, show colors similar to those of the circumnuclear regions of NGC 3310, substantially bluer than

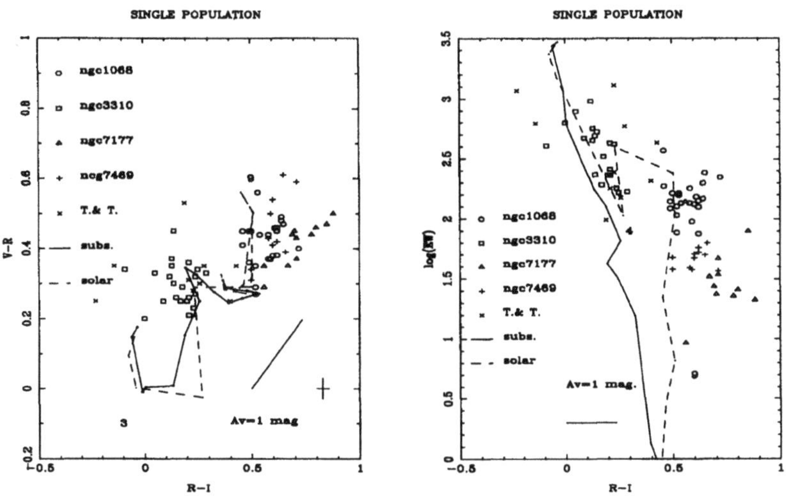

Figure 1. Two-color diagram and Hα equivalent width *versus* continuum color for the sample regions together with data on HII galaxies from Telles & Terlevich (1997) and theoretical models for single stellar populations of solar and 0.25 solar metallicities. The effect of 1 mag extinction as well as a typical error bar are shown.

in the rest of the sample galaxies. Regarding Mayya's observations, only V-R colors are available for comparison and are, in average, similar to those found for our circumnuclear regions.

Some information about the ages of the stellar populations present in these regions can be extracted with the use of evolutionary synthesis models. In Figs 1 and 2 we show the theoretical predictions from the models of Leitherer & Heckman (1995) for single stellar populations (Fig 1) and composite two-burst populations (Fig 2) of different metallicities.

The broad-band colors of the regions in NGC 1068, NGC 7177 and NGC 7469 can be reproduced with single population models with ages of 10 to 100 Myr and solar metallicities but they are also compatible with younger ages if an internal extinction of at least 1 mag is assumed. On the other hand, ages of about 6 Myr for the region in NGC 1068 and between 6 and 9 Myr for those in NGC 7177 and NGC 7469 are needed in order to reproduce their observed Hα equivalent widths. There is no need to invoke high values of extinction if a two-burst model is assumed (Fig 2). Two bursts of 3 and 10 Myr for NGC 1068, and somewhat older (8 and 15 Myr) for NGC 7177 and NGC 7469 and a moderate amount of extinction adequatelly reproduce the observations. In all cases, the age difference between bursts is consistent with the evolutionary time for supernova production from the first star formation episode.

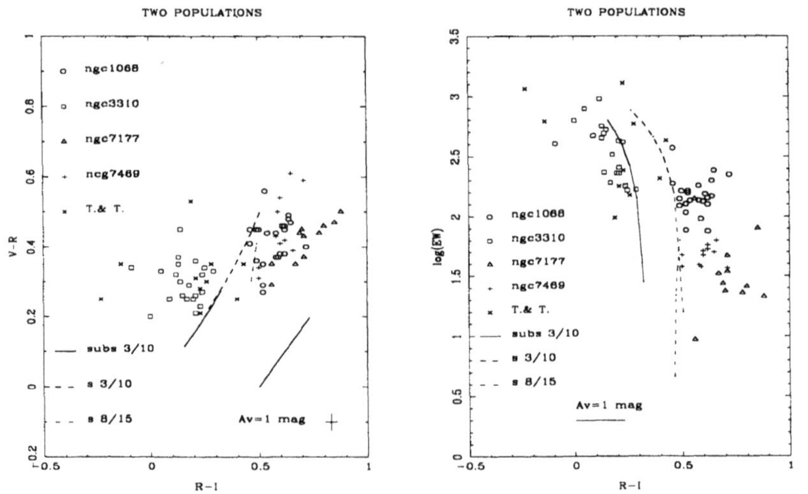

Figure 2. Same data as in Fig 1 but showing theoretical two-burst models. The contribution by the young burst to the total mass varies from 1% to 50% from bottom to top in the left side panel and from top to bottom in the right side one. Each tic represents a 10% increment.

In the case of NGC 3310, both the broad-band colors and the Hα equivalent widths of the observed regions are consistent with single populations of subsolar metallicity and an age of about 4 Myr. The two-burst population model is unable to reproduce the blue colors observed. The circumnuclear star formation in this galaxy seems to be coeval and could be related to an interaction or merging process occurred previously.

References

Arsenault, R., 1989, A&A, 217, 66
Bottinelli, L., et al., 1984, A&A 56, 381
de Vaucouleurs et al., 1991, *Third Reference Catalog of Bright Galaxies*, New York:Springer
Díaz, A.I., Álvarez-Álvarez, M., Terlevich, E. , Terlevich, R., Sánchez-Portal, M., Aretxaga, I,. 1999, in preparation
Leitherer, C., Heckmann, T., 1995, ApJS, 96, 9
Mayya, Y.D., 1994, AJ, 108, 1276
Sandage, A., Tammann, G.A., 1975, ApJ, 196, 313
Telles, E., Terlevich, R., 1997, MNRAS, 286, 183
Terlevich, R., Sánchez-Portal, M., Díaz, A.I., Terlevich, E., 1991 MNRAS, 249, 36

FUNCTIONAL AND PHYSICAL PARAMETERS OF HII GALAXIES

ÁNGELES I. DÍAZ
Dpto. de Física Teórica, Universidad Autónoma de Madrid
Cantoblanco, 28049-Madrid, Spain

1. Introduction

The term "HII galaxy" currently denotes dwarf emission line galaxies undergoing violent star formation. At optical wavelengths, their observable properties are dominated by the young component of their stellar population. The gas is photoionized by hot, recently formed, stars and therefore shows a characteristic emission line spectrum whose features are controlled by three fundamental parameters: the shape of the ionizing continuum, the degree of ionization and the gas chemical composition. The gas density also affects the emission line spectrum but to a lesser extent, as long as it remains below the critical density for line deexcitation.

If simple photoionization models are used in which a single O-B star is assumed as the ionization source, these fundamental parameters can be represented by what we call the *functional parameters*: effective temperature of the ionizing star – T_{eff} –, ionization parameter – U – and metallicity – Z –. The functional parameters can be derived from observed emission line ratios and translated into physical properties of the HII galaxies by applying more realistic photoionization models in which the central ionizing star is substituted by a young evolving cluster.

2. Determination of Functional Parameters

I have determined the functional parameters of a large sample of HII galaxies whose data have been collected from the literature and then I have tried to interpret these parameters in terms of the physical properties of the ionizing clusters. Most studies of HII galaxies made so far have been restricted to objects for which measurements of the weak temperature sensitive line of [OIII]λ 4363 Å are reported, since in these cases an accurate determi-

Astrophysics and Space Science is the original source of publication of his article. It is recommended that this article is cited as: *Astrophysics and Space Science* **263**: 143–146, 1999.

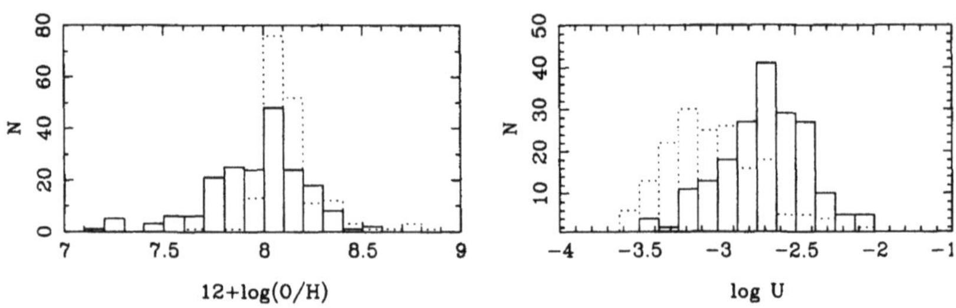

Figure 1. Distributions of a) oxygen abundance and b) ionization parameter for the sample galaxies. Solid and dotted lines for galaxies with and without measurements of the [OIII] λ 4363 Å line respectively.

nation of the oxygen abundance is possible. It has been implicitly assumed that the subsample thus defined is representative of the total HII galaxy sample. I have included also HII galaxies without measurements of this line in order to test if this assumption is valid. Details concerning the sample selection and characteristics can be found in Díaz (1998). The functional parameter determination is based on an extensive grid of single ionizing star photoionization models (CLOUDY, Ferland 1990) in the range of physical conditions appropriate to represent HII galaxies (Díaz, 1998 for a full discussion). The global metallicity of a given galaxy is represented by its oxygen abundance. The rest of elements vary with it in solar proportions. For galaxies with observed [OIII] λ 4363 Å line, the oxygen abundances have been derived following standard techniques (*e.g.* Pagel *et al.* 1992). For the rest of the objects two empirical calibrations of oxygen abundance *versus* the R_{23} parameter (Pagel *et al.* 1979), have been used for the upper branch (12+log(O/H) $>$ 8.00) and the lower branch(12+log(O/H)$\leq$ 8.00) respectively. The objects have been assumed to lie on one of the two branches according to their observed $[NII]\lambda6584/[OII]\lambda3727,29$ line ratio. The ionization parameter – defined as the ratio of the density of ionizing photons to the particle density – has been derived from: the intensity of the [OII]λ 3727,29 Å line relative to /Hβ; the intensity of the [SII]λ 6717,6731 Å line relative to Hα and [OII] λ 3727,29 /[OIII] λ 5007, looking for consistency between the three indicators. The T_{eff} of the ionizing radiation has been derived using multiregression formulae involving the [OIII] /Hβ, R_{23} and [OII]/Hβ ratios and are calculated for different metallicity ranges.

Figure 1 shows a) the metallicity and b) the ionization parameter distributions for the sample galaxies. It can be seen from this figure that HII galaxies without measurements of the [OIII] λ 4363 Å line have, in average, higher oxygen abundances and lower ionization parameters than HII galaxies for which measurements of this line exist. The two subsamples seem to be substantially different and the no inclusion of one of them in any statistical study introduces an important bias. Regarding T_{eff}, most of the galaxies show values between 35000 and 55000 K with an almost gaussian distribution, although there are some objects with an estimated T_{eff} higher than 55000 K. For the seven objects with derived oxygen abundance greater than 0.5 times solar, no T_{eff} estimation could be made.

3. Physical Properties of the Ionizing Clusters

Some properties of the ionizing populations of the studied HII galaxies can be deduced with the use of evolutionary synthesis models (Cerviño & Mas-Hesse 1994; García-Vargas *et al.* 1995 (GBD95); Leitherer & Heckman 1995). Various combinations of stellar evolution and atmosphere models can then be used which, together with the numerical methods employed in the different codes, can give rise to somewhat different predictions that can be compared to observations. There are nevertheless some general results which are common to all computed models. At the start of their evolution the stars in the cluster are distributted along a ZAMS, according to their assumed IMF. The total number of ionizing photons per unit mass, $Q_m(H)$ can be computed for each ionizing cluster. This number depends on metallicity and assumed IMF and increases for flatter IMF and/or higher upper mass limit, and decreasing metallicity. Once the cluster starts its evolution it loses ionizing photons as the more massive stars end their lives and, as a consequence, $Q_m(H)$ decreases with increasing age of the cluster. On the other hand, the computed equivalent width of the Balmer lines in emission decreases as the cluster ages due both to the loss of ionizing photons and the increase of the stellar continuum at longer wavelengths. Their computed values are rather independent of both IMF parameters and metallicity and therefore can be considered as good indicators of the age of the cluster in the absence of underlying stellar populations. Since both the number of ionizing photons per unit mass and the equivalent width of Hβ decrease with the age of the cluster, a correlation between these two quantities is expected on the basis of the computed models. This correlation is shown in Figure 2a). A linear regression fit gives: $log(Q(H)/M_{\odot}) = 0.86 logEW(H\beta) + 44.48$. Using this relation and the number of ionizing photons for each sample galaxy, derived from its observed Hβ fluxes using the distance derived from its measured redshift, the masses of the ionizing clusters can be calculated. No effects due to the presence of dust are included and therefore these values represent lower limits to the actual cluster masses. The cluster masses derived are between 2×10^3 and 2×10^8 $M_{\odot}$.

The decrease of the number of ionizing photons with the age of the cluster also makes the ionization parameter decrease. This can be seen in

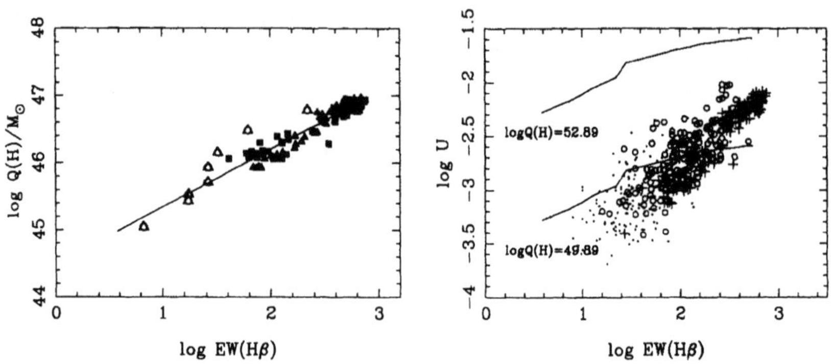

Figure 2. a) Relation between the number of ionizing photons per unit mass and the equivalent width of Hβ for ionizing star cluster models. Filled symbols: GBD95; open symbols SL96. b) Relation between the ionization parameter and the equivalent with of Hβ for the sample HII galaxies.

Figure 2b) where the ionization parameter *vs* the equivalent width of Hβ is plotted for the galaxy sample (open circles and dots refer to HII galaxies with and without measurements of the [OIII] λ 4363 Å line respectively). A positive correlation is readily apparent. The trend is well reproduced by the models by GBD95 (crosses in the plot) in which the HII galaxies evolve at a constant radius, thus decreasing the ionization parameter and the filling factor. If our models are correct, HII galaxies without measurements of the [OIII] λ 4363 Å line are, in average, older that those for which measurements of this line exist. The solid lines in the plot represent two different model sequences from Stasińska and Leitherer (1996;SL96). The one at the top corresponds to the clusters which provide a number of ionizing photons in the range of that observed in HII galaxies (logQ(H)=52.88). However this sequence has too large ionization parameters. Furthermore, the sequence which lies in the region occupied by the data does not adequatelly reproduce the observed trend. This is probably due to the fact that keeping the filling factor constant, as is made in these models, makes the galaxies evolve as effectively reducing the radius of the ionized region.

References

Cerviño, M., Mas-Hesse, M. 1994 A&A, 284, 749
Díaz, A.I. 1998 in preparation
Ferland, G.J. 1996 *HAZY: A Brief Introduction to CLOUDY*, Univ. Kentucky
García-Vargas, M.L., Bressan, A., Díaz, A.I. 1995 A&ASS, 112, 13
Leitherer, C., Heckman, T. 1995 ApJS, 96, 9
Pagel, B.E.J., Edmunds, M.G., Blackwell, D.E., Chun, M.S., Smith, G. 1979 MNRAS, 189, 95
Stasińska, G., Leitherer, C. 1996 ApJS, 107,661

NIR PROPERTIES OF THE UCM STAR-FORMING GALAXIES

A. GIL DE PAZ, J. GALLEGO AND J. ZAMORANO
Dpto. de Astrofísica, Universidad Complutense de Madrid
Avda. Complutense s/n, 28040, Madrid, Spain
A. ALONSO-HERRERO
Steward Observatory, Tucson, AZ 85721, USA
AND
A. ARAGÓN-SALAMANCA
Institute of Astronomy, Madingley Road, Cambridge, England

Abstract. We have obtained near-infrared (JK) imaging for 67 star-forming galaxies from the Universidad Complutense de Madrid survey. Aperture photometry at one and three disk-scale lengths have been performed on the J, K and r-band images. By comparing the $r - J$, $J - K$ colors and equivalent widths of Hα with evolutionary synthesis models we derive the mass of newly formed stars, which is typically a 1 per cent of the galaxy stellar mass. The galaxy stellar masses obtained range between 3×10^9 and 3×10^{11} M$_\odot$. The distribution of burst strengths and specific star formation rates indicate a low level of star formation activity.

1. Introduction

The Universidad Complutense de Madrid (UCM hereafter) survey constitutes a well defined sample of local ($z \leq 0.045$) star-forming galaxies selected by their Hα emission [1][2], which provided the first estimate of the Local Universe star-formation rate density [3]. A meaningful comparison between their ongoing and past star-formation will give important clues about the evolution of the star formation rate (SFR) density with redshift [4][5]. In particular, the different scenarios for this star formation evolution will be tested by using the galaxy stellar masses, as derived from the K-band luminosity [6], along with the properties of the ongoing star formation, deduced from the optical-nIR colors and Hα equivalent widths (EWHα).

Astrophysics and Space Science is the original source of publication of his article. It is recommended that this article is cited as: *Astrophysics and Space Science* **263**: 147–150, 1999.

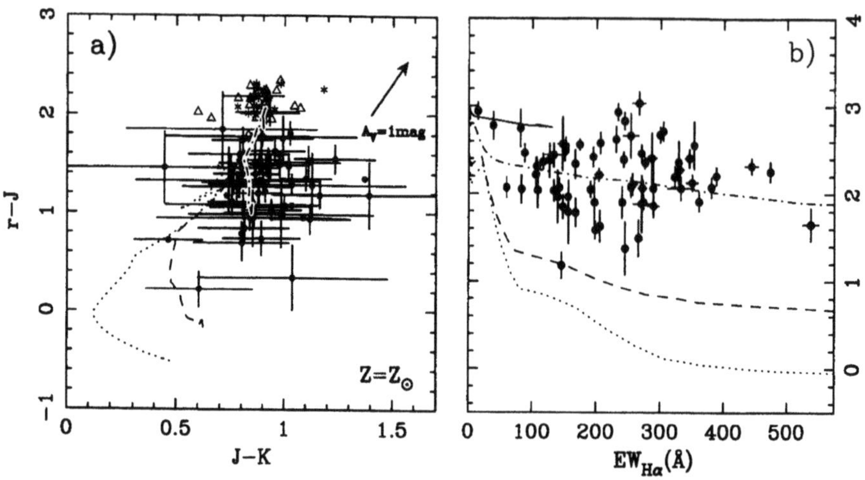

Figure 1. **a)** $r-J$ - $J-K$ color-color diagram using three disk-scale length apertures. Solar metallicity models with burst strength, b=1 (*dotted*), 0.1 (*dashed*), 0.01 (*dash-dotted*) and 10^{-3} (*solid*) are also drawn. Asterisks and triangles are bulge and disk data of relaxed spirals [7]. **b)** The same for $r-K$ vs. Hα equivalent width.

2. Observations and Analysis

The J and K-band near-infrared images for the 67 galaxies analized were obtained during three observing runs on the 1m telescope at Lick Observatory and one run on the 2.2m telescope at Calar-Alto Observatory, both using 256×256 NICMOS3 array cameras. This sample represents the same fraction (≃30%) of the UCM complete sample for every physical property, M_r, $L_{H\alpha}$ and EWHα [8]. Aperture photometry, at one and three disk-scale lenght apertures [9], were performed for the whole sample. The same apertures were chosen for the previously obtained Gunn-r images [9].

The optical-nIR colors, together with the equivalent widths of Hα [8] were extinction corrected (see Figure 1) assuming a gas to continum extinction relation of $E(B-V)_{cont}=0.44\times E(B-V)_{gas}$ [10], and using prior $E(B-V)_{gas}$ data [8]. The $r-J$ and $J-K$ errors shown are associated both to intrinsic variations and calibration errors. Finally, we computed the errors in the equivalent widths of Hα from the signal-to-noise at the continuum, Hα line and equivalent widths [11][8].

3. Models

Evolutionary synthesis models including both stellar continuum [12] and nebular line and continuum emission [13] were developed. Our models cover a wide range in metallicity ($1/50\,Z_\odot < Z < 2\,Z_\odot$), age (0–1 Gyr), and burst strength (b=10^{-4}–1), defined as the fraction of stellar mass in the burst over total mass in the galaxy. The metallicity, age and burst strength (Z,t,b) that better match the observed quantites ($r-J$, $J-K$, EWHα) were determined by using a maximum likelihood method, with the following estimator [14],

$$\mathcal{L}(Z,t,b) = \prod_{n=1}^{3} \frac{1}{\sqrt{2\pi}\Delta C_n} \exp\left(-\frac{(c_n - C_n)^2}{2\Delta C_n{}^2}\right) \tag{1}$$

where C_n are the measurements ($r-J$, $J-K$ and $2.5\log \mathrm{EWH}\alpha$), ΔC_n are the associated errors, and c_n=$c_n(Z,t,b)$ are the model predictions.

Then, from the K-band luminosities per unit solar mass predicted by the *best-fitting* models, and the galaxy K-band luminosities, we can derive the galaxy total stellar masses, avoiding the uncertainties present in standard mass-to-light ratios.

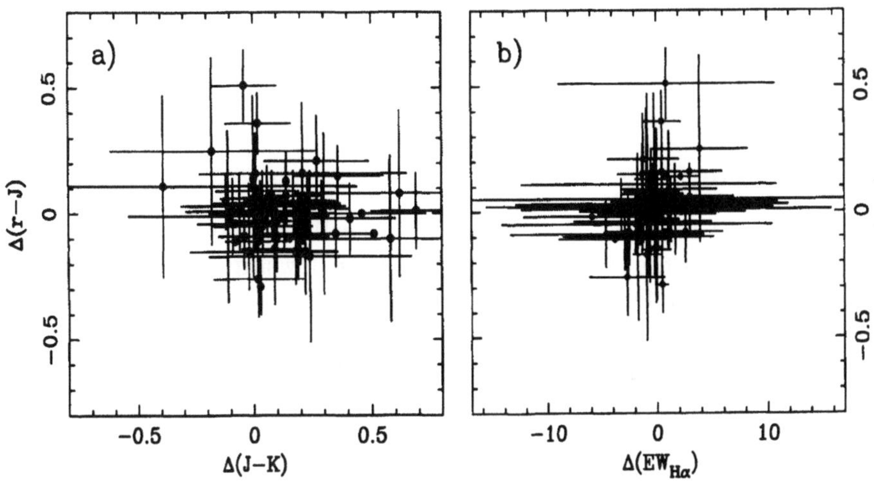

Figure 2. **a)** Differences between the measured $r-J$ and $J-K$ colors and those of the *best-fitting* model, Δ=$C_n - c_n$. **b)** The same for $r-J$ and the Hα equivalent width (Å).

4. Results and Conclusions

Figure 2 shows the differences between our measurements and the model predictions. The small residuals found demonstrate the significance of our results, except perhaps for those galaxies with large photometric errors. We have derived the burst strengths, galaxy stellar masses, star formation rates [13], SFR, and specific star formation rates [4], SFR/M, for the whole sample. The measured burst strengths, galaxy stellar masses and specific SFR are shown in Figure 3. Both the burst strengths (b=0.01) and specific star formation rates (SFR/M ~7×10^{-11} yr^{-1}) indicate moderately low local star formation activity. These two quantities should be most relevant for further works studying the evolution of the density of star formation rate as a function of redshift [3][4][5].

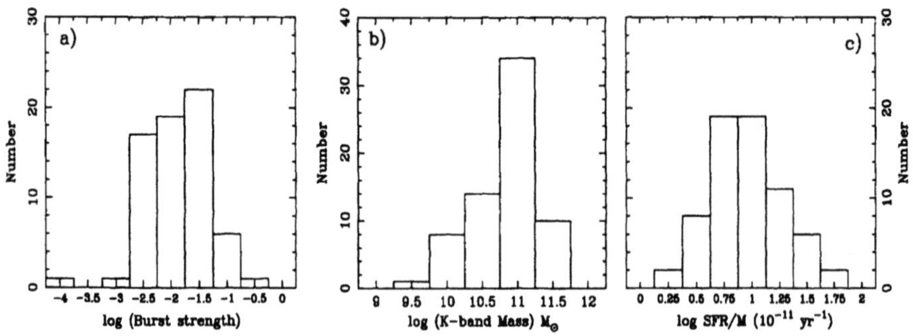

Figure 3. **a)** Burst strength histogram. **b)** Galaxy total stellar mass histogram. **c)** Specific star formation rate histogram.

Acknowledgements

We would like to thank G. Kauffmann for her collaboration in this project. AGP acknowledges the receipt of a FPU/MEC fellowship. AGP also thanks the hospitality and facilities provided by the IoA Cambridge. This research was supported by the Programa Sectorial de Promoción General del Conocimiento under grant PB96-0610. AAH was supported by NASA on grant NAG 5-3042.

References

1. Zamorano, J., Rego, M., Gallego, J., Vitores, A., et al., 1994, ApJS, 95, 387
2. Zamorano, J., Gallego, J., Rego, M., Vitores, A., Alonso, O., 1996, ApJS, 105, 343
3. Gallego, J., Zamorano, J., Aragón-Salamanca, A., Rego, M., 1995, ApJ, 455, L1
4. Guzmán, R., Gallego, J., Koo, D.C., Phillips, A.C., Lowenthal, J.D., Faber, S.M., Illingworth, G.D., Vogt, N.P., 1997, ApJ, 489, 559
5. Gallego, J., 1998, this volume
6. Charlot, S., 1998, in NGST – Science Drivers & Tech. Challenges, astro-ph/9810408
7. Peletier, R.F., Balcells, M., 1996, AJ, 111, 2238
8. Gallego, J., Zamorano, J., Rego, M., Alonso, O., Vitores, A., 1996, A&AS, 120, 323
9. Vitores, A., Zamorano, J., Rego, M., Alonso, O., Gallego, J., 1996, A&AS, 118, 7
10. Calzetti, D., 1997, in The Ultraviolet Universe at Low and High z, astro-ph/9706121
11. Gil de Paz, A. et al., 1998, in preparation
12. Bruzual, G. and Charlot, S., 1996, ftp://gemini.tuc.noao.edu/ (GISSEL96)
13. Alonso-Herrero, A., Aragón-Salamanca, A., Zamorano, J., Rego, M., 1996, MNRAS, 278, 417
14. Abraham, R.G. et al., 1998, MNRAS, in press, astro-ph/9807140

MRK86: A 'NOT-SO-DARK' GALAXY

A. GIL DE PAZ, J. ZAMORANO AND J. GALLEGO
Dpto. de Astrofísica, Universidad Complutense de Madrid
Avda. Complutense S/N, 28040, Madrid

Abstract. We have studied the velocity field of the Blue Compact Dwarf galaxy Mrk86 using an Hα Fabry-Perot image and 14 long-slit optical spectra. From the ionized gas velocities measured we have found that Mrk86 follows a solid-body rotation law with a central angular velocity of 34 ± 5 km s^{-1} kpc^{-1}. The comparison of the modeled rotation curve with our velocity data indicates a clear dominance of the stellar mass component over the dark matter. Related with the galaxy most intense star forming knot, a very strong gradient steeping is observed. Finally, we give kinematic evidence for the Mrk86–C expanding bubble.

1. Introduction

The kinematical properties of dwarf galaxies have been widely studied in the past (see [1] and references therein). From the study of their global velocity fields, [2] have shown that they are dark matter dominated, although the dark matter density profiles deduced [3] are significantly different from those predicted by the CDM simulations [4]. In particular, [3] suggests the existence of a constant density core, meanwhile the simulated CDM profiles grow rapidly at short radii. The existence of this core would imply a higher contribution of the luminous matter in the central regions of dwarf galaxies. This contribution should be more important in Blue Compact Dwarf galaxies (BCDs hereafter), since they show higher central surface brightnesses [5] and HI column densities [1] than other dwarf galaxies.

Typical solid-body velocity fields of BCDs are commonly disturbed, with evidence of extended gas not kinematically connected to the main system [1] and recent mergers [6][7]. Finally, an increasing number of works have shown that galactic supernova driven winds are relatively frequent in dwarf

Astrophysics and Space Science is the original source of publication of his article. It is recommended that this article is cited as: *Astrophysics and Space Science* **263**: 151–154, 1999.

star forming galaxies [8][9]. These structures are thought to be crucial for the mass and metallicity evolution of dwarf galaxies [10].

We have studied all these points in Mrk86, a prototypical iE type BCD [11], the most numerous BCD galaxy type ($\sim$ 70 %; [11]).

2. Global velocity field

We have fitted the Hα profiles along the spectral dimension of the Fabry-Perot datacube, obtaining a high spatial resolution velocity field (see figure 1b). We have also obtained velocity information from 14 long-slit optical spectra of different resolutions ($1000 < R < 7000$). In figure 2a we plot the velocities measured along the slit positioned closest to the galactic center and with position angle perpendicular to the rotation axis (slit #1 in figure 1a). Then, we have parametrized all the mass components. The following mass density profiles were assumed for the underlying stellar population (ρ^s; 7 Gyr old [12], M/L$_B$=2.6 M$_\odot$/L$_\odot$ [13]), *starburst* (ρ^{sb}; 1 Gyr old [12], M/L$_B$=0.5 M$_\odot$/L$_\odot$ [13]), HI ($\rho^{\rm HI}$) and dark matter components ($\rho^{\rm DM}$).

$$\rho^s(r) = \rho_0^s \exp(-r/R_s) \quad [5]$$

$$\rho^{sb}(r) = \rho_0^{sb} \exp(-r/R_{sb})^2 \quad [5]$$

$$\rho^{\rm HI}(r) \simeq \rho_0^{\rm HI} \exp(-r/R_{\rm HI}) \quad [1]$$

$$\rho^{\rm DM}(r) = \frac{\rho_0^{\rm DM}}{1+(r/R_{\rm DM})^2} \quad [14]$$

Param.	Value	Refs.
ρ_0^s	20 at$_{\rm H}$/cm^3	[5][13]
R_s	1.0 kpc	[5]
ρ_0^{sb}	31 at$_{\rm H}$/cm^3	[5][13]
R_{sb}	0.14 kpc	[5]
$\rho_0^{\rm HI}$	0.7 at$_{\rm H}$/cm^3	[15]
$R_{\rm HI}$	2.0 kpc	[15]
$\rho_0^{\rm DM}$	1.00 at$_{\rm H}$/cm^3	free
$R_{\rm DM}$	1.14 kpc	[14]

Due to $\rho_0^{\rm DM}$ and $R_{\rm DM}$ are not independent [14], only $\rho_0^{\rm DM}$ and the inclination of the rotation axis in the sky (i) are free parameters. Considering dark matter models with different $\rho_0^{\rm DM}$ we observe that the velocity curve can be reproduced up to galactocentric distances of 1 kpc and $i \leq 50°$ just using the stellar component[1] (see figure 2a for $\rho_0^{\rm DM}$ = 1 at$_{\rm H}$/cm^3). If we measure the velocity gradient perpendicular to the rotation axis at different galactocentric distances we observe a progressive shallowing towards outer galaxy regions from 34$\pm$5 km s^{-1} kpc^{-1} to 8$\pm$3 km s^{-1} kpc^{-1}. This change in the velocity gradient agrees with the global velocity field expected for a $i \sim 40°$ inclined rotating disk [7].

[1]Physical quantities have been converted to H_0=50 km s^{-1} Mpc^{-1} and q_0=0.5

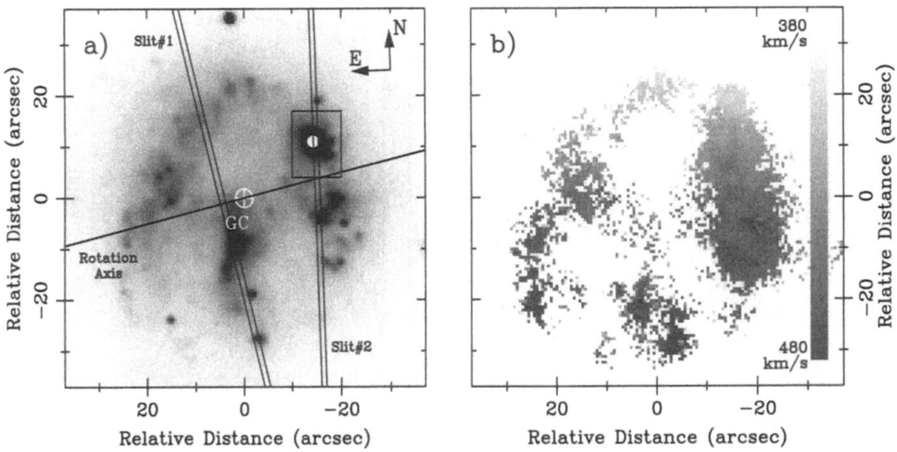

Figure 1. **a)** Mrk86 B-band image. The white dot shows the position of peak of the K-band brightness (see Figure 2a) . Slits (see text) and rotation axis orientations are also shown. **b)** Hα velocity field.

3. Evidence for merging and the Mrk86–C bubble

The slit #2 (figure 1a) velocity profile shows a rapid velocity gradient steeping (figure 2b). This profile suggests that the steep gradient component of 70 ± 5 km s^{-1} kpc^{-1} is due to a kinematically independent object, probably a recently fused dwarf galaxy (box in figure 1a). Meanwhile, the shallower component of 19 ± 2 km s^{-1} kpc^{-1} should be related with the galaxy main body. This scenario has been also suggested for IIZw40 [1] and IZw18 [6].

Near this K-band peak, 7 arcsec (=0.3 kpc) north, a local deep velocity minimum is observed. Its velocity profile is compatible with the approaching side of an expanding structure of about 110 pc of radius and expansion velocity of 17 km s^{-1}. These expanding bubbles are driven by collective supernova winds originated in very active star forming regions. We call this bubble Mrk86–C (following the sequence of the previously discovered Mrk86–A [9] and B [7] bubbles). The observation only of the approaching side of the bubble is probably due to the higher extinction suffered by the bubble receding side emission (E$(B-V)_{\rm gas}$=0.73$^{\rm m}$; [7]).

4. Conclusions

In this work we have shown that dwarf star forming galaxies, such as Mrk86, are not necessarily dark matter dominated at short radii. In the case of the BCDs, this fact is probably related with a peculiar, more active, star formation history. Therefore, it is necessary to extend these studies to dwarf galaxies with different star formation histories.

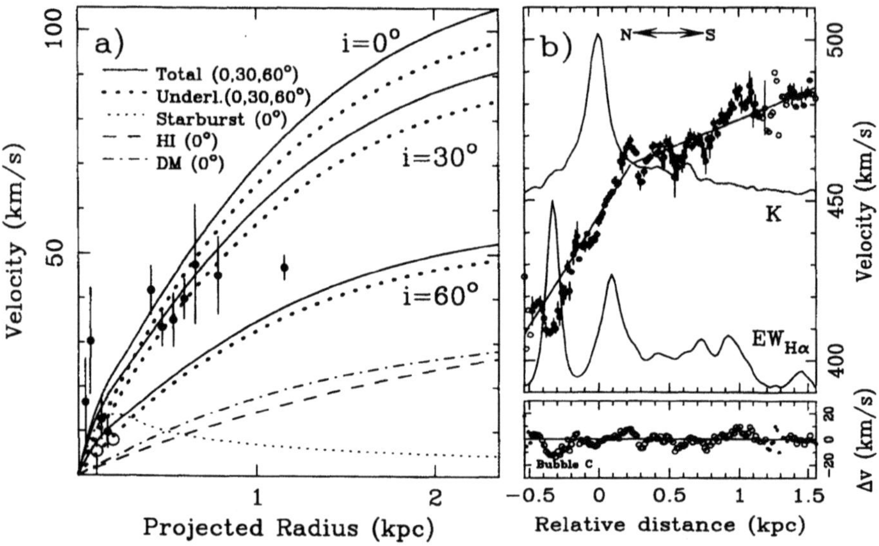

Figure 2. **a)** Radial velocity curve. **b)** *Upper panel:* Velocity curve at PA=12° and 0.85 kpc from the galactic center (slit #2 in figure 1a). *K*-band and EW(Hα) profiles are also shown. *Lower panel:* Residuals of the fitted gradients.

Acknowledgements

We would like to thank C. Sánchez Contreras and L. F. Miranda for obtaining the high resolution spectra. AGP acknowledges the receipt of a FPU/MEC fellowship. This research was supported by the Programa Sectorial de Promoción General del Conocimiento under grant PB96-0610.

References

1. Van Zee L., Skillman E.D. and Salzer J.J. (1998), *AJ*,**Vol. no. 116**, pp. 1186–1204
2. Carignan C. and Freeman K.C. (1988), *ApJ*, **Vol. no. 332**, pp. L33–L36
3. Moore B. (1994) *Nature*, **Vol. no. 370**, pp. 629–631
4. Navarro J.F., Frenk C.S. and White S.D.M. (1997), *ApJ*, **Vol. no. 490**, pp. 493–508
5. Papaderos P., Loose H.-H., Thuan T.X. and Fricke H.J. (1996), *A&AS*, **Vol. no. 120**, pp. 207–228
6. Skillman E.D. and Kennicutt R.C. (1993), *ApJ*, **Vol. no. 411**, pp. 655–666
7. Gil de Paz A., Zamorano J. and Gallego J. (1999), *MNRAS*, accepted
8. Marlowe A.T. et al. (1995), *ApJ*, **Vol. no. 438**, pp. 563–589
9. Martin, C.L. (1998), *ApJ*, **Vol. no. 506**, pp. 222–252
10. Dekel A. and Silk J. (1986), *ApJ*, **Vol. no. 303**, pp. 39–55
11. Loose H.-H., and Thuan T.X. (1985), in *Star-Forming Dwarf Galaxies*, eds. Kunth D. et al., Editions Frontiéres
12. Gil de Paz A., Zamorano J. and Gallego J. (1998), in *Dwarf Galaxies and Cosmology*, XVIIIth Moriond meeting, in press
13. Bruzual G. and Charlot S., (1996), ftp://gemini.tuc.noao.edu/, (BC96)
14. Mac Low M.-M. and Ferrara A. (1998), *ApJ*, astro-ph/9801237
15. WHISP survey; Sancisi R. (1994), *NFRA*, **3**, http://thales.astro.rug.nl/~whisp/

DWARF SPHEROIDAL GALAXIES: DYNAMICAL EVOLUTION

M.A. GÓMEZ-FLECHOSO*
Observatoire de Genève
Ch. des Maillettes 51, CH-1290 Sauverny, Switzerland

Abstract. We show the results of numerical simulations of dwarf spheroidal galaxies (DSphs), which allow to understand their evolution in the framework of no internal virial equilibrium. Under this hypothesis, we reproduce with a low dark matter (DM) model the high values of the virial M/L ratio measured observationally for these objects. In this work, we present a model of the Sagittarius (Sgr) DSph, with all its observational features.

1. Introduction

The main questions about the DSphs as satellites of the Milky Way (MW) are their actual content of DM and their implications in the hierarchical formation of the Galactic halo.

The present studies about DSphs are based on the hypothesis of its internal virial equilibrium. Under this assumption, the mass of the DSph, M, can be inferred by measuring the structural parameters of the system: central velocity dispersion, σ_o, half brightness radius, r_{hb}, central surface brightness, μ_o, and the luminosity, L, and using the expression:

$$\left(\frac{M}{L}\right)_o \propto \frac{\sigma_o^2}{\mu_o r_{hb}} \tag{1}$$

The values of the total M/L of the DSphs, obtained by this method, vary from 8 $(M/L)_\odot$ for Leo I to 100 $(M/L)_\odot$ for Sgr in the literature (see Mateo 1997 for a review). However, the satellites with high M/L ratio are more frequent. If the virial hypothesis were true, the mass contribution of the DSphs to the Galactic halo could be very important and the DSphs could play a fundamental role in the hierarchical evolution of the MW.

*Fundación Ramón Areces Fellow

Astrophysics and Space Science is the original source of publication of his article. It is recommended that this article is cited as: *Astrophysics and Space Science* **263**: 155–158, 1999.

However, the virial equilibrium condition in the DSphs is not fully proved. Furthermore, observational evidences show that most of the DSphs present tidal tails and other interaction features that prove the existence of strong internal tidal forces. This argument questions the viability of using the Eq. (1) for calculating the mass content of the DSphs.

In this paper, we present simulations of the Sgr DSph. This galaxy is supposed to be DM dominated. It is the closest satellite to the center of the MW, $R_{GC} \sim 16$ kpc[1]. Recent observations (Mateo *et al.*, 1998) show that Sgr presents a large extension ($68^o \times 8$–10°) including a long tidal tail. New calculations of the M/L based on these observations and assuming virial equilibrium drop this value from 50 to 10 $(M/L)_\odot$ at the center of Sgr. However, it is still larger than the expected value for its stellar population (Sgr is not much more luminous than a typical globular cluster). Previous numerical simulations of Sgr assuming either a low DM content (Ibata & Lewis, 1998) or a large amount of DM (Johnston, Spergel & Hernquist, 1995; Velázquez & White, 1995) are not able to reproduce its observational features and, additionally, fail in obtaining a long lifetime satellite . The previous condition is required to explain the long and complex evolution of Sgr (Montegriffo *et al.*, 1998). The only simulations able to reproduce long lifetime DSphs are those of Klessen & Kroupa (1998), but these authors obtained satellites fainter than the observations.

The main differences between our simulations and the previous ones are the initial configurations. Whereas other authors assume that the DSph is in virial equilibrium at the beginning of the simulation and, therefore, it has not been modified by the tidal forces, we consider that the DSph has been orbiting into the tidal potential of the MW and it had time to readjust its internal configuration to a new *no-virial* equilibrium with the tidal forces (for details see Gómez-Flechoso & Domínguez-Tenreiro 1998).

2. The model

We have performed simulations of N-body models of a Galaxy+satellite system, using the treecode algorithm provided by Barnes & Hut (1986).

2.1. THE GALAXY: THE MILKY WAY

The MW is represented by a multicomponent model, described by Fux (1997) as m04 model. The disc, spheroid and halo masses are 4.6×10^{10} $M_\odot$, 3.65×10^{10} $M_\odot$ and 4.11×10^{11} $M_\odot$, respectively, and the number of particles is 8739 (disc), 6932 (spheroid) and 29648 (halo).

[1]The parameters of Sgr can be found in Ibata *et al.* (1997) and Mateo *et al.* (1998)

TABLE 1. Parameters of the satellites (modified King's spheres)

	r_o (kpc)	σ_o (km/s)	M_t ($10^7 M_\odot$)	r_t (kpc)	W_o
Model 1	0.1	15.0	5.27	1.93	5.42
Model 2	0.1	11.0	2.33	1.48	4.87
Model 3	0.3	15.0	6.04	2.02	2.98

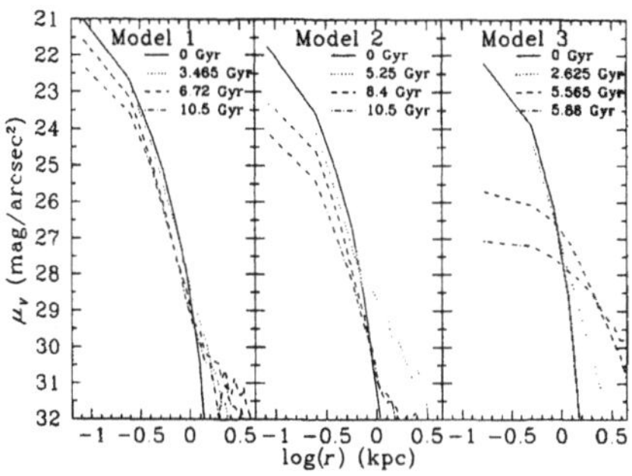

Figure 1: Surface brightness of the models, assuming $(M/L)_V = 2$ (M/L)$_\odot$.

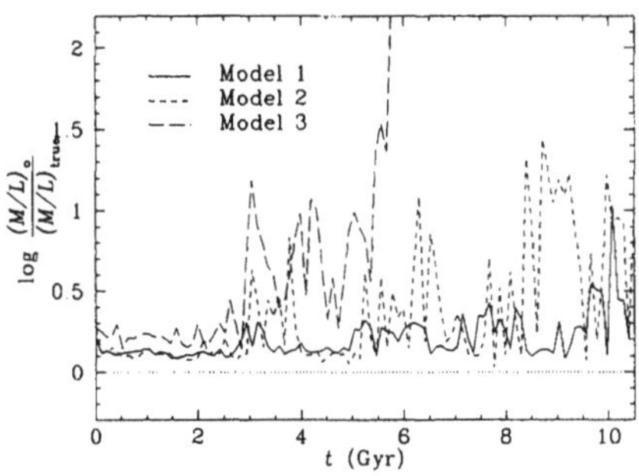

Figure 2: Evolution of the log of the ratio between the virial and the true M/L.

2.2. THE SATELLITE: SAGITTARIUS

The satellites are represented by N-body ($N = 8000$ particles) modified King's spheres described by Gómez-Flechoso & Domínguez-Tenreiro (1998). These models have two free parameters: the core radius, r_o, and the central velocity dispersion, σ_o, which have been selected within the observational range. The other parameters of the models (total mass, M_t, tidal radius, r_t, and dimensionless central potential, W_o) are automatically determined by the method (the values are listed in Table 1). The orbit of the DSph reproduces the observational position and velocity of Sgr (Ibata *et al.*, 1997), with the apocenter and pericenter at 74 and 10 kpc, respectively[2].

3. Results and discussion

In contrast with other simulations (e.g. Velázquez & White 1995; Ibata & Lewis 1998), our models show a lifetime long enough to explain the evolution of Sgr. This difference is a consequence of the initial configuration of the models. Whereas we assume that the DSph is initially in equilibrium with the environment (as result from either a slow accretion or its formation inside the Galactic halo), other authors use isolated configurations, which do not represent the real state of the DSph when it orbits inside the Galactic halo. In the latter case, the simulations do not only reflect the evolution

[2]Other models are analyzed in detail by Gómez-Flechoso, Fux & Martinet (1998)

of the DSph, but also the numerical effects due to the erroneous initial conditions.

In Fig. 1, we have plotted the surface brightness of the models at several timesteps (assuming $M/L = 2$ $(\mathrm{M/L})_\odot$). We can see that the evolution of the satellites depends on its concentration, defined as $c = \log(r_t/r_o)$. The least concentrated satellite (model 3) suffers the strongest evolution and it is finally destroyed at $\sim$ 6 Gyr. However, model 3 at $\sim$ 5.6 Gyr (right before destruction) reproduces the structural characteristics of Sgr (for more details see Gómez-Flechoso, Fux & Martinet 1998). It supports the hypothesis that Sgr is an almost disrupted DSph and that it has suffered strong tidal interactions with the Galactic potential.

The evolution of the ratio between the M/L obtained from virial arguments (Eq. (1)), $(M/L)_o$, and the true M/L value, $(M/L)_{true}$, for the models is plotted in Fig. 2. This plot displays the *no virial situation* of the DSphs. The horizontal dotted line shows the evolution of a virialized system. If we assume virial conditions for the DSphs when they are suffering tidal forces (and therefore they are not in virial equilibrium), we always obtain values of the M/L larger than the real ones. This is also true for the DSphs which are not close to their disruption (e.g. model 1). However, the more disrupted the satellite is, the larger $(M/L)_o/(M/L)_{true}$ we obtain. So, the large $(M/L)_o$ of Sgr measured by Ibata *et al.* (1997) can be reproduced with a DSph close to the disruption (e.g. the model 3 at 5.6 Gyr).

Finally, special attention should be paid to the situations in which the virial theorem cannot be applied, because it could lead to erroneous conclusions about the DM content. We also advise the use of initial configurations of the models which reflect the true state of the system, otherwise the numerical effects can mask the effective evolution of the DSphs.

References

Barnes, J.E., Hut, P., 1986, Nature, 324, 446
Fux, R., 1997, A&A, 327, 983
Gómez-Flechoso, M.A., Domínguez-Tenreiro, R., 1998, MNRAS, submitted
Gómez-Flechoso, M.A., Fux, R., Martinet, L., 1998, A&A, submitted
Ibata, R., Wyse, F.G., Gilmore, G., Irwing, M.J., Suntzeff, M.B., 1997, AJ, 113, 634
Ibata, R., Lewis, G.F., 1998, ApJ, 500, 57
Johnston, K.V., Spergel, D.N., Hernquist, L., 1995, ApJ, 451, 598
Klessen, R.S., Kroupa, P., 1998, ApJ, 498, 143
Mateo, M., 1997) in *The Nature of Elliptical Galaxies*, Arnaboldi, M., Da Costa, G.S., Saha, P. (eds.), in press
Mateo, M., Olszewski, E.W., Morrison, H.L., 1998, astroph/9810015
Montegriffo, P., Bellazzini, M., Ferraro, F.R., Martins, D., Sarajedini, A., Fusi Pecci, F., 1998, MNRAS, 294, 315
Velázquez, H., White, S.D.M., 1995, MNRAS, 275, L23

STELLAR POPULATIONS IN DWARF ELLIPTICAL GALAXIES

S. PEDRAZ, J. GORGAS AND N. CARDIEL

Departamento de Astrofísica. Facultad de Físicas. Universidad Complutense de Madrid, 28040, Madrid, Spain

AND

R. GUZMÁN

Dept. of Astronomy, Yale University, USA

1. Introduction

At first sight, the stellar populations of the different families of elliptical galaxies (E's) share some global properties that make them remarkably similar. In particular, dwarf ellipticals fit extremely well in the extrapolation of the Mg_2–σ and the color–magnitude relations of the giant ellipticals towards lower luminosities. In spite of these similarities, important structural differences (luminosity profiles, positions in the fundamental plane) suggest different galaxy formation and/or evolutionary processes for giant and dwarf ellipticals. To get a deeper insight into these problems we are carrying out a spectroscopic survey devoted to study in detail the stellar populations of dwarf elliptical galaxies.

Our basic sample was observed during two runs with the 2.5 m INT at La Palma (April 94) and the 3.0 m. Shane Telescope at Lick (April 97). It comprises 7 compact ellipticals (cE, with $r_{\rm eff} \leq 0.5$ kpc), 6 bright spheroidals (Sph, according to their exponential luminosity profiles. Note that these galaxies are sometimes called "dwarf" ellipticals) and 6 low luminosity E's (with $M_B > -20.5$). As a comparison sample of giant E's we have used the data from González (1993). The analysis is based on the measurement of central line–strength Lick indices (namely Hβ, Mgb, Mg_2, Fe5270 and Fe5335). The indices presented in this paper correspond to a 4 arcsec. central aperture and have been fully corrected to the Lick system.

Astrophysics and Space Science is the original source of publication of his article. It is recommended that this article is cited as: *Astrophysics and Space Science* **263**: 159–162, 1999.

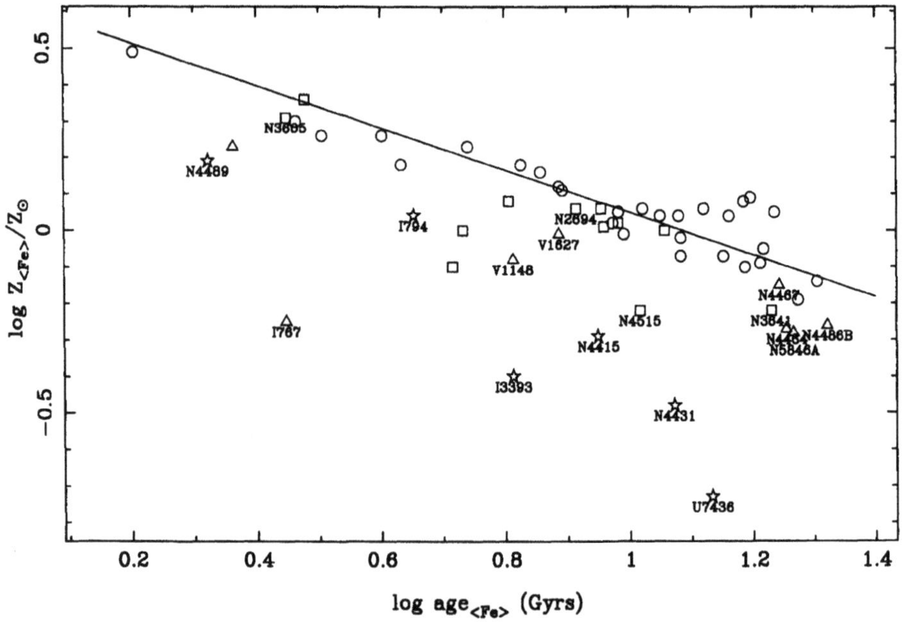

Figure 1. Age–metallicity diagram for the galaxy sample. The galaxies observed in this work are labelled with their names. Circles correspond to bright E's ($M_B < -20.5$) and squares to low–luminosity E's ($M_B > -20.5$). Triangles represent cE's, whereas stars are used for bright Sph galaxies. The straight line is a least–squares fit to the bright ellipticals.

2. Ages and metallicities

To investigate the stellar populations of dwarf and giant ellipticals we have compared their line-strength indices with the predictions of single–burst stellar populations models from Worthey (1994). In particular, we have employed the Hβ–⟨Fe⟩ and Hβ–[MgFe] diagrams to get estimates of luminosity–weighted mean ages and metallicities (see González, 1993, for the indices definitions). Given the discrepancies in the derived ages when using different models (typically $\sim$ 4 Gyr for old populations and metallicities around solar; e.g. Worthey, 1994, versus Vazdekis et al., 1996) absolute values are quite uncertain, although relatives differences among galaxies should be reliable. In Fig. 1 we plot the metallicities and ages for the galaxy sample as derived from the Hβ–⟨Fe⟩ diagram (see also Trager 1997).

From Figure 1 it is evident that bright ellipticals ($M_B < -20.5$) span a wide range in mean age (González 1993; Worthey et al. 1996). Furthermore, ellipticals that look younger tend to be more metal–rich. Concerning the dwarf ellipticals, cE's also extend along this range in age and follow the E's trend but at lower metallicity. The bright Sph sample seems to populate different regions of the diagram, suggesting a steeper trend at much lower metallicities.

In order to interpret these results in terms of star formation histories, one must keep in mind the well–known Mg overabundance problem. Peletier (1989) and Worthey et al. (1992) showed that the metallicities inferred in E galaxies from indices based on the strength of Mg lines are higher than those derived from Fe indices. This implies an enhancement of the [Mg/Fe] ratio in bright E's compared to the solar ratio (employed in the calibration stellar library). It is important to note that our results (see also Gorgas et al. 1997) clearly indicate that bright Sph's and isolated cE's (i.e. except NGC 4486B and NGC 5846A) exhibit solar [Mg/Fe]. Since Mg is created rapidly by Type II SN while Fe is produced in a longer timescale by Type Ia SN, these differences between bright and dwarf E's clearly imply a difference in their star formation histories. Following these arguments, the bulk of star formation in giant E's could have taken place in a short timescale (with a high efficiency). Subsequent bursts (induced by pre-enriched gas infall or mergers) would then be needed to explain the age–metallicity relation shown in Fig. 1. On the other hand, the star formation in bright Sph's might have proceeded in a longer timescale (with a lower efficiency), producing a solar [Mg/Fe] ratio.

3. Do we still have a mass-metallicity relation?

The standard picture of a monolithic dissipative collapse for the formation of galaxies establishes that massive E's are able to retain the enriched gas against SN–driven winds and, therefore, they attain higher metallicities than low-mass E's, in which further star formation is inhibited by the lost of the processed gas (e.g. Arimoto & Yoshii 1987). Note, however, that this is clearly in contradiction with the arguments we have given above. From Fig. 2, in which we plot a mass–metallicity diagram for our galaxy sample, it is apparent that massive E's do not exhibit the relation predicted by the standard scenario. In fact, the metallicity of a massive E seems to be determined by the mean age rather than by the mass. However, for masses below a threshold value of $\sim 10^{10}$ $M_{\odot}$ a clear difference emerges. Interestingly, most bright Sph's attain much lower metallicities than cE's or the extrapolation of bright E's, confirming that the dichotomy observed in the structural properties of spheroidal and elliptical galaxies is also found in their stellar populations.

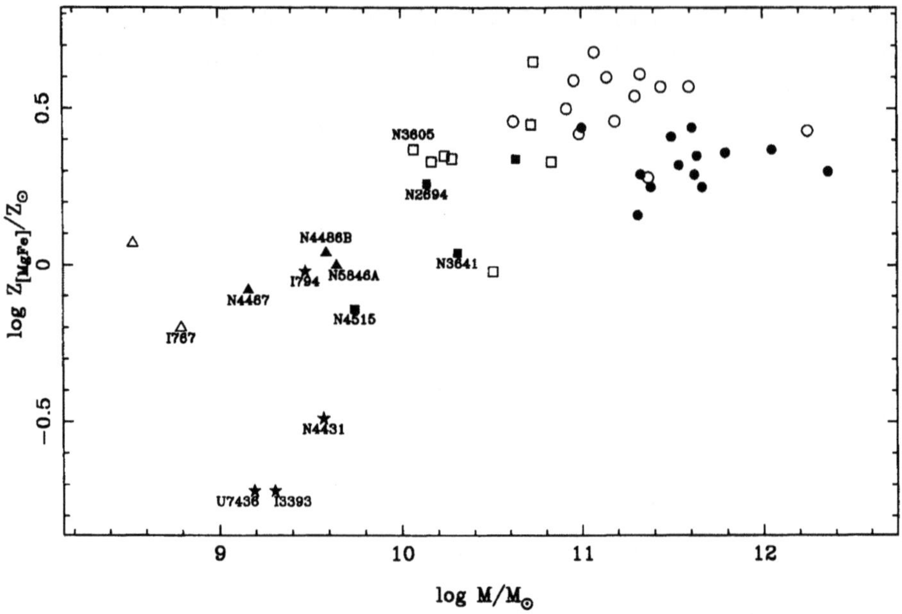

Figure 2. Mass–metallicity relation for galaxy sample. The metallicities are determined from the Hβ–[MgFe] diagram using Worthey (1994) models. Masses are from Bender et al. (1992). Symbol types are the same as in Fig 1. Filled and empty symbols correspond to galaxies with mean ages above and below 5 Gyr respectively.

Acknowledgements

This work was supported by the Spanish "Programa Sectorial de Promoción del Conocimiento" under grant No. PB96–610. We are grateful to Jesús González for fruitful discussions.

References

Arimoto, N, Yoshii, Y., 1987, A&A, 173, 23
Bender, R., Burstein, D., Faber, S.M., 1992, ApJ, 399, 462
González, J.J., 1993, Ph. D. Thesis, University of Santa Cruz, California
Gorgas, J., Pedraz, S., Guzmán, R., Cardiel, N., González, J.J., 1997, ApJ, 481, L19
Peletier, R.F., 1989, Ph. D. Thesis, Rijksuniversiteit Groningen
Trager, S.C., 1997, Ph. D. Thesis, University of California, Santa Cruz
Vazdekis, A., Casuso, E., Peletier, R.F., Beckman, J.E., 1996, ApJS, 106, 307
Worthey, G., Faber, S.M., González, J.J., 1992, ApJ, 398, 69
Worthey, G., 1994, ApJS, 95, 107
Worthey, G., Trager, S.C., Faber, S.M., 1996, in *Fresh Views on Elliptical Galaxies*, eds. A. Buzzoni, A. Renzini, & A. Serrano, ASP Conf. Ser, 86, 203

KINEMATICS OF GAS AND STARS IN SPIRAL GALAXIES

J.C. VEGA BELTRÁN AND E. PIGNATELLI
Osservatorio Astronomico di Padova,
Vicolo dell'Osservatorio 5, I-35122 Padova, Italy
W.W. ZEILINGER
Insitut für Astronomie, Universität Wien,
Türkenschanzstrasse 17, A-1180 Wien, Austria
A. PIZZELLA
European Southern Observatory
Alonso de Corodova 3107, Casilla 19001, Santiago 19, Chile
E.M. CORSINI AND F. BERTOLA
Dipartimento di Astronomia, Università di Padova
vicolo dell'Osservatorio5, I-35122 Padova, Italy
AND
J.E. BECKMAN
Instituto Astrofísico de Canarias
Calle Via Láctea s/n, E-38200 La Laguna, Spain

In order to investigate the interplay between the morphological and kinematical properties of disk galaxies we measured and analyzed the kinematics of ionized gas and stars along the major axis of a sample of 20 unbarred S0 and spiral galaxies. In the outer regions of all the sample galaxies the stellar rotation velocity is comparable to that of the ionized gas ($\Delta V_\star \approx \Delta V_{\rm gas}$), while in the inner regions the following kinematic features are noteworthy.

In central regions of the late-type spirals we observe almost the same velocity gradient for both stars and gas ($\Delta V_\star/\Delta r \approx \Delta V_{\rm gas}/\Delta r$). The same is true for their velocity dispersion ($\sigma_\star \approx \sigma_{\rm gas}$), which are characterized by low values (~ 50 km s^{-1}) over all the observed radial range. As an example of this kind of kinematics we show in Figure 1 the Scd galaxy NGC 1160. This is also the case of NGC 949 (Sc) and NGC 2541 (Scd). We deduce that in most of the late-type spirals we studied the stars and the ionized gas are virtually moving at the circular velocity.

Astrophysics and Space Science is the original source of publication of his article. It is recommended that this article is cited as: *Astrophysics and Space Science* **263**: 163–166, 1999.

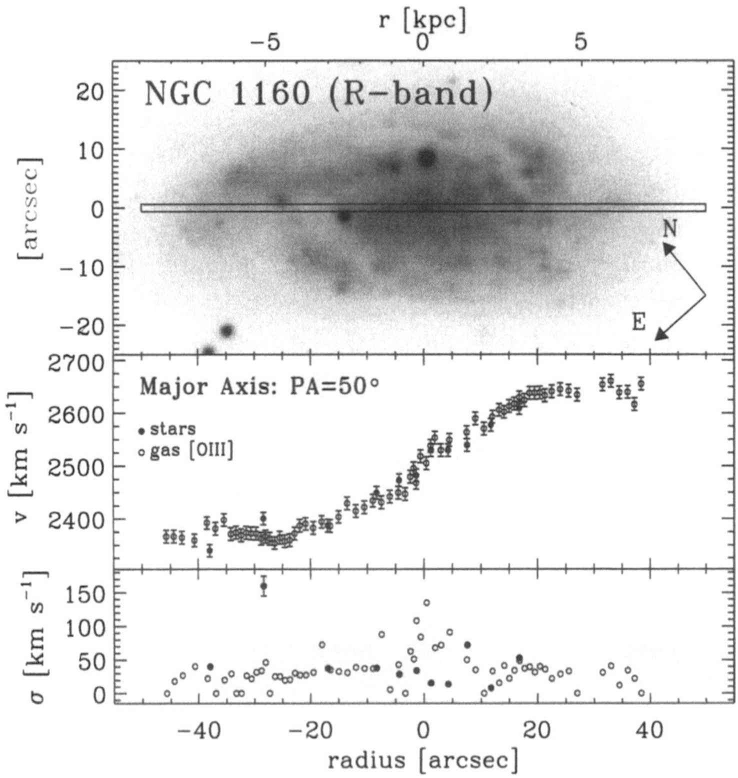

Figure 1. Ionized gas and stellar kinematics along the major axis of the late-type spiral NGC 1160. In the upper panel the position of the slit of the spectrograph is plotted superimposed to an R−band image of the galaxy The velocity curves (middle panel) and the velocity dispersion profiles (lower panel) of the gaseous (open circles) and stellar (filled circles) components are plotted

The central parts of early-type disk galaxies reveal a wider variety of different behaviors of stars and gas. We usually measured for stars and gas $\Delta V_\star/\Delta r < \Delta V_{\rm gas}/\Delta r$ and $\sigma_\star > \sigma_{\rm gas} \sim 50$ km s^{-1} (e.g. NGC 5064, NGC 7782). This can be easily explained by considering that the stellar and gaseous kinematics are dominated by dynamical pressure and rotation respectively. The observed stellar rotation can be corrected into that corresponding to the circular velocity traced by the gas rotation taking into account the asymmetric drift effect (e.g. Binney & Tremaine 1987). In the early-type disk galaxies NGC 772, NGC 980 and NGC 3898, we derived $\Delta V_\star/\Delta r \approx \Delta V_{\rm gas}/\Delta r$ and we found $\sigma_\star \approx \sigma_{\rm gas} \geq 120$ km s^{-1} over an extended radial range. In Figure 2 we show the kinematics of the Sa NGC 3898. The presence of the gaseous component with a high central velocity dispersion has been previously detected in the nucleus of

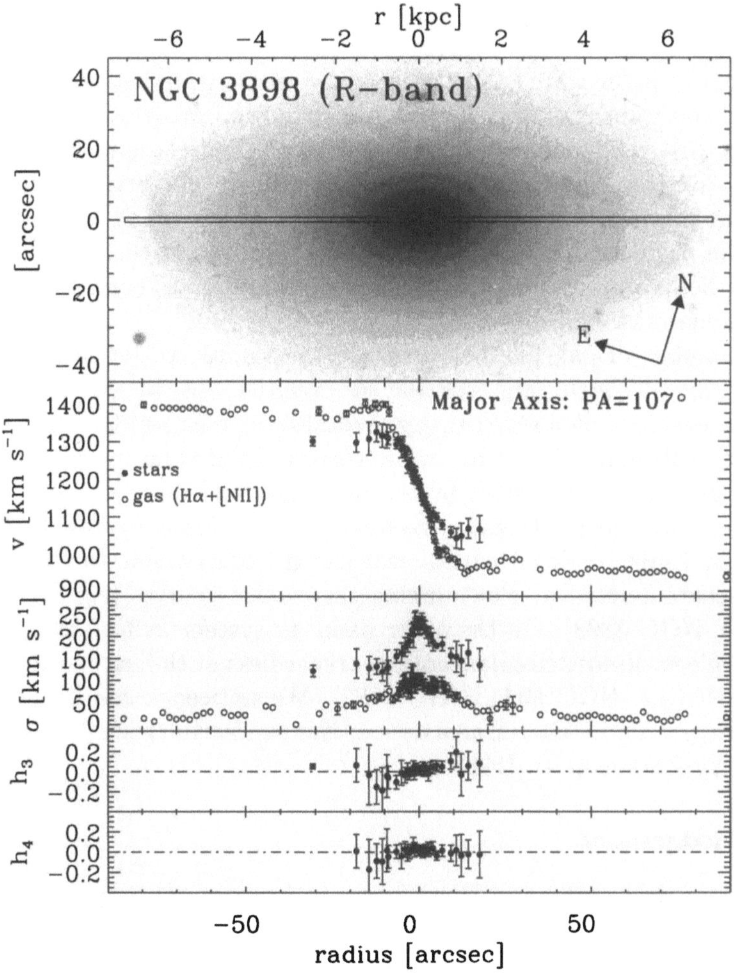

Figure 2. Same a Fig. 1 for the early-type spiral NGC 3898. In the two additional panels the radial profiles of Gauss-Hermite coefficients h_3 and h_4 are plotted

some lenticulars and early-type spirals by different authors (Fillmore et al. 1986, Kent 1988, Kormendy & Westpfahl 1989). It has been explained by Bertola et al. (1995) showing that random motions are also crucial for the dynamical support of the gas (e.g. see the detailed model for NGC 4036 by Cinzano et al. 1998). The peculiar stellar and/or gaseous kinematics of the remaining early-type spiral of the sample is due to the presence of triaxial (e.g. NGC 3521, NGC 4419), counterrotating (e.g. NGC 2841, NGC 7331) or kinematically decoupled (e.g. IC 4889, NGC 4698) components. These components have been detected (even when they were photometrically unresolved) due to the hallmarks they leave in the observed kinematics (e.g. "wavy pattern" rotation curves, "figure-of-eight" rotation curves "M-shaped" velocity dispersion profiles, h_3 peculiar radial profiles).

In order to investigate quantitatively the relationship between the kinematics of stars and gas and the mass distribution of spirals we adopted the self-consistent dynamical models by Pignatelli & Galletta (1998). The models derive from the stellar data the gravitational potential in which the ionized gas is expected to orbit. They take into account the asymmetric drift effects, the projection effects along the line-of-sight and the non-Gaussian shape of the line profiles due to the presence of different components with distinct dynamical behavior.

The number of galaxies belonging to the sample is not large enough to draw general conclusions. Anyway we found a possible correlation between the presence of slowly-rising gas rotation curves and the ratio of the bulge/disk half luminosity radii, while there is no obvious correlation with the key parameter represented by the morphological classification, namely the bulge/disk luminosity ratio. Systems with a diffuse dynamically hot component (bulge or lens) with a scale length comparable to that of the disk are characterized by slowly-rising gas rotation curves (e.g. NGC 772, NGC 980, NGC 3898). On the other hand, in systems with a small bulge the gas follows almost circular motions, regardless of the luminosity of the bulge itself (e.g. NGC 5064, NGC 7782). We noticed a similar behavior also in the gas and stellar kinematics of the two early-type spiral galaxies modeled by Corsini et al. (1998).

Acknowledgements

The data of NGC 1160 and NGC 3898 are based on observations carried out at the MMT telescope, Tucson (Arizona), at the 2.5-m INT telescope, La Palma (Canary Islands) and at VATT: the Alice P. Lennon Telescope and the Thomas J. Bannan Astrophysics Facility, Tucson, Arizona.

References

Bertola, F., Cinzano, P., Corsini, E.M., Rix, H.-W., Zeilinger, W., 1995, ApJ, 448, L13
Binney, J., Tremaine, S., 1987, Galactic Dynamics. Princeton University Press, Princeton
Cinzano, P., Rix, H.-W., Sarzi, M., Corsini, E.M., Zeilinger, W.W., Bertola, F., 1998, MNRAS, submitted
Corsini, E.M., Pizzella, A., Sarzi, M., Vega Beltran, J.C., Funes, J.G., Bertola, F., Persic, M., Salucci, P., 1998, A&A, in press (astro-ph/9809366)
Fillmore, J.A., Boroson, T.A., Dressler, A., 1986, ApJ, 302, 208
Kent, S.M., 1988, AJ, 96, 514
Kormendy, J., Westpfahl, D.J., 1989, ApJ, 338, 752
Pignatelli, E., Galletta, G., 1998, A&A, submitted (astro-ph/9705042)

TOWARDS AN UNDERSTANDING OF THE λ4000Å BREAK BEHAVIOUR IN OLD STELLAR POPULATIONS

J. GORGAS, N. CARDIEL AND S. PEDRAZ
Departamento de Astrofísica. Facultad de Físicas
Universidad Complutense de Madrid, 28040 Madrid, Spain

Abstract. We present new empirical fitting functions describing the behaviour of the discontinuity at λ4000 Å in terms of the stellar atmospheric parameters: effective temperature, metallicity and surface gravity. These calibrations can be easily incorporated into stellar population models, providing, for the first time, accurate predictions of the break amplitud for, relatively old, composite systems.

1. Introduction

The λ4000Å break is a sudden onset of absorption features bluewards of λ4000Å clearly noticeable for spectral types cooler than G0. The amplitude of the break is therefore quite sensitive to the stellar effective temperature, but also must depend on metallicity. Due to this behaviour, the break is an important tool to investigate the properties of composite stellar populations. In fact, a number of authors have already employed this discontinuity to get insight into the star formation histories of early–type galaxies (see references in Gorgas et al. 1999)

In this work we are using the definition of the break adopted by Bruzual (1983), which can be rewritten as:

$$\mathrm{D}_{4000} = \int_{4250\,(1+z)}^{4050\,(1+z)} \lambda^2 f_\lambda \, \mathrm{d}\lambda \Bigg/ \int_{3750\,(1+z)}^{3950\,(1+z)} \lambda^2 f_\lambda \, \mathrm{d}\lambda \,, \tag{1}$$

where f_λ is the flux density per unit wavelength, and z is the redshift.

Compared with other features usually used in the blue spectral region, the D_{4000} has the clear advantage of being insensitive to spectral resolution (i.e. velocity dispersion). Furthermore its measurement does not require a high signal-to-noise (SN) ratio (Cardiel et al., 1998, have shown that

Astrophysics and Space Science is the original source of publication of his article. It is recommended that this article is cited as: *Astrophysics and Space Science* **263**: 167–170, 1999.

a SN/Å = 1 translates into a 10% uncertainty in D_{4000}). Obviously, this advantage, due to the large wavelength interval included in its definition, translates into an important drawback: many absorption features are included in the break bandpasses and, therefore, its behaviour is expected to be complex. Several authors have tried to calibrate the break in terms of the stellar atmospheric parameters using either observed star samples (Wildey et al. 1962, van den Bergh & Sackmann 1965, Bruzual 1983, Hamilton 1985) or theoretical stellar atmospheres (Poggianti & Barbaro 1997). However, due to several reasons (incomplete stellar libraries, poor resolution in the parameter space, uncertainties in the theoretical calibrations) these previous works do not provide conclusive D_{4000} predictions, in particular for old stellar populations. In this paper, we present empirical fitting functions, based on an ample stellar library, which can be used to model the break behaviour as a function of the age and metallicity of relatively old stellar systems.

2. The sample

In order to derive a confident empirical calibration of the D_{4000} we have measured it in the 430 stars of the Lick/IDS stellar library (Gorgas et al. 1993 –G93–, Worthey et al. 1994 –W94–), which spans the following ranges in atmospheric parameters: $T_{\rm eff}$: 2747 – 21860 K, $\log g$: 0 – 5.12, and [Fe/H]: −2.70 – 0.46. The observations were performed in 13 different observing runs from 1991 to 1996 using the JKT, INT and WHT (La Palma), the 3.5 m. telescope at Calar Alto, and the 2.12 m. telescope from San Pedro Mártir (Mexico). Although the observational configuration changes from run to run, most of the spectra have a resolution of 3 Å (FWHM) in the spectral range from 3600 to 4500 Å. The reduction of data was performed with REDUCEME[1]. This allowed a careful treatment of the observational random and systematic errors. A final table with individual D_{4000} indices and associated errors for the whole library is available upon request.

3. The fitting functions

Following the previous works of G93 and W94, the fitting functions are expressed as polynomial functions in the atmospheric parameters (using $\theta \equiv 5040/T_{\rm eff}$ as the temperature indicator), with two different forms:

$$D_{4000}(\theta, [\mathrm{Fe/H}], \log g) = p(\theta, [\mathrm{Fe/H}], \log g) \quad (2)$$

and

$$D_{4000}(\theta, [\mathrm{Fe/H}], \log g) = \mathrm{const.} + \exp\left[p(\theta, [\mathrm{Fe/H}], \log g)\right], \quad (3)$$

[1]See http://www.ucm.es/OTROS/Astrof/reduceme/reduceme.html

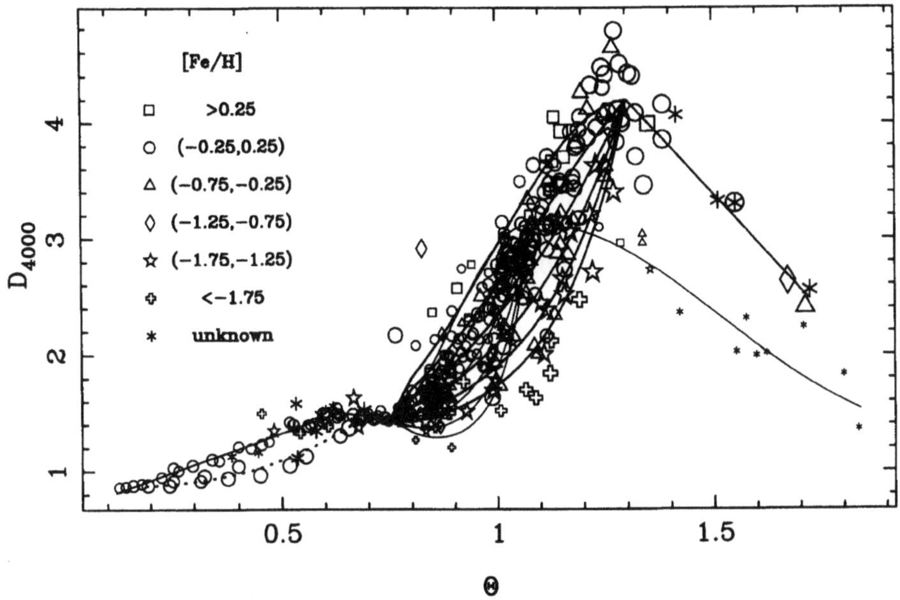

Figure 1. D_{4000} as a function of $\theta \equiv 5050/T_{\rm eff}$ for the sample together with the derived fitting functions. In the low θ range, the solid line corresponds to dwarf and giant stars, whereas the dashed line is used for supergiants. For $\theta > 0.75$ thick and thin lines show the functions for giant ($\log g < 3.25$) and dwarf ($\log g > 3.25$) stars respectively. In the region around $\theta \approx 1$ different lines correspond to the metallicities $[{\rm Fe/H}] = 0.5, 0, -0.5, -1, -1.5, -2$, from top to bottom. The stars are shown with different symbols depending on the metallicity and with sizes inversely proportional to surface gravity.

where p is a polynomial with terms up to the third order, including all possible cross-terms among the parameters.

The polynomial coefficients were determined from a least squares-fit where all the stars were weighted according to the D_{4000} observational errors. The strategy followed to determine the final fitting function is based on the succesive inclusion of terms, starting with the lower powers. At each step, the term which yielded a lower new residual variance was tested. To do this, the significance of the new term, as well as those of all the previously included coefficients, was computed using a t-test and only terms statistically significant were retained. No cross-terms between $\log g$ and [Fe/H] were necessary. Unfortunately, the behavior of the break can not be reproduced by a single function in the whole temperature interval, and four regions were used with smooth transitions among them. In addition, cold giants and dwarfs were found to follow different trends. In Figure 1 we plot the final fitting functions as a function of temperature. In the whole, the unbiased residual standard deviation of the fitted functions is $\sigma_{\rm std} = 0.160$.

This is significantly larger than the typical error in the D_{4000} measurements ($\sigma_{\rm typ} = 0.064$). Since these latter errors are quite realistic (from a subsample of multiply observed stars), the extra scatter must arise from uncertainties in the input atmospheric parameters. The determination coefficient of the fit is 0.96, i.e. 96% of the observed variance is explained with the present fitting functions. More details about the derived functions are presented in Gorgas et al. (1999). A FORTRAN subroutine for computing predicted D_{4000} indices, including errors in the predictions, is also available upon request.

4. Conclusions

The empirical calibration presented in this paper can be easily incorporated into stellar populations models to provide accurate predictions of the D_{4000} for composite systems. These predictions should be valid for populations with ages between 0.1 and 20 Gyr, and [Fe/H] ranging from -1 to $+0.5$ dex. The combination of these predictions with those for spectral features such as the CaT in the near-infrared (Gorgas et al., in preparation) will provide new simple methods to disentangle age and metallicity effects in old stellar populations.

Acknowledgements

The JKT, INT and WHT are operated on the island of La Palma by the Royal Greenwich Observatory at the Observatorio del Roque de los Muchachos of the Instituto de Astrofísica de Canarias. The Calar Alto Observatory is operated jointly by the Max-Planck-Institute für Astronomie, Heidelberg, and the Spanish Comisión Nacional de Astronomía. This wok was supported by the Spanish 'Programa Sectorial de Promoción del Conocimiento' under grant No. PB96-610.

References

Bruzual, A.G., 1983, ApJ, 273, 105
Cardiel, N., Gorgas, J., Cenarro, J., González, J.J., 1998, A&AS, 127, 597
Gorgas, J., Faber, S.M., Burstein, D., González, J.J, Courteau, S., Prosser, C., 1993, ApJS, 86, 153
Gorgas, J., Cardiel, N., Pedraz, S., González, J.J., 1999, A&AS, submitted.
Hamilton, D., 1985, ApJ, 297, 371
Poggianti, B.M., Barbaro, G., 1997, A&A, 325, 1025
van den Bergh, S., Sackmann, I.J., 1965, AJ, 70, 353
Wildey, R.L., Burbidge, E.M., Sandage, A.R., Burbidge, G.R., 1962, ApJ, 135, 94
Worthey, G., Faber, S.M., González, J.J., Burstein, D., 1994, ApJS, 94, 687

STELLAR α ELEMENTS ABUNDANCE GRADIENT IN M33

M.I. MONTEVERDE AND A. HERRERO
Instituto de Astrofísica de Canarias
E-38200 La Laguna, Tenerife, Spain

Abstract. In previous papers we have published preliminary results on the stellar oxygen-abundance gradient in M33 by using B-type supergiants. In the present study, we include one new B-supergiant and derive effective temperatures and gravities for these stars. The determination of atmospheric parameters is based on analyses of silicon ionisation balance.

Detailed non-LTE model atmospheres and non-LTE line formation calculations were used for the abundance determination of Si and O. Careful differential analyses are used to quantify the differences in metal abundances between M33 stars and MK spectral standards in our Galaxy.

We compare our results with the abundances expected from studies of H II regions.

1. Introduction

Radial abundance gradients in spiral galaxies are an important diagnostic in the study of galaxy evolution. As we know from studies in our Galaxy, the spectroscopic analysis of B-type stars is an excellent way of determining abundances – see the very recent work on the abundance gradient in the Galaxy by Smartt & Rolleston (1997), and references therein.

Here, we are dealing with M33, for which there are several abundance studies based on H II surveys. These studies have been summarized by Vila-Costas et al. (1992) and Henry & Howard (1995), and all of them establish a well determined abundance gradient for oxygen in M33. It is therefore important to check if the gradient obtained from H II regions is echoed by stellar spectroscopy.

In Monteverde et al. (1997) we have published first results concerning the stellar oxygen abundance gradient in M33 from B supergiants and have compared them to results from emission-line spectra of H II regions. That technique relies in a precise classification of the M33 stars againts its galactic counterparts. Now we try to confirm our preliminary results by obtaining stellar parameters and then differential abundances of sim-

Astrophysics and Space Science is the original source of publication of his article. It is recommended that this article is cited as: *Astrophysics and Space Science* **263**: 171–174, 1999.

ilar galactic and M33 supergiants. Details of the observational data and spectral classification can be found in Monteverde et al. (1996).

2. Method of Analysis

Our studies are based on non-LTE, plane-parallel geometry in hydrostatic and radiative equilibrium, which are not completely realistic physical models for B supergiants. Then, we will concentrate on a non-LTE differential analysis relative to Galactic standards. Such a differential analysis relies on comparing each star to the appropriate Galactic standard, ensuring that both have similar parameters.

3. Determination of Stellar Parameters

In our preliminary studies on the oxygen-abundance gradient the effective temperature was adopted from the spectral type and the microturbulence, ξ, was assumed to be that of the Galactic standard counterpart. In the present work we have improved the stellar parameter determination by using silicon ionization balance, which are available for use as an indicator of T_{eff} in B-type supergiants. Silicon lines were also used to derive ξ values.

The atmospheric parameters T_{eff}, log g, [He], [Si] and ξ are not independent and a degree of iteration is required. The analysis of the hydrogen and silicon spectra allows a consistent determination of the effective temperature, the gravity, the microturbulence and the unknown silicon abundance. We have assumed a solar helium abundance for all our supergiants.

Preliminary values of log g-T_{eff} were first determined. As Balmer lines are almost unaffected by ξ, H_δ and H_γ have been used as indicators of log g. For each grid value of T_{eff} we have determined an unique log g, for which observed and theoretical line profiles agreed best. For each of those (log g, T_{eff})-pairs we calculated silicon line profiles at different ξ. A solar silicon abundance is initially assumed. From the diagram of theoretical equivalent width ratios for each pair of silicon lines we determine log g, T_{eff} and ξ by comparison with the observed equivalent width.

TABLE 1. Non-LTE Atmospheric parameters for the M33 stars and Galactic standards derived from the silicon ionisation balance.

Star	Spec. Type	T_{eff}	log g	ξ	Ion. balance
M33 - 110A	B1 Ia^+	19000	2.20	19	Si III/Si IV
M33 - ob21-108	B1.5 Ia^+	20700	2.37	20	Si III/Si IV
M33 - 1054	B2 Ib	21000	2.55	9	Si II/Si III
M33 - B133	B2.5 Ia	21300	2.53	15	Si III/Si IV
HD13854	B1Iab	21700	2.52	14	Si III/Si IV
HD14956	B1.5Ia	19300	2.28	15	Si III/Si IV
HD206165	B2Ib	20000	2.45	15	Si II/Si III
HD198478	B2.5Ia	18500	2.20	8	Si II/Si III

Then, we considered the preliminary values of log g-T_{eff} fixed and the microturbulence and Si abundance are freely varied in order to ensure that the derived abundances from individual lines are not a function of strength. Si III lines, which come from a single multiplet, were chosen to create the diagram of Si abundance against observed equivalent width. The new value of microturbulence is that one which produces a zero slope in the diagram and the corresponding Si abundance comes from the value indicated by that zero slope.

The so defined Si abundance served as a basis for a new (log g - T_{eff}, ξ) determination. These steps are repeated until the change of T_{eff}, log g, ξ and Si abundance was within the error in the equivalent width measurements.

4. Metal Abundances

We generated non-LTE model atmospheres for each star with the atmospheric parameters listed in Table 1. The subsequent non-LTE line-formation calculations were performed using the programs DETAIL and SURFACE. Regarding the oxygen and silicon models atoms, we used the model for O II set up by Becker & Butler (1988) and for Si II , Si III and Si IV by Becker & and Butler (1990). Non-LTE curves of growth were constructed for the lines of interest for each element using a range of abundances. Table 2 gives the differential abundance for the M33 stars.

5. Oxygen and Silicon Abundance Gradients in M33

The Si and O non-LTE differential abundances for the M33 B-type supergiants compared to galactic standards with similar atmospheric parameters are listed in Table 3, along with their projected galactocentric distance in arcminutes, ρ.

A comparison of our O II results with those from H II studies is made in Fig. 1 (left). The oxygen differential abundances for M33 stars obtained here are represented by asterisks. The line represents the fit of Henry & Howard (1995) to H II region data from different authors, and diamonds give individual points by Vilchez et al. (1988). The triangle represents one more point from Monteverde (1998).

Fig. 1 (right) shows the differential silicon abundances for the pro-

TABLE 2. Silicon and oxygen non-LTE differential abundances.

M33 Star	Standard	ρ	$\log(Si/H) - \log(Si/H)_\odot$	$\log(O/H) - \log(O/H)_\odot$
1054	HD206165	1′.0	+0.25±0.14	+0.27±0.18
B133	HD13854	9′.4	-0.55±0.26	-0.80±0.24
ob21-108	HD14956	11′.7	-0.67±0.20	-0.40±0.23
110A	HD14956	17′.1	-0.59±0.20	-0.34±0.33

gramme stars (asterisks) and the solid line represents the silicon gradient from the fit to the points.

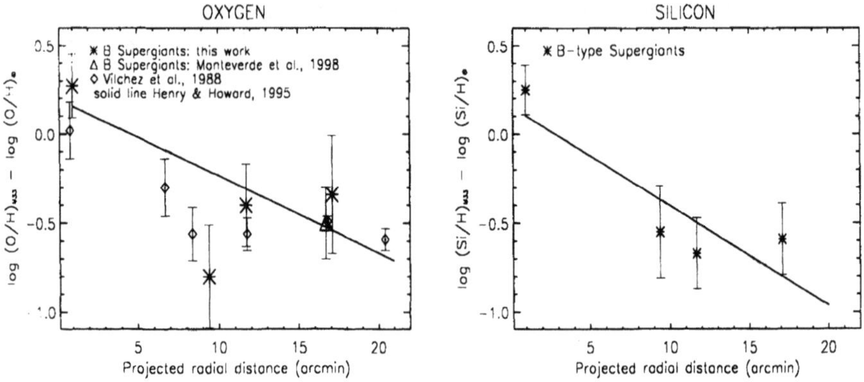

Figure 1. Oxygen and silicon differential abundance gradient for M33 stars

6. Conclusions

We have confirmed here, with a more realible stellar parameter determination, first results by Monteverde et al. (1997) and Monteverde (1998) concerning the oxygen abundance gradient in M33 from B supergiants.

We find consistency for old and new oxygen abundances in B supergiants (except for B133). We also confirm that our oxygen abundances are in good agreement with the H II regions results. A comparison of differential silicon and oxygen abundances within each star indicates they are well correlated and also the O and Si gradients obtained are compatible with each other. We obtain -0.17 ± 0.36 dex/Kpc for O and -0.27 ± 0.25 dex/Kpc for Si.

We are planning to complete this work with the analysis of two more M33 B-type supergiant along with a magnesium abundance determination. Also new, better observations with 10m-class telescopes will help to reduce the large errors, dominated by the equivalent width measurements.

References

Becker, S. R., Butler, K., 1988, A&AS, 201, 232
Becker, S. R., Butler, K., 1990, A&AS, 235, 326
Henry, R. B. C., Howard, J. W., 1995, ApJ, 438, 170
Monteverde, M. I., Herrero, A., Lennon, D. J., Kudritzki, R. P., 1996, A&A, 312, 24
Monteverde, M. I., Herrero, A., Lennon, D. J., Kudritzki, R. P., 1997, ApJ, 474, L107
Monteverde, M. I., 1998, in *Boulder-Munich II: Properties of Hot, Luminous Stars*, ed. I.D.Howarth, San Francisco, ASP Conf. Series, 131, 188
Smartt, S. J., Rolleston, W. R. J., 1997, ApJ, 481, L47
Vila-Costas, M. B., Edmunds, M. G., 1992, MNRAS, 259, 121
Vilchez, J. M., Pagel, B. E. J., Díaz, A. I., Terlevich, E., 1988, MNRAS, 235, 633

INFRARED ASTROPHYSICS IN THE SPACE WITH ISO

From ISO to FIRST

J. CERNICHARO
CSIC. Intituto de Estructura de la Materia.
Dpt. Molecular Physics. Serrano 121, 28006 Madrid, Spain

Abstract. I present some of the results we have obtained with the four instruments on board of the ISO satellite. The determination of water vapour abundance, the study of star forming regions and the determination of the physical and chemical conditions in evolved stars are discussed. The implication of these results for future space-based observatories like FIRST and large ground-based infrared telescopes are analyzed.

1. Introduction

In spite of the important astrophysical output that could be obtained from infrared observations, the wavelength range 2-200 μm has remained poorly studied due to the absorption of radiation produced by the Earth's atmosphere. The instruments on board the Infrared Space Observatory (ISO) provide a unique opportunity to study the dust emission and important molecular species through transitions which are inaccessible from the ground or airborne platforms. The ISO spectrometers offer the possibility to detect, with medium spectral resolution, the ro-vibrational transitions of important molecules. The SWS spectrometer covers the region 2-45 μm and it is well adapted to the study of ro-vibrational transitions (stretching and bending modes) while the LWS spectrometer provides a powerful tool to observe the pure rotational transitions of light species like H_2O. In this review we present the main results on gas phase molecules obtained with the ISO-SWS and ISO-LWS spectrometers and also the low resolution CVF-ISOCAM spectrometer.

2. The Water Abundance in Interstellar Clouds

The determination of the abundance of water vapour is a long standing problem in astrophysics. Because water vapour is predicted to be an abundant molecule in the gas phase, a determination of its spatial extent, its distribution, and its abundance is crucial in modeling the chemistry and the physics of molecular clouds. In warm molecular clouds, water vapour

Astrophysics and Space Science is the original source of publication of his article. It is recommended that this article is cited as: *Astrophysics and Space Science* **263**: 175–192, 1999.

could play a critical rôle in the gas cooling (Neufeld & Kaufman 1993) and, hence, in the evolution of these objects. Unfortunately, due to its high abundance in the terrestrial atmosphere, observations from the ground of the rotational, vibrational and electronic transitions of water vapour are particularly difficult. Some ground-based and airborne observations of water vapour have been performed, but most of the observed transitions are maser in nature making very difficult the analysis of the data (Cernicharo et al. 1990; Menten & Melnick 1991). Even the observation of the $H_2{}^{18}O$ isotope gives only poor estimates on x(H_2O) due to contamination of the observed transitions by other molecular lines (Jacq et al. 1988, Gensheimer et al., 1996), or to the observation of only one line which avoids the correct determination of the population of the rotational levels of the ^{18}O isotopic species of water vapour (Zmuidzinas et al. 1995). HDO has been also observed (Jacq et al. 1990; Helmich et al., 1996b), but it is also difficult to derive the H_2O abundance from this rare isotopic species.

The 6_{16}-5_{23} masing transition of H_2O at 22 GHz, which arises from levels around 700K, has been used since its detection by Cheung et al. (1969), to trace high excitation gas around star forming regions and evolved stars. The size of the emitting regions at that frequency is typically of the order of a few milliarcseconds (a few 10^{13} cm at the distance of Orion). Hence, no information has been obtained on the rôle of water vapour at large spatial scales from this line. Other H_2O lines have been detected from ground or airborne based telescopes like the 3_{13}-2_{20} transition at 183 GHz (Waters et al., 1980; Cernicharo et al. 1990, 1994, 1996a; González-Alfonso et al. 1994, 1998a), the 4_{14}-3_{21} transition at 380 GHz (Phillips, Kwan & Huggins 1980), the 10_{29}-$9_{3,6}$ transition at 321 GHz (Menten, Melnick & Phillips 1990a) and the 5_{15}-4_{22} transition at 325 GHz (Menten et al. 1990b). Among these lines only the 3_{13}-2_{20} transition at 183 GHz, for which the terrestrial atmosphere has good transmission under excellent weather conditions (see Cernicharo 1985, 1988; Cernicharo et al. 1990; Pardo 1996), have been used to map the emission of water at very large spatial scale (Cernicharo et al. 1994, hereafter referred to as Cer94). The map of the Orion molecular cloud shown in Cer94 is 6 orders of magnitude larger than the size of the spots detected at 22 GHz and for the first time a water abundance estimation was derived for the different large scale components of the Orion molecular cloud.

The ISO satellite has provided the opportunity to observe thermal lines of water in the medium and far infrared (see the reviews by van Dishoeck 1997; Cernicharo 1997; and Cernicharo et al 1997a). Water emission has been detected towards young stellar objects by Liseau et al. (1996) and Saraceno et al. (1996). The physical conditions derived from the observed water vapour lines and the high excitation lines of CO in these objects cor-

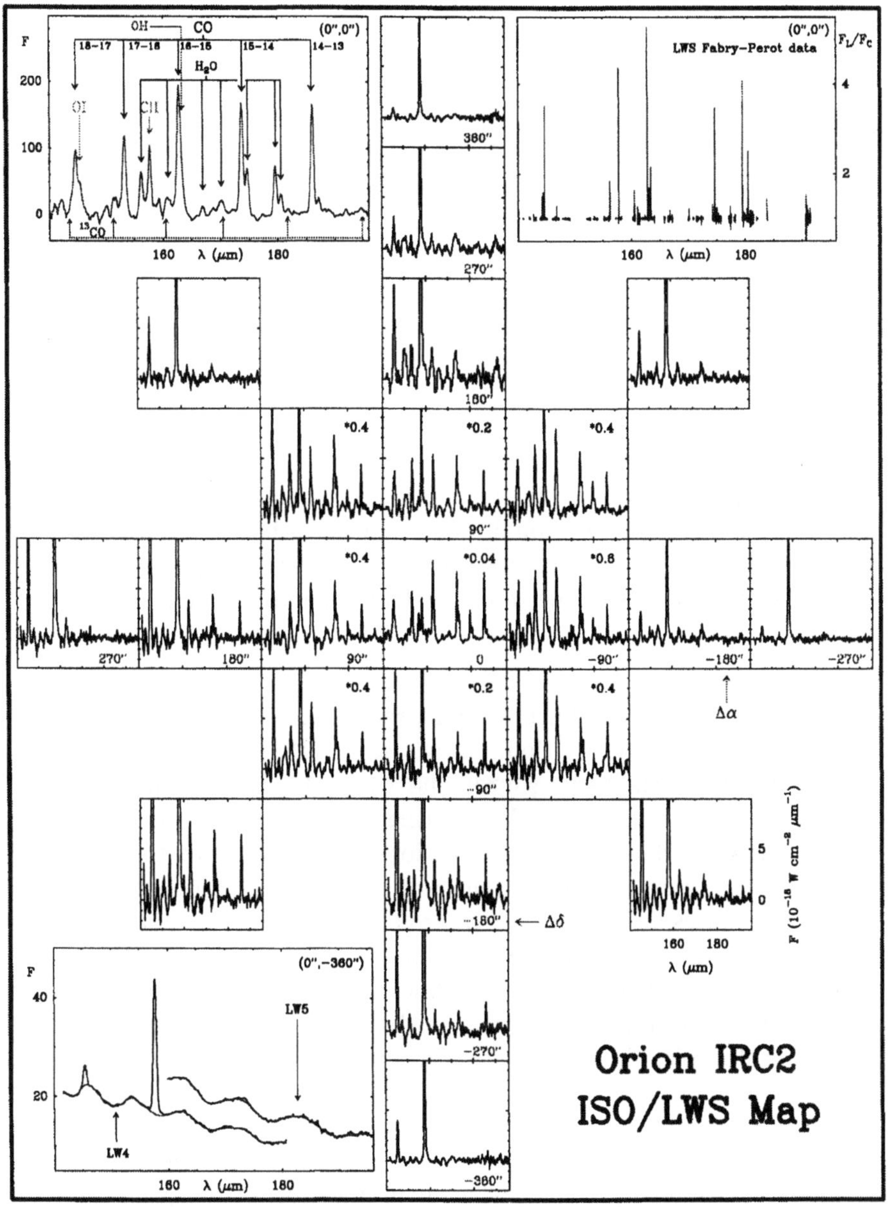

Figure 1. LWS raster of the Orion molecular cloud between 140 and 197 μm. Offsets, in arcseconds, are indicated in each box (in declination for the vertical strip and in right ascension for the horizontal one). Line identification are given in the zoom to the central position in the top-right panel.

respond to shocked gas with $T_K \simeq 300$ K and $x(H_2O) \simeq 10^{-5}$. Water vapour is probably produced in these shocked regions through evaporation of water ice from dust grain mantles. The SWS spectrometer has been used by van Dishoeck and collaborators to study the ro-vibrational transitions of the bending mode of H_2O in the direction of the bright infrared sources associated with AFLG2591, NGC7538, AFLG2136 and AFGL4176 (van Dishoeck & Helmich 1996., Helmich et al. 1996). Mapping of the SgrB2 molecular cloud by Cernicharo et al. (1997b) has definitely shown that water vapour is an ubiquitous molecule in molecular clouds with an abundance of 10^{-5}, i.e., similar to that derived by Cer94 from the masing line at 183 GHz. Maps of the emission of several H_2O lines in Orion IRc2 have been obtained by Cernicharo et al. (1997a, 1999a). Observations of the central position have been also obtained by van Dishoeck et al. (1998), González-Alfonso et al. (1998) and Harwit et al. (1998). However, the ISO observations of H_2O have to be complemented with ground based observations, such as those obtained by Cernicharo et al (1994) at 183.3 GHz, in order to disantagle the different spatial components of water vapour and to model correctly the emission of the different rotational lines. We discuss here the available data at all wavelengths and we give an estimate of the water vapour abundance from model calculations of the radiative transfer in molecular clouds.

2.1. ORION-IRC2

Cernicharo et al. (1997a) have shown preliminary results of their study of H_2O in the direction of Orion. Harwit et al. (1998) have also observed a few H_2O lines in the same source and conclude that the water vapour abundance is $\simeq 5\ 10^{-4}$. The LWS grating data taken in this source are shown in Figure 1. Some line identifications are given in the top panels of Figure 1. This Figure shows that the water vapour emission extends over several arc minutes around IRc2. Towards the central position all the lines of H_2O involving levels with energies below 2000 K have been detected in the grating spectrum. The water lines are typically 1/3 less stronger than the CO lines. In order to study in detail the excitation conditions of H_2O in a prototypical high-mass star forming region as Orion, we observed a large number of transitions of H_2O and its ^{18}O and ^{17}O isotopes (see Cernicharo et al., 1998a, 1999b). The lines involving rotational levels at high energy appear in emission. However, those arising from levels of low and intermediate energy show P-Cygni profiles for $\lambda < 100$ μm, while for longer wavelengths they appear in emission.

As described by Cernicharo et al. (1994), the water vapour emission is very extended in Orion. They detected at least four different components : the extended molecular ridge, the high velocity wind, the "plateau" or wind of moderate velocity, and many narrow and very strong features which are

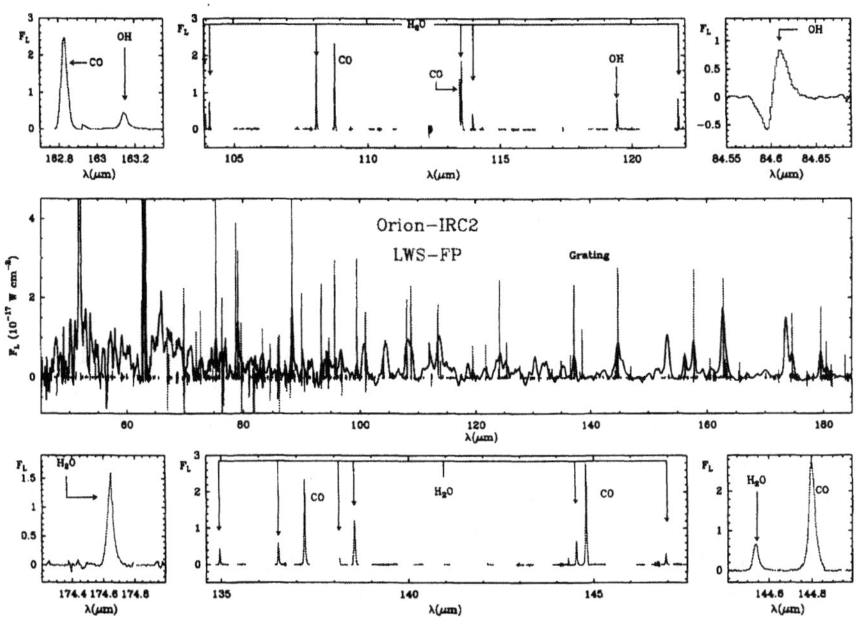

Figure 2. The central panel shows the FP data together with the grating spectrum of Figure 1. The panels at top and bottom show selected ranges of the FP data. All strong lines in the FP are due to CO and water vapour. Some weak lines from NH_3, H_3O^+, and other species (see text) have been also detected.

probably associated with the small bullets observed at 22 GHz by Genzel et al (1981).

In order to model the H_2O emission in Orion we have used the data provided by the grating and FP LWS observations on CO and ^{13}CO. These data are useful to constraint the kinetic temperature, the CO column density and the size of the emitting regions. With these data as input we derived the H_2O column density and the H_2 volume density. An additional input was the dust temperature. The contribution of the extended emission of H_2O in the Orion molecular cloud has been also taken into account in the models. Dust and molecules are treated simultaneously (see, e.g., González-Alfonso et al. 1998). The models, which account for two regions of different temperature and density, reproduce qualitatively the P-Cygni profiles of H_2O and the fully absorbed lines observed with the SWS spectrometer. However, the results are very dependent on the assumed geometry and spatial distribution of the dust. Our calculations indicate that some lines will be redshifted due to self-absorption in the emitting layers and in the foreground gas, an effect that is observed for some H_2O lines. The comparison of the CO and H_2O emission allows to estimate an abundance ratio, $x(CO)/x(H_2O)$, $\simeq$2-5 in the low velocity wind, and $\simeq$10-50 in the extended molecular cloud. The maser lines at 183.3 and 325 GHz (Cernicharo et al. 1994, 1999a) are well reproduced in our calculations. A detailed description of the models and the main results will be published elsewhere (Cernicharo et al 1999b; see also Cernicharo et al. 1998a).

2.2. SGR B2

The grating ISO/LWS spectrum towards the central position of Sgr B2 shows a strong absorption at 179.5 μm corresponding to the water vapour 2_{12}-1_{01} transition (see Cernicharo et al., 1997b). The grating spectra in different positions of the cloud are shown in Figure 3. The most striking result from the raster map is the presence of the water vapour 179.5 μm line in absorption in all directions (at least up to $\pm$830 arcsec, i.e., over 80 pc; see Figure 3). At the spectral resolution of the grating (i.e., $\simeq$ 1000 km s^{-1}), the absorption remains strong and roughly constant through the map. The absorption depth of $\simeq$15%, is consistent with saturated water vapour lines with a line width of $\simeq$150 km s^{-1}. Other H_2O transitions are detected in absorption: the 3_{03}-2_{12} H_2O line at 174.5 μm is also present in all the raster positions with an absorption depth of 1%. The absorption at 180.5 μm (also detected in all positions) could be a blend of the 2_{21}-2_{12} H_2O transition at 180.488 μm, the CH line at 180.478 μm and the 2_{12}-1_{01} $H_2{}^{18}O$ line at 181.05 μm. Sensitive observations with the LWS-FP spectrometer of the lines of CH towards the central position of SgrB2 show that only the fundamental transitions of this species, around 149.5 μm, are detected. Therefore, the grating feature at 181 μm is not contaminated by CH and probably arises from $H_2{}^{18}O$. Nevertheless, we note that the inversion transition of H_3O^+ is exactly at the same frequency than the $H_2{}^{18}O$ line, and some contamination can not be excluded. A preliminary analysis of other H_3O^+ lines seems to indicate that an important fraction of the absorption is due to $H_2{}^{18}O$. Finally, note that the C[II] line is very strong in all observed positions but it decreases in intensity when approaching the center where it appears in absorption. Baluteau et al (1997) have also found the O[I] line in absorption towards the central position while it is detected in emission in other positions through the cloud (see Figure 7).

The main result from the present data is that the far-infrared lines of water vapour are seen in absorption towards SgrB2, instead of in emission (Cernicharo et al., 1997b). This suggests that in SgrB2, where the continuum emission in the far-infrared is optically thick, the H_2O lines arise in a region where the excitation temperatures are smaller than the dust temperature ($<$ 30 K). Nevertheless, the problem of H_2O line excitation is not simple because, apart from self-absorption and the possibility of collisional excitation, the excitation temperatures will also be affected by the continuum emission of the dust. The rôle of dust is essential in the excitation

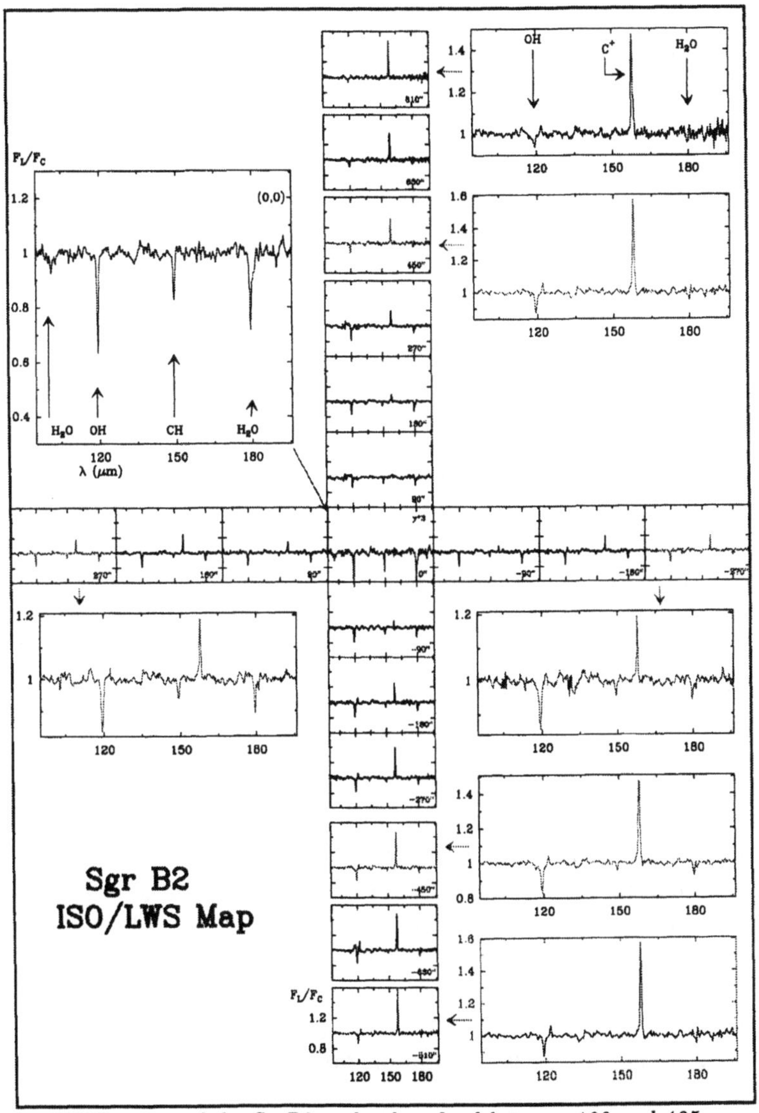

Figure 3. LWS raster of the SgrB2 molecular cloud between 100 and 195 μm. Offsets, in arcseconds, are indicated in each box (in declination for the vertical strip and in right ascension for the horizontal one). Line identification are given in the zoom to the central position in the top-right panel. The adjacent panels show the spectra observed at selected positions (indicated by arrows). The line around 122 μm corresponds to N[III].

of molecules such as H_2O which have transitions at far- and near-infrared wavelengths (see Cernicharo et al. 1997b). In addition to the H_2O lines, the LWS grating spectra show the OH and CH lines across the cloud. The lines of OH arising from the J=3/2 $^2\Pi_{3/2}$ level and connecting with the J=5/2 $^2\Pi_{3/2}$ and J=1/2 and J=3/2 of the $^2\Pi_{1/2}$ level at 54, 80 and 119 μm are observed at many positions. The strength of the OH features, as for the water vapour lines, indicates that the lines are saturated.

Water vapour will be one of the main scientific goals of the FIRST satellite. It will be the only instrument allowing a detailed study (with the HIFI instrument) of the emission of H_2O in interstellar clouds. ISO has revealed how abundant H_2O could be in star forming regions. However, its limited angular resolution and sensitivity, together with its poor velocity resolution has left unrevealed the most important part of the "iceberg". FIRST will provide the angular and velocity resolution to study the innermost layers of protostellar objects.

3. Star Formation in Molecular Clouds

The formation of stars rarely appears as isolated events. There is ample evidence that, in star-forming regions, once stars are formed the process spreads further into the parent molecular cloud. HII regions are particularly interesting as they form from the photoionisation of the molecular gas surrounding the massive newly-born stars. The densest condensations of the clouds are left behind the ionisation front where they are exposed to the stellar radiation. These condensations are known as Cometary Globules (CGs). The Trifid nebula is a nearby galactic H[II] region with an estimated distance of 1.68 kpc. The main exciting star is an O-type star of 30 $M_\odot$, emitting Ly-c photons at a rate 5 10^{48} s^{-1}. The size of the optical nebula is only 5 arcmin (or 8 10^{18} cm) indicating that the nebula is rather small and hence relatively young as compared to the Rosette nebula, for instance, which is located at the same distance. The Rosette nebula is about 40 times bigger and has dissipated most of its parental gas. On the contrary, the molecular clouds in the Trifid are still close to the recently formed central star, exposed to its strong radiation field. Thereby, they provide excellent sites to search for the youngest generation of stars, to study the earliest stages of star formation and quantify the impact of the ionising field. For these reasons we have started an observing program using radio and optical telescopes and the four instruments on board ISO (see Cernicharo et al., 1998b).

Complete maps of the Trifid have been obtained in the continuum emission at 20 cm (VLA), 1.3 mm (30-m IRAM telescope), 200, 105, 60 μm (ISOPHOT) and several broad-band filters with ISOCAM. The radio free-free emission, which traces the low-density (200-300 cm^{-3}) ionised gas, shows that the exciting star has created a bubble of radius $\simeq$ 4 10^{18} cm which is limited by a well-delineated ionisation front along the border of the adjacent molecular cloud (see Figures in Cernicharo et al., 1998b). The thermal, optically thin dust emission at 1.3 mm shows low-emissivity extended emission and five point- like sources with peak fluxes ranging between 0.10 and 0.36 Jy/beam (see Cernicharo et al., 1998b). At far-infrared

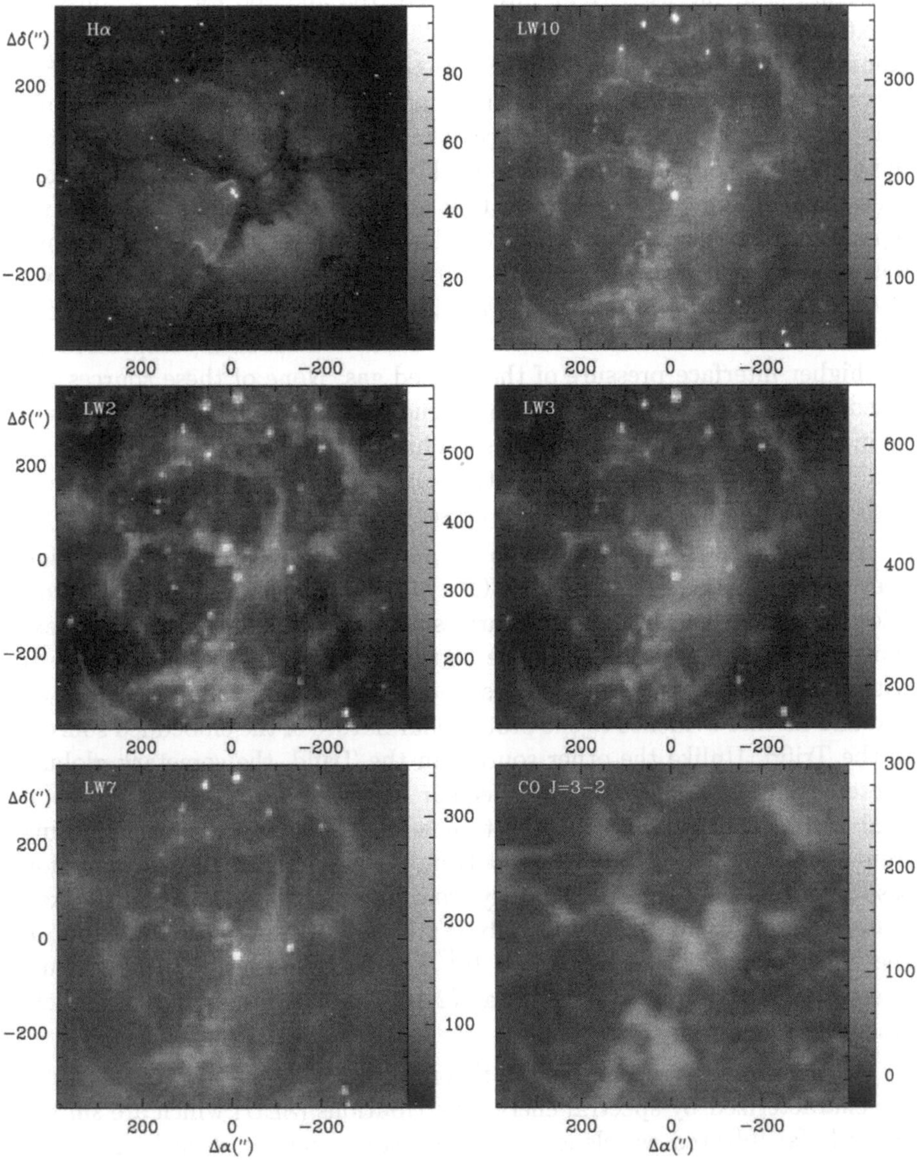

Figure 4. The top left panel shows a Hα image of the Trifid Nebula taken with the IAC80 telescope (in arbitrary units). The panels from top right to bottom left show the dust emission as observed with ISOCAM in different filters (units are in mJy/pixel). The angular resolution of the maps is 6 arcsec, except for filter LW10 (12 μm) which is 3 arcsec. The 12 μm image looks very similar, at first glance, to the optical picture shown in the top left panel. However, a detailed comparison of both images show that the 12 μm emission arises from the photon- dominated region (PDR) associated to the neutral gas seen in absorption in the optical picture. These regions have strong emission in the CO J=3-2 line (see bottom-right panel). Note the large number of embedded infrared sources detected in the filter LW2.

wavelengths in the ISOPHOT filters at 60, 105 and 200 μm, the emission is extended and poorly contrasted. At 200 μm dust emission is detected around the millimeter sources found in the south-west part of the Trifid. However, due to the poor spatial resolution of the 200 μm observations, it is not clear if this emission traces the extended dust emission around the embedded source or the source itself. It is only in the dust continuum emission at millimeter wavelengths that all sources are clearly detected and separated from the extended emission. The sensitivity of these observations was good enough to detect sources of 0.5 $M_\odot$ in one telescope beam. Two of these sources are located inside the ionised bubble. The other ones are found well beyond the ionisation front, in the neutral shell compressed by the higher interface pressure of the ionised gas. None of these sources are listed in the IRAS Point Source Catalogue which indicates that they must be in a very young evolutionary stage. This is further strengthened by their non detection in the high-sensitivity ISOCAM image at 12 μm (Figure 4) with upper limits of 45 mJy in a 6 arcsec pixel.

The sources are also well seen in molecular transitions tracing high density gas such as SiO, CS or HCO^+. Maps obtained at the IRAM 30m-telescope show that all the sources are associated with high velocity gas (except in the cometary globule in the South-East of the Trifid; see Figure 4) with velocities up to 60 kms^{-1}. These motions characterize bipolar outflows and add further evidence to the protostellar nature of the embedded sources in the Trifid. Unlike the other sources in the Trifid, the cometary globule quoted above does not display molecular line wings. However, it is associated with a conspicuous optical jet (now classified as HH399) emanating from the head of the globule (see bottom panel of Figure 6). The emission associated with the jet arises mostly from lines probing dense and shocked material propagating from the globule's head into the ionised medium of the HII region. The jet extends 5 10^{17} cm from the head of the globule and reappears 65 arcsec further away (2 10^{16} cm from the origin of the jet) where it traces a large bow-shock.

All the protostars discovered in this multiwavelength study of the Trifid are characterized by spectral energy distributions (SED) which are shifted towards far-infrared wavelengths and, for some sources, even longward of 200 μm. This suggests that all of them must be very young and could possibly be members of the Class 0 objects. Given the sensitivity of our millimeter survey, it appears that this second episode of star formation in the Trifid resulted in either massive or clusters of low/intermediate-mass stars. As shown by the present results, this episode has started just after the birth of the nebula, as soon as the ultraviolet photons from the central star propagated through the surrounding neutral gas.

The dust emission shown in Figure 4 seems to follow the dust lanes of

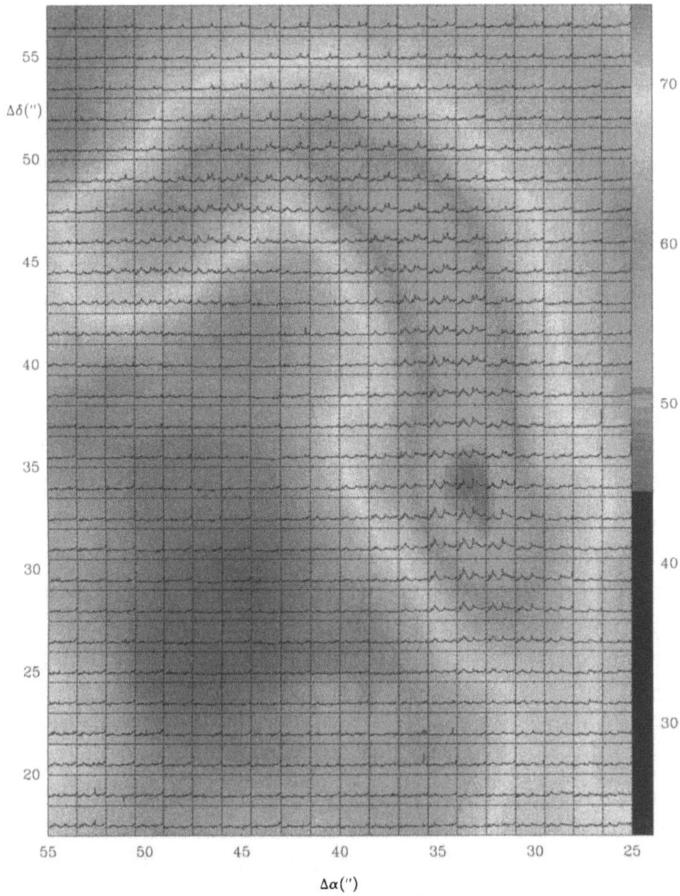

Figure 5. Infrared emission observed with the LW10 ISOCAM filter with a pixel size of 1.5" in the direction of the cometary globule from which the optical jet HH399 is arising (grey scale). The CVF spectra between 5 and 16 μm are shown in the small boxes. The infrared emission is dominated by the vibrational bands of PAHs; a reference spectrum has been removed.

the optical Hα image. However, a detailed comparison of the ISOCAM images and the optical one indicates that the infrared emission is arising from the PDRs of the Trifid nebula. Figure 5 shows one of the cometary globules in the Trifid as observed with the LW10 ISOCAM filter and a pixel size of $1.5''$ (the one associated to the HH399 optical jet). This Figure also shows the CVF spectra observed around the same globule (a reference spectrum, containing the contribution of the zodiacal light and the systematic contribution of the H[II] region has been removed). All the emission, which is dominated by the infrared bands of large carbon-rich macromolecules (the so called PAHs), arises from the PDR associated to the bright rim of the cometary globule. Important differences in the spatial distribution of each band can be observed suggesting that different types of molecules are contributing to the infrared emission.

The FIRST satellite will provide a broad wavelength coverage with two of its instruments, PACS and SPIRE. They will have large arrays of IR detectors covering the 80-600 μm range. The high angular resolution of FIRST (as high as $7''$), together with the high sensitivity of PACS and SPIRE, will permit detailed studies of the dust emission from protostellar objects as those found with ISO in the Trifid nebula. Although ground-based instruments, like the LSA/MMA, or airborne telescopes, like SOFIA, will provide important information in this field, FIRST will have a much larger frequency coverage and sensitivity.

4. Evolved Stars

4.1. AGB STARS

IRC+10216 is the brightest C-rich evolved object in the sky. It has an extended envelope where more than 50 molecular species have been detected and it is probably one of the best studied objects at all wavelengths. This object has a particularly rich carbon chemistry and most of the species detected in its molecular envelope are carbon chain radicals which are formed in the external layers of the molecular envelope (Guélin, Lucas & Cernicharo 1993). Some of the detected species are metal-bearing molecules as the metal halides NaCl, AlCl, AlF, KCl, first detected in this envelope by Cernicharo & Guélin (1987). The spatial distribution of these molecules peaks on the central star. The innermost regions of the envelope are dominated by a chemistry at thermodynamical equilibrium (Tsuji, 1973). Most of the stable diatomic and triatomic species form there efficiently (CO, HCN, C_2H_2, SiO, CS, SiS, ...). The most abundant of these molecules are observed at radio wavelengths, but most species are too weak to be detected. They could nevertheless be detected in absorption against the strong IR flux of the central source. Hence, the molecular envelope of IRC+10216 constitutes an interesting laboratory for the study of the circum- and interstellar chemical processes; it is also the best evolved star for near-, mid- and far infrared studies.

We have observed with ISO (SWS full grating full resolution and LWS grating spectrometers) the medium and far infrared spectrum of IRC+10216, CRL2688, CRL618, and NGC7027 with the aim to obtain a complete study of the molecular content of these objects and to relate it to their degree of evolution. The proposed surveys will give us a detailed spectroscopic information that will allow to identify new molecular species, some of them playing an important role in the chemistry of these objects. Molecules like CO_2, CS_2, NCCN, CH_3, CH_4, carbon chains, complex hydrocarbons, H_3^+, ..., will be easily detected in the SWS and LWS spectral scans. Among

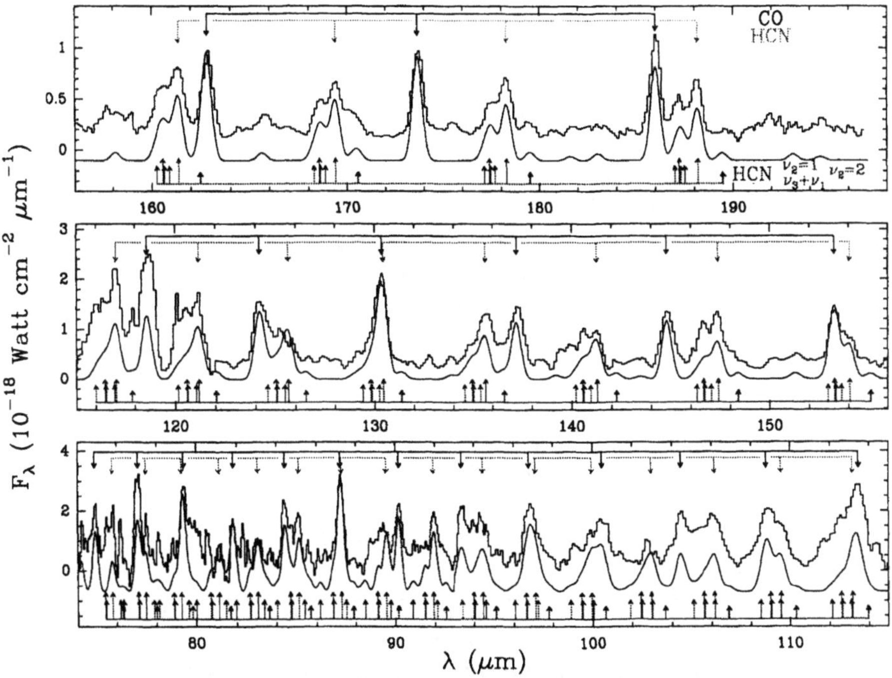

Figure 6. Continuum subtracted LWS grating spectrum of IRC+10216 from 75 to 197 μm (thick line). An offset has been introduced to the continuum to show simultaneously the computed emission of the CO ν=0, 1, ^{13}CO ν=0, HCN and H^{13}CN ν=0, and HCN ν_2=1,2 $\nu_{1,3}$ =1 (thin line); see Cernicharo et al 1996b for details. The rotational transitions of CO, HCN, and HCN ν_2=1,2 and $\nu_{1,3}$ =1 are indicated at the top and bottom of each panel. The weak lines in the model correspond to H^{13}CN, ^{13}CO and to the stretching modes of HCN (from Cernicharo et al. 1996b).

the expected new molecules it is worth mentioning the N-atomic carbons C_3, C_4, C_5, C_6, C_7, which are efficiently formed in the external layers of the circumstellar envelope. The detection, identification and spectroscopic characterization of these species will bring up key information for the chemistry of these objects. On the other hand, the interaction of the strong winds from the central objects of PPNe with their molecular envelope produced in the AGB phase will be studied in detail. The ISO data will allow a determination of the physical conditions of the high velocity gas and an accurate determination of the abundances of key molecular species.

The far-infrared spectrum of IRC+10216 has been analyzed by Cernicharo et al. (1996b); see Figure 6. The near and medium infrared spectrum of IRC+10216 is dominated by absorption bands of C_2H_2 and HCN. The 15 μm spectrum of IRC+10216 (see Figure 7) is dominated by hot and combination bands of C_2H_2 (small vertical lines indicate the position of individual lines in the R and P branches of the ν_5 bending mode) plus a lot of emission bands from HCN, the most conspicuous being the 02^00-01^10

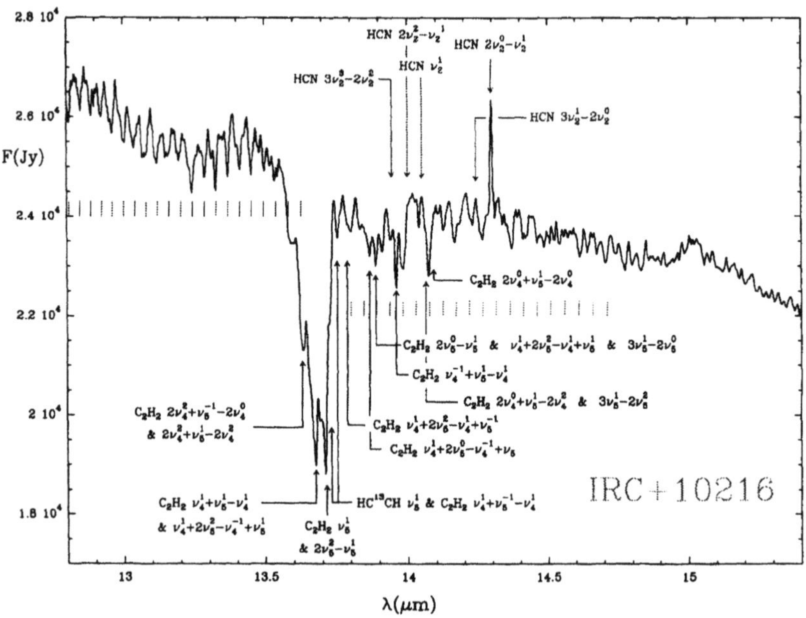

Figure 7. *The SWS spectrum of IRC+10216 around 15 μm. Band identification are indicated. Vertical lines indicate the position of individual R and P lines of the bending mode of C_2H_2.*

(see González-Alfonso & Cernicharo 1998). Radiative pumping affects considerably the population of the vibrational levels of HCN, not only around the central star, but even far away where collisions are inefficient to pump these levels.

4.2. CH^+ IN PLANETARY NEBULAE

CH^+ was one of the first molecules detected in Astrophysics through its $^1\Pi$-$^1\Sigma$ electronic transition at optical wavelengths. This molecule is abundant in the diffuse interstellar medium and during many years the chemical reactions involved in its formation have been subject of important controversy. Cernicharo et al (1997c) have reported the discovery of several pure rotational transitions of CH^+ in the LWS spectrum of NGC 7027. The strong line at 179.62 μm and the lines at 119.90 and 90.03 μm (reported as unidentified by Liu et al. 1996), whose frequencies are in harmonic relation 2:3:4, are shown to arise from the J=2-1, 3-2, and 4-3 rotational transitions of CH^+. This identification is strengthened by the detection in the LWS spectra of the next two rotational lines of CH^+, at 72.17 and 60.22 μm; see Figure 8. Cernicharo et al (1997c) have derived a rotational temperature for the CH^+ lines of 160 K, a volume density of a few 10^7 cm^{-3}, and a CH^+/CO abundance ratio of 0.2-1 10^{-3}. Given the carbon-rich nature of

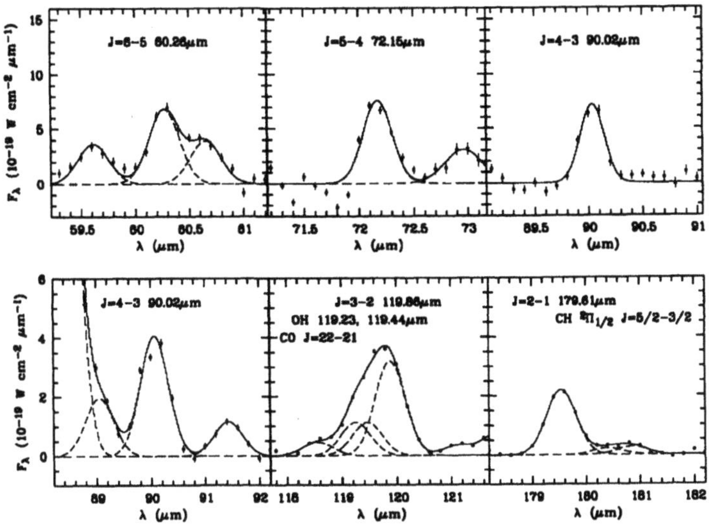

Figure 8. Pure rotational lines of CH^+ detected in NGC 7027 with the LWS grating spectrometer. The wavelengths of the lines are given in each panel (from Cernicharo et al. 1997c).

NGC 7027 and the strong UV radiation field in its inner regions it is thus not too surprising to find CH^+ in this prototypical PN. However, this is the first time that CH^+ has been seen through its pure rotational spectrum. Taking into account the dipole moment of CH^+ (1.7 D) and the structure of its rotational levels (the J=10 level is at $\simeq$ 2200 K), the rotational transitions of this molecular species constitute a unique tool to trace the physical conditions of the PDRs of the interstellar and circumstellar medium.

4.3. O-RICH EVOLVED STARS

In O-rich stars, Barlow et al (1996) and Neufeld et al (1996) have observed several rotational lines of water vapour. Taking into account the large variation of the temperature and density profiles across the envelopes of evolved stars, these lines, which cover a very large range of energies and Einstein coefficients will provide the best tool to derive the physical conditions of the gas in these objects. We have modelled the H_2O emission in O-rich stars with the use of non-ETL and non-Local codes that allow to treat simultaneously dust and molecules. The details of these models are given by González-Alfonso & Cernicharo (1998, 1999). Figure 9 shows the results for the $\nu_2 = 1 \rightarrow \nu = 0$ band of water vapour in two CSEs which differ in the mass loss rate, the terminal velocity of the gas (hereafter v_t) and/or the inner radius of the water envelope (hereafter R_{int}). All other model parameters are the same for the two models: the star has a radius of $3.7\,10^{13}$ cm, an effective temperature of $2.5\,10^3$ K, and the H_2O abundance

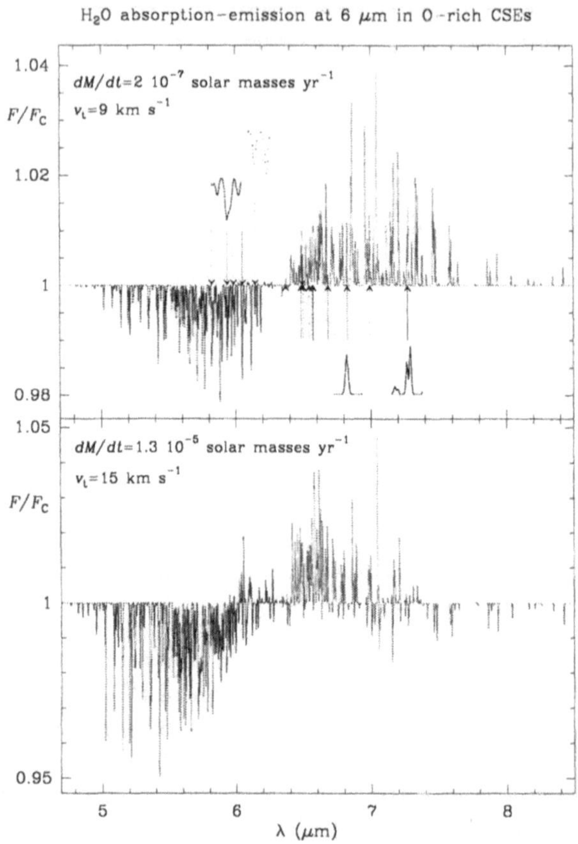

Figure 9. *The predicted $\nu_2 = 1$ H_2O band for two models which differ in the mass loss rate and/or the inner radius of the H_2O envelope (R_{int}) (from González-Alfonso & Cernicharo 1998).*

is $3\,10^{-4}$, according to the value inferred from the 183 GHz maser emission in O-rich evolved stars (González-Alfonso & Cernicharo, 1999). The J_{k_1,k_2} levels of the $\nu_2 = 1$ state are populated mainly via the strong transitions $\nu_2 = 1, J_{k_1,k_2} \leftarrow \nu = 0, J{-}1_{k_1\pm1,k_2\pm1}$ and $\nu_2 = 1, J_{k_1,k_2} \leftarrow \nu = 0, J_{k_1-1,k_2+1}$. These lines, which lie at $\lambda < 6.3$ μm, are observed in absorption. The $\nu_2 = 1$ levels depopulate mainly through $\nu_2 = 1, J_{k_1,k_2} \rightarrow \nu = 0, J{+}1_{k_1\pm1,k_2\pm1}$ and $\nu_2 = 1, J_{k_1,k_2} \rightarrow \nu = 0, J_{k_1+1,k_2-1}$. If dust absorption of line photons is negligible, and star blocking can be ignored, these lines, which lie at $\lambda > 6.3$ μm, are expected to be observed in emission against the continuum. Dust absorption through the envelope is negligible in stars with the lowest mass loss rate (panel b) so that the above pattern of absorption/emission lines will most probably arise in CSEs that are optically thin in the continuum at 6 μm.

Although the CSE with dM/dt= $1.3\,10^{-5}$ $M_\odot$ yr^{-1} is still optically thin at 6 μm, dust absorption does increase the intensity of the absorption lines over that of the emission lines. Furthermore, the effect of star blocking, which reduces the intensities of the lines observed in emission, becomes more important when line opacities increase, i.e., in case of stars with high mass loss rate. Panel c of Figure 9 corresponds to a model which differs from the model of panel b only in the value of R_{int}. In panel c, the density and velocity dispersion in the region comprised between the photosphere and $r = 3R_*$ are $2\,10^9$ cm^{-3} and 5 km s^{-1}. These values are consistent with those associated to the innermost regions of CSEs where the SiO masers arise. The presence of water vapour in these zones changes the expected appearance of the 6 μm band because of the blocking by the star. Hence, depending mainly on the inner radius of the H_2O envelope, we expect that most lines in the $\nu_2 = 1$ band with $\lambda > 6.3$ μm will be observed in absorption or emission against the continuum.

Unlike the spectrum of IRC+10216, the far-infrared spectrum of O-rich stars is dominated by the emission of the rotational lines of H_2O. Even the CO lines are much weaker than those of H_2O for the mass loss rate of W Hya. The cooling of the gas is dominated in these objects by the emission of water vapour lines. A detailed explanation of these results is given by González-Alfonso & Cernicharo (1998, 1999). The same models have been applied to the case of molecular clouds to explain the 6 μm H_2O absorption/emission features in Orion (González-Alfonso et al 1998).

The far-infrared spectrum of evolved stars as observed by ISO has revealed the importance of molecular emission in this wavelength domain for stars having large mass loss rates. However, the line profile of the transitions of HCN, CO, H_2O and other molecular species have not been resolved with the ISO spectrometers. The information carried by the line profile on the acceleration region (the dust formation zone), and the possibility to observe evolved stars with low mass loss rate will be provided by the FIRST satellite. FIRST will be the only instrument allowing a detailed study of water vapour transitions covering a very large range of energies and excitation conditions. The frequency coverage of the heterodyne receiver (HIFI) will permit to reveal the chemical complexity of C-rich objects through the observation of ro-vibrational transitions of the low energy bending modes of heavy molecules (most of them without permanent dipole moment). Rotational transitions of light species, like CH^+, will be observed with FIRST towards a large variety of objects allowing detailed chemical studies. Hence, FIRST will be a unique tool in all fields of submillimeter and infrared astrophysics. The ISO's view of the Universe will be an excellent input for the studies that FIRST will be able to perform.

Acknowledgments

J. Cernicharo acknowledges DGYCIT grants PB96-0883 and ESP98-1351.

References

Baluteau, J.P., Cox, P., Cernicharo, J. et al., 1997, A&A, 322, L33
Barlow, M., Nguyen-Rieu, Truong-Bach, 1996, A&A, 215, L241.
Cernicharo, J., 1985, 'ATM a program to compute atmospheric transmision between 0-1000 GHz', IRAM internal report
Cernicharo, J., Guélin, M., 1987, A&A 183, L10
Cernicharo, J., 1988, Ph.D. Thesis, Université de Paris VII
Cernicharo, J., Thum, C., Hein et al., 1990, A&A, 231, L15
Cernicharo, J., González-Alfonso, E., Alcolea, J. et al., 1994, (Cer94), ApJ, 432, L59
Cernicharo, J., González-Alfonso, E., Bachiller, R., 1996a, A&A, 305, L5
Cernicharo, J., et al., 1996b, A&A, 315, L201
Cernicharo, J., 1997, 'The Far Infrared and Submillimeter Universe', 15-17 April 1997, Grenoble, France. ESA SP-401, page 91
Cernicharo, J., González-Alfonso, E., Lefloch, B., 1997a, 'First Workshop on Analytical Spectroscopy', 6-8 October 1997, ESA SP-419, page 23
Cernicharo, J., Lim, T., Cox, et al., 1997b, A&A, 323, L25
Cernicharo, J., X-Liu, González-Alfonso, E., et al. 1997c, ApJ, 483, L65
Cernicharo, J., González-Alfonso, E., et al., 1998a, in 'The Universe as seen by ISO', Paris, October 1998
Cernicharo, J., Lefloch, B., et al., 1998b, Science, October 16, page 462
Cernicharo, J., Pardo, E., Serabyn et al., 1999a, submitted to the ApJ Letters
Cernicharo, J., González-Alfonso, E., et al., 1999b, in preparation
Cheung, A.C., Rank, D.M., Townes, C.H. et al., 1969, Nature, 221, 626
Genzel, R., Reid, M.J., Moran, J.M., Downes, D., 1981, ApJ, 244, 884
González-Alfonso, E., Cernicharo, J., Bachiller, R., Fuente, A. ,1994, A&A, 293, L9
González-Alfonso, E., 1995, Ph.D., Universidad de Valencia (Spain)
González-Alfonso, E., Cernicharo, J., Alcolea, J., Orlandi, M.A., 1998, A&A, 334, 1016.
González-Alfonso, E., Cernicharo, J., van Dishoeck, E.F., et al., 1998, ApJ, 502, L169
González-Alfonso, E. and Cernicharo, J., 1998, in 'The Universe as seen by ISO', Paris, October 1998
González-Alfonso, E., Cernicharo, J., ApJ, 1999, submitted.
Green, S., Maluendes, S., McLean, A.D., 1993, ApJS, 85, 181
Guélin, M., Lucas, R., Cernicharo, J., 1993, A&A, 280, L19
Harwit, M. Neufeld, D.A., Melnick, G.J., Kaufman, M.J., 1998, ApJ, 497, L105
Helmich, F.P., van Dishoeck, E.F., Black, J.H., et al., 1996, A&A 315, L173
Jacq, T., Jewell, P.R., Henkel, C., Walmsley, C.M., Baudry, A., 1988, A&A, 199, L5
Jacq, T., Walmsley, C.M., Henkel, C., et al., 1990, A&A, 228, 447
Liseau, R., Ceccarelli, C., Larsson, B., et al., 1996, A&A, 315, L181
Liu, X.-W., Barlow, M.J., Nguyen-Q-Rieu, et al., 1996, A&A, 315, L257
Menten, K.L, Melnick, G.J., Phillips, T.G. 1990a, ApJ, 350, L41
Menten, K.L, Melnick, G.J., Phillips, T.G., Neufeld, D.A., 1990b, ApJ, 363, L27.
Menten, K., Melnick, G.J. 1991, ApJ, 377, 647
Neufeld, D.A., Kaufman, M.J., 1993, ApJ, 418, 263
Neufeld, D.A., Chen, W., Melnick, G.J. et al., 1996, A&A, 315, L237
Pardo, J.R., 1996, Ph.D. Thesis, Université Pierre et Marie Curie-Universidad Complutense de Madrid
Phillips, T.G., Kwan, J., Huggins, P.J. 1980, in 'Interstellar Molecules', IAU Symposium 87, ed. B.H. Andrew (Dordrecht: Reidel), page 21
Saraceno, P., Ceccarelli, C., Clegg, P., et al., 1996, A&A, 315, L293
Tsuji, T., 1973, A&A, 23, 411
van Dishoeck, E., Helmich, F.P., 1996, A&A, 315, L177
van Dishoeck, F., Wright, C.M., Cernicharo, J., et al., 1998, ApJ, 502, L173
Waters, J.W., Gustincic, J.J., Kakar, et al., 1980, ApJ, 235, 57
Zmuidzinas, J., Blake, G.A., Carlstrom, J., et al., 1995, ASP Conference Series, Vol. 73. Ed. M.R. Haas, J.A. Davidson and E.F. Erickson

TEMPERATURE FLUCTUATIONS IN GALACTIC HII REGIONS

C. ESTEBAN
Instituto de Astrofísica de Canarias
38200 - La Laguna, Tenerife, SPAIN

1. Introduction

The determination of confident values of the chemical composition of Galactic and extragalactic nebulae is an essential task for studies on chemical evolution of galaxies and stellar and primordial nucleosynthesis. The traditional methods for deriving chemical abundances in ionized nebulae are based largelly on the analysis of collisionally excited lines, which are strongly dependent on the electron temperature. If spatial temperature fluctuations are present in the gas (which are not predicted by photoionization models) the chemical abundances could be underestimated.

The presence of temperature fluctuations in gaseous nebulae was proposed by Peimbert (1967). There have been several estimations of these fluctuations in H II regions (see compilation by Peimbert 1993) and planetary nebulae, PNe, (see Mathis et al. 1998 and references therein) based mainly on: a) the comparison of the electron temperature derived from the ratio of the [O III] lines, and the temperature derived from the ratio of the Balmer continuum to the intensity of Hβ; b) the comparison of the abundance of a given ion determined from collisionally excited lines and recombination lines.

2. Temperature Fluctuations in three Galactic H II Regions

We have obtained a set of homogeneous echelle observations taken with the 2.1m telescope at San Pedro Mártir Observatory (Baja California, Mexico) for the bright Galactic H II regions M42, M8 and M17. The spectral range covered is from 3500 Å to 7060 Å in the case of M42 (two slit positions) and from 3500 Å to 10300 Å in the cases of M8 and M17 (one and two slit positions, respectively). The spectral resolution is of 0.234 Å pixel^{-1} at

Astrophysics and Space Science is the original source of publication of his article. It is recommended that this article is cited as: *Astrophysics and Space Science* **263**: 193–196, 1999.

Hα. The number of lines identified in each object is about 200, a third of these correspond to permitted lines of heavy element ions.

TABLE 1. Temperature fluctuations in Galactic H II regions

			Abundance 12+log(X^+/H^+)			
Object	Position	Ion	Coll. excited lines	Recomb. lines	t^2	Reference
M42	1	O^{++}	8.39	8.52	0.018	1
		C^{++}	8.32	8.47	0.022	1
		Average			**0.022**	1
	2	O^{++}	8.39	8.59	0.028	1
		C^{++}	8.29	8.49	0.028	1
		Average			**0.028**	1
M8	HGS	O^{+}	8.18	8.28	0.022	2
		O^{++}	8.02	8.32	0.040	2
		C^{++}	7.95	8.28	0.032	2
		Average			**0.032**	2
M17	3	O^{++}	8.49	8.75	**0.033**	3
	14	O^{++}	8.40	8.74	**0.044**	3

1.- Esteban, Peimbert, Torres-Peimbert, & Escalante (1998)
2.- Esteban, Peimbert, Torres-Peimbert, García-Rojas, & Rodríguez (1999)
3.- Esteban, Peimbert, Torres-Peimbert, & García-Rojas (1999)

We have determined abundances of several ions making use of both, collisionally excited lines and recombination lines (which are almost independent of the temperature structure of the nebula). The comparison of these abundances is shown in Table 1. As we can see in the table the ionic abundances obtained from recombination lines are *in all cases* between 0.10 to 0.34 dex (factors between 1.25 and 2) larger than those obtained from collisionally excited lines. In column 6 we show the value of the mean square temperature fluctuation, t^2, that accounts for the difference of the abundances given in columns 4 and 5. It is remarkable the consistency of the values of t^2 obtained for different slit positions and ions for a given object, and for the three H II regions. This result suggests strongly that temperature fluctuations are present in the nebulae. Similar values of t^2 have been obtained for these objects from the comparison of the electron temperature derived from the ratio of collisionally lines and the temperature derived from the Balmer continuum (see Peimbert 1995 and references therein).

We have found differences in the total abundances derived for each of the three nebulae, fact that could be addressed to the presence of the Galactic abundance gradient. Distance determinations of M17, M8 and M42 indicate that they are located at galactocentric distances of 5.9, 6.5 and 8.4 kpc. In Table 2 we show the computed radial abundance gradients derived from the data of the three H II regions for the cases of $t^2 = 0.00$ and $t^2 > 0.00$ (the value derived for each nebula). It is remarkable the similarity of the radial gradients found for N, O, Ne, S, and Cl when the observed temperature fluctuations are taken into account. In fact, the dispersion decreases dramatically when $t^2 > 0.00$ is considered. In this case, the average value of the radial gradient for those elements is -0.046 dex kpc^{-1}. This value is about 0.03 dex lower than the mean gradient obtained by the other works compiled in column 4. It must be considered that the values of the literature are computed for larger samples of objects with a wider coverage of galactocentric distances.

TABLE 2. Radial gradients for the three H II regions

Gradient (dex kpc^{-1}	t^2=0.00	t^2>0.00	Other works
Δlog(He/H)/ΔR	−0.004	−0.004	−0.001[a]
Δlog(C/H)/ΔR	−0.133	−0.133	−0.080[b]
Δlog(N/H)/ΔR	−0.033	−0.048	−0.090[a]
Δlog(O/H)/ΔR	−0.032	−0.049	−0.070[a]
Δlog(Ne/H)/ΔR	+0.031	−0.045	−0.080[c]
Δlog(S/H)/ΔR	+0.006	−0.055	−0.070[c]
Δlog(Cl/H)/ΔR	+0.021	−0.031	...

[a] Shaver et al. (1983).
[b] Peimbert et al. (1992).
[c] Simpson et al. (1995).

Values of t^2 ranging from 0.00 to 0.09 have been found for PNe, with a typical value around 0.04 (see Peimbert 1995, and references therein). Mathis et al. (1998) present a compilation of results from the literature finding that, in PNe, the C^{++} and O^{++} abundances are typically up to 10 and 5 times larger, respectively, when determined from the recombination lines rather than collisionally excited lines; differences much larger that those obtained for our sample H II regions. This fact indicates that the temperature fluctuations found for PNe are, in general, larger than those we have derived for Galactic H II regions. Although the results are suggestive, we need careful and homogeneous sets of determinations of t^2 for more objects to settle definitively whether these differences are real or not and the ultimate origin of these fluctuations. In particular, we need data for Galactic H II regions at different galactocentric distances and PNe of different morphological types and metallicities. On the other hand, it would be highly desirable to have t^2 determinations for giant extragalactic H II regions of different metallicity and stellar content. There are very few estimations available for these objects (González-Delgado et al. 1994; Terlevich et al. 1996) and they give different results. Moreover, Esteban & Peimbert (1995) and Pérez (1997) suggest the presence of important temperature fluctuations in the ionized gas associated with young burts of star formation. This could be specially important in the case of H II galaxies, whose chemical abundances are essential for the comparison of models of chemical evolution and primordial nucleosynthesis.

References

Esteban, C., Peimbert, M., 1995, A&A, 300, 78
Esteban, C., Peimbert, M., Torres-Peimbert, S., Escalante, V., 1998, MNRAS, 295, 401
Esteban, C., Peimbert, M., Torres-Peimbert, S., García-Rojas, J., 1999, RMAA, in press
Esteban, C., Peimbert, M., Torres-Peimbert, S., García-Rojas, J., Rodríguez, M., 1999, AJSS, in press
González-Delgado, R. M. et al., 1994, ApJ, 437, 239
Mathis, J. S., Torres-Peimbert, S., Peimbert, M., 1998, ApJ, 495, 328
Peimbert, M., 1967, ApJ, 150, 825
Peimbert, M., 1993, RMAA, 27, 9
Peimbert, M., 1995, in *The Analysis of Emission Lines*, ed. R. E. Williams & M. Livio, Cambridge University Press, Cambridge
Peimbert, M., Torres-Peimbert, S., Ruíz, M. T., 1992, RMAA, 24, 155
Pérez, E., 1997, MNRAS, 290, 465
Shaver, P. A., McGee, R. X., Newton, L. M., Danks, A. C., Pottasch, S. R., 1983, MNRAS, 204, 53
Simpson, J. P., Colgan, S. W., Rubin, R. H., Erickson, E. F., Haas, M. R., 1995, ApJ, 444, 721
Terlevich, E., Díaz, A. I., Terlevich, R. J., González Delgado, R. M., Pérez, E., García-Vargas, M. L., 1996, MNRAS 279, 1219

SPECTROSCOPIC STUDY OF THE CYGNUS SUPERBUBBLE

J. TRAPERO
Laboratorio de Astrofísica Espacial y Física Fundamental-INTA, P.O. Box 50727, E-28080 Madrid, Spain;
Instituto de Astrofísica de Andalucía-CSIC, P.O. Box 3004, E-18080 Madrid, Spain;
and Universidad SEK, c/ Cardenal Zúñiga s/n, 40003 Segovia, Spain

E.J. ALFARO
Instituto de Astrofísica de Andalucía-CSIC, P.O. Box 3004, E-18080 Madrid, Spain

AND

D. DE MIGUEL
Laboratorio de Astrofísica Espacial y Física Fundamental-INTA, P.O. Box 50727, E-28080 Madrid, Spain

Abstract.
The Cygnus Superbubble is a ring-like structure asociated with atomic and molecular gas, dust and objects related to recent star forming activity. Its origin and nature has been interpreted both as a single structure originated in OB asociations or as an arrangement of unrelated objects. We present a high resolution spectroscopic study of this region designed to probe the velocity and density structure of the gas asociated with the Superbubble.

1. Introduction

The Cygnus Superbubble is a ring-like structure of about 15 by 25 degrees detected in X rays (Cash et al. 1980) that includes 8 associations of young stars, different kind of nebulosities, filaments, and a number of radio and X ray sources (including Cyg X and Cygnus Loop). It was interpreted that all these objects were physically related and its origin was attribute to the

Astrophysics and Space Science is the original source of publication of his article. It is recommended that this article is cited as: *Astrophysics and Space Science* **263**: 197–200, 1999.

combined action of some 100 supernovae in the dense association Cyg OB2 (Cash et al. 1980) or the result of a single supermassive star (Blinnikov et al. 1982). However, Bochkarev and Sitnik (1985) showed that the Cygnus Superbubble can be explained as an arrangement of unrelated objects.

The Cyg OB1, OB2, OB3, OB8 and OB9 associations are in the western part of the Superbubble in a region structured with filaments of gas (Garmany & Stencel 1992, Odenwald & Schwartz 1993). According to Comerón et al. (1993) these structures could be fragments of the shell generated by the expanding bubble originated in Cyg OB2. We have focused our spectroscopic study in this region of the Cygnus Superbubble, one of the best laboratories for the study of the propagation of stellar formation (see e.g. Comerón, Torra and Gómez, 1998).

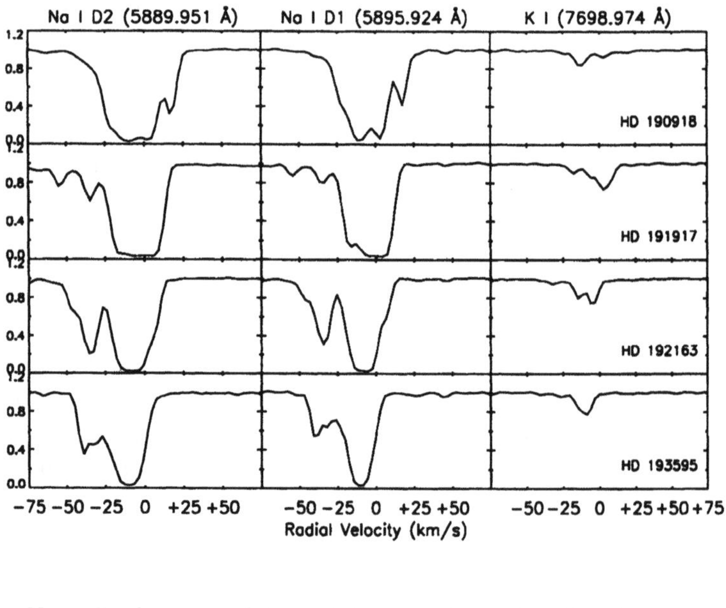

Figure 1. Normalized spectra of three of the stars observed. Left and center panel show the Na I doublet, and the right panel the K I line.

2. Observations

We obtained high resolution spectrum of 41 stars of the Cyg OB1, OB3 and OB9 associations distributed on the sky in order to form a three dimensional grid of the region. The observations qere carried out with the UES spectrograph at the William Herschel Telescope at the Observatorio del Roque de los Muchachos. We observed absorption interstellar spectrum of the Na I (5890 and 5896 Å), Ca II (3934 and 3968 Å) and K I (7699 Å) lines. The Na I line measure the weak high velocity lines while the K I line allow us to measure the stronger lines without saturation effects. The Ca II line (at the moment in analysis process) will give information of the warmest gas, usually associated with violent processes.

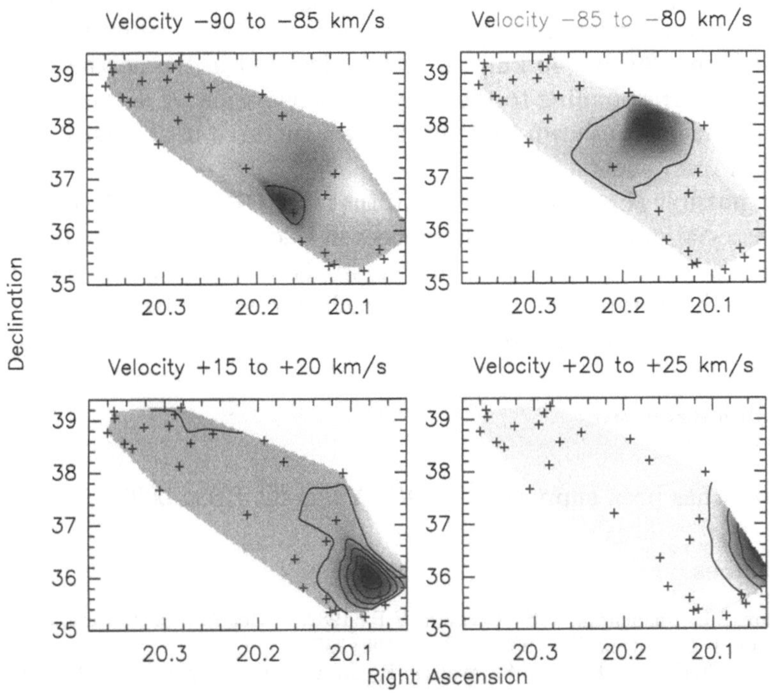

Figure 2. Na I column density maps at the most negative and positive velocities detected. Note that gas with negative and positive velocities is present at different parts of the region explored. Crosses indicate the position of the stars observed.

3. Results

We used a profile fitting method in the data analysis. We first obtained the column density, line width and radial velocity for each component present in the K I spectrum. Then we can use these parameters to fit the Na I lines because both lines are produced by gas with the same physical conditions. Using this method we obtained reliable column densities even for the most saturated absorptions. More than 200 components have been detected, with a velocity range from −90 to +25 km/s.

We have produced Na I column density maps for the different radial velocities detected. In Fig. 2 we present the maps obtained for the most negative and positive velocities. They show the region where gas at the given velocity is presnt. It is clear that gas with more negative velocity is present at the central part of the explored region, corresponding to the Cyg OB1 association. Positive velocities rise at the southern part, corresponding to the Cyg OB3 association. The stronger absorptions measured correspond to radial velocities of about −10 km/s. However, we find this absorption in stars at only 200 pc from the Sun showing that this component is originated in gas not related to the Superbubble. Gas with radial velocities in a range from −30 to −40 km/s is clearly detected in the northern part of the region, corresponding to the Cyg OB9 association. A detailed study of this component has shown that this gas is structured in at least 2 spatially separated components.

It is possible (Comerón et al. 1993) that Cyg OB1 and OB9 constitute a single association, so that the differences in velocity that we are measuring correspond to the geometry of the gas in the association. As soon as we introduce the distance information for each star we will be able to produce three dimensional maps and test models for expanding bubbles.

Acknolowdgements

This work has been supported by DGCYT grant PB94-0007.

References

Blinnikov, S.I., Imshennik, V.S., Utrobin, V.P., 1982, Soviet Astron. Letters, 8, 361
Bochkarev, N.G., Sitnik, T., 1985, Ap&SS, 108, 237
Cash, W., Charles, P., Bowyer, S., Walter, F., Garnire, G., Riegler, G., 1980, ApJ, 238, L71
Comerón, F., Torra, J., Jordi, C., Gómez, A.E., 1993, A&ASS, 101, 37
Comerón, F., Torra, J., Gómez, A.E., 1998, A&A, 330, 975
Garmany, C.D., Stencel, R.E.,1992, A&ASS, 94, 211
Odenwald, S.F., Schwartz, P.R., 1993, ApJ, 405, 706

OBSERVATIONS TOWARDS STAR FORMATION REGIONS: H_2 NEAR–INFRARED EMISSION LINES

JAVIER PALACIOS AND CARLOS EIROA
Universidad Autónoma de Madrid
Ciudad Universitaria de Cantoblanco
Madrid 28049 (Spain)

Abstract. We present near-infrared images of HH93 and L1251. The detection of H_2 emission allows us to study deeply embedded shocks, produced by the strong winds of young stars.

1. Introduction

During their earlier stages of evolution, young stars develope strong winds, which propagate far from the central protostar usually with a bipolar morphology. The interaction of these supersonic winds with the interstellar medium produces shock-waves, easily traced by means of the de-excitation emission lines of the molecular and atomic species present in the medium. Traditionally studied in the visible range with atomic lines, these shocks can be traced closer to the central object using near-IR emission lines.

Intermediate velocity shocks present in the wings of the bow-shock, or in the bulk of the flow produce the excitation of molecular hydrogen, without dissociating it. Some of the most prominent vibro-rotational lines of this molecule, at IR wavelengths, allow us a direct determination of physical parameters as excitation temperatures and extinction.

In this contribution, we report on our observations of the Herbig-Haro object HH93 and of the star formation region L1251.

2. Observations and data reduction

Observations were carried at the Calar Alto Observatory 3.5m telescope, in Almería (Spain), with the imaging mode of the near-infrared camera MAGIC, equipped with a NICMOS array of 256 × 256 pixels. The filters

Astrophysics and Space Science is the original source of publication of his article. It is recommended that this article is cited as: *Astrophysics and Space Science* **263**: 201–204, 1999.

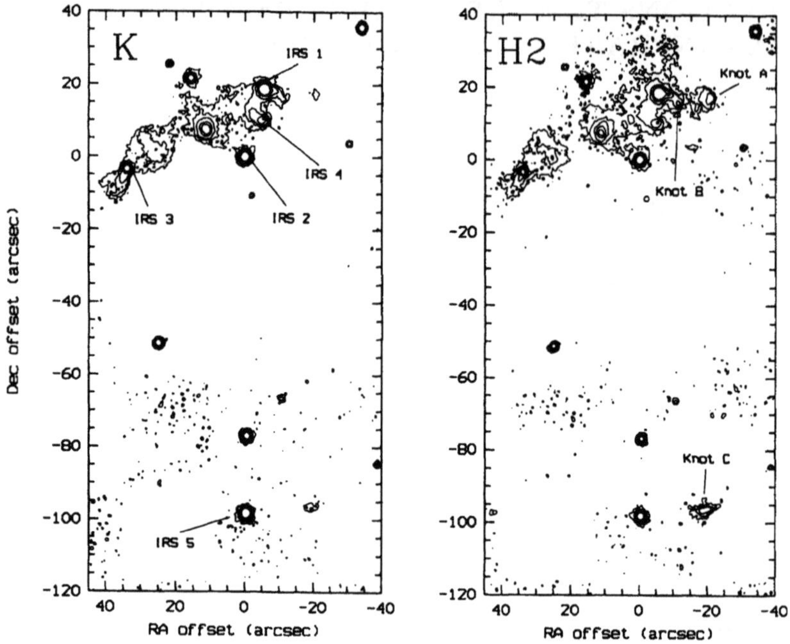

Figure 1. Contour plots of the near-infrared images of IRAS 22376+7455. Some objects mentioned in text has been labelled. First contour is 3σ. Offsets relative to IRS 2 ($\alpha_{1950} = 22^h37^m37.3^s$ $\delta_{1950} = 74°55'49''$). North is up, East is left.

used were K ($\lambda_c = 2.2\mu m, \Delta\lambda = 0.2\mu m$) and a $H_2S(1)$ narrow filter ($\lambda = 2.122\mu m, 1\%$).

Data reduction was achieved in the usual way in infrared imaging: flat fielding and substracion of a mean sky calculated from five dithered images taken around each object.

3. Results

3.1. IRAS 22376+7455 (L 1251)

The first indication of star formation activity was the detection of a compact CO outflow (Sato and Fukui 1989). More sensitive works of this region (Sato et al. 1994) showed a bipolar outflow 5' long ($PA \approx 30°$) and a H^3CO^+ core with 1-3 $M_\odot$. Optical images from Eiroa et al. (1989) revealed the presence of a strong reflexion nebula, as well as three Herbig-Haro emission knots (HH 189).

Our K image reveals the presence of at least five red stars not seen at the I images, altought most of them will probably be background objects. One of these stars (labelled IRS 3 in Figure 1) is associated with a biconical

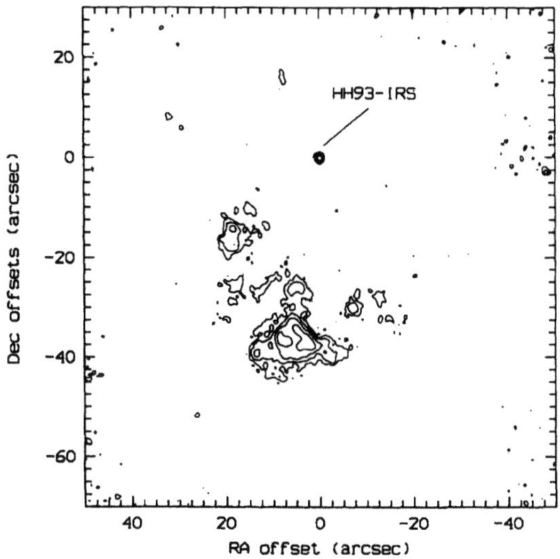

Figure 2. Contour plot of the H_2 image of HH 93. Except the labelled source, all the emission is from Hydrogen. First contour is 3σ. North is up, East is left.

reflection nebulae oriented NW and (rougly) aligned with the CO outflow from Sato et al. (1994) . The reflection emission is brighter (both in K and I) in NW lobe (the blueshifted one) than in the SW. IRS 3 is the likely powering source of the molecular flow. Regarding the other stars, IRS 1 and 4 present smalls arms of nebulosity emanating from each of them.

A comparison of the K image with the 2.12μm image reveals the presence of three strong H_2 knots. Condensation C its the near-IR counterpart of HH 189 C, remaining undetected the other optical knots of HH 189. The other H_2 knots seems related to the stars embedded in the main nebulosity, most likely to IRS 1.

3.2. HH 93

HH93 is a Herbig-Haro object first observed by Reipurth (1985), who described it as "a loose clustering of HH knots". Centimetric observations by Rodríguez and Reipurth (1994) discovered a radio source located at the northern edge of the optical object. Its spectral index upper limit ($\alpha \leq 0.2$) is compatible with that of a thermal radio jet. The radio source possibly indicates the posicion of the powering source of HH93.

Only one star labelled labelled HH93-IRS, is seen at K in the MAGIC field of the region. Its K magnitude is 14.8. Due to the absence of any other star in the image, an astrometric position of HH93-IRS is only possible relying on the pointing accuracy of the 3.5m telescope. Thus HH93-IRS could be the K counterpart of the radio source HH93-VLA2

References

Eiroa, C., Torrelles, J.M., Miranda, L.F., Anglada, G., Estalella, R., 1994, A&AS, 108, 73

Reipurth, B., 1985, A&AS, 61, 319

Rodríguez, L.F., Reipurth, B., 1994, A&A, 281, 882

Sato, F., Fukui, Y., 1989, ApJ, 343, 773

Sato, F., Mizuno, A., Nagahama, T., Onishi, T., Yonekura, Y., Fukui, Y., 1994, ApJ, 435, 279

FAR–INFRARED CO ROTATIONAL LINES IN THE ORION MOLECULAR CLOUD

M.J. SEMPERE, J. CERNICHARO, E. GONZÁLEZ- ALFONSO,
B. LEFLOCH AND S. PÉREZ-MARTÍNEZ
Instituto de Estructura de la Materia (CSIC)
Serrano 121, E–28006 Madrid. Spain

AND

S.J. LEEKS
Queen Mary & Westfield College. London, E1 4NS, U.K.

Abstract.

We present a high signal–to–noise grating spectrum between 43–196.9 μm of the Orion molecular cloud towards the massive star–forming region IRc2, obtained with the Long Wavelength Spectrometer (LWS) on board the Infrared Space Observatory (ISO).

CO lines up to J=20–19 have been detected around Orion-IRc2, while in the central position higher quantum numbers have been found. Lines of the ^{13}CO isotopic species have also been observed in several directions. In addition, high quality LWS-FP observations of some CO lines have been performed towards IRc2.

The data analysis suggest that at least two regions of Orion-IRc2 contribute to the observed CO emission: the ridge, responsible of the spatial extension, and the *plateau*, dominating the line flux observed towards the center of the map.

CO emission through the Orion molecular cloud has been studied in terms of temperature, column density and H_2 volume density, using and Large Velocity Gradient (LVG) model. We find that the flux ratio of the several CO lines can not be explained in terms of an homogeneous source, but a gradient in temperature and density must be involved.

Besides the CO lines, several molecular and fine–structure atomic lines have been detected in all observed positions. A detailed discussion of other molecular species rather than CO (H2O, OH...) can be found in the contribution by Cernicharo et al (1998).

Astrophysics and Space Science is the original source of publication of his article. It is recommended that this article is cited as: *Astrophysics and Space Science* **263**: 205–208, 1999.

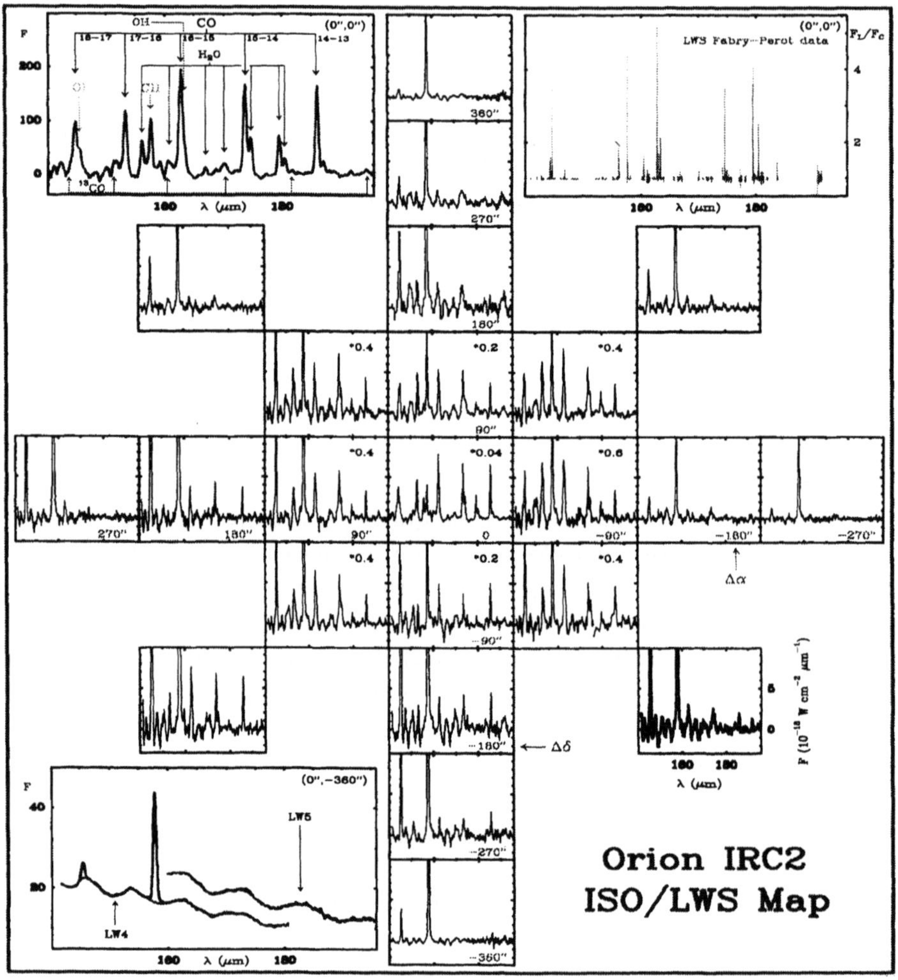

Figure 1. Full LWS grating rasters around IRc2 from 140 to 197μm. Observations of the central point in the LWS-FP mode are displayed at the top right corner. The main molecular features are identified at the top left corner.

1. Results

A total of 23 positions around IRc2, with an angular separation of 90″ (~ beam size), were observed. The obtained spectra are shown in Fig. 1.

TABLE 1. Integrated CO lines intensities in Orion–IRc2

Levels	J=14–13	J=15–14	J=16–15	J=17–16	
$\lambda(\mu m)$	185.999	173.631	162.812	160.305	
Position		Flux ($W\,cm^{-2}$)			$T(K°)$
(0″, −270″)	$4.9\,10^{-18}$	$4.6\,10^{-18}$	$1.7\,10^{-18}$	$2.2\,10^{-19}$	–
(0″, −180″)	$7.5\,10^{-18}$	$1.0\,10^{-17}$	$6.5\,10^{-18}$	$2.2\,10^{-18}$	80
(0″, −90″)	$4.3\,10^{-17}$	$4.2\,10^{-17}$	$4.0\,10^{-17}$	$3.0\,10^{-17}$	150
(0″, 90″)	$3.2\,10^{-17}$	$4.8\,10^{-17}$	$3.8\,10^{-17}$	$4.0\,10^{-18}$	120
(0″, 180″)	$2.8\,10^{-18}$	$4.5\,10^{-18}$	$2.7\,10^{-18}$	$1.3\,10^{-18}$	80
(−270″, 0″)	$8.0\,10^{-19}$	$2.2\,10^{-18}$	$3.3\,10^{-19}$	$3.0\,10^{-19}$	–
(−180″, 0″)	$1.9\,10^{-18}$	$1.3\,10^{-18}$	$1.3\,10^{-19}$	–	60
(−90″, 0″)	$1.6\,10^{-17}$	$1.7\,10^{-17}$	$1.9\,10^{-17}$	$1.9\,10^{-18}$	120
(90″, 0″)	$1.6\,10^{-17}$	$1.8\,10^{-17}$	$1.5\,10^{-17}$	$3.9\,10^{-18}$	100
(180″, 0″)	$7.1\,10^{-18}$	$7.0\,10^{-18}$	$4.5\,10^{-18}$	$2.3\,10^{-18}$	80
(270″, 0″)	$2.7\,10^{-18}$	$3.0\,10^{-18}$	$6.4\,10^{-19}$	–	–
(−90″, 90″)	$1.6\,10^{-17}$	$2.3\,10^{-17}$	$2.5\,10^{-17}$	$6.2\,10^{-18}$	100
(90″, −90″)	$1.8\,10^{-17}$	$1.9\,10^{-17}$	$2.5\,10^{-17}$	$3.8\,10^{-18}$	120
(90″, 90″)	$1.7\,10^{-17}$	$1.8\,10^{-17}$	$6.7\,10^{-18}$	$2.5\,10^{-18}$	80
(−90″, −90″)	$1.9\,10^{-17}$	$2.9\,10^{-17}$	$2.7\,10^{-17}$	–	100

The central position has been also observed in the LWS FP mode. The FP spectra are displayed at the top right corner of Fig 1. Also, the corresponding grating observations are plotted at the top left margin, where the most noticeable features are identified.

Although some components of Orion-IRc2 are practically point-like sources for the LWS beam, residual fringes due to the extended Orion molecular cloud have been observed and removed using special data processing algorithms (see bottom left corner of Fig. 1).

All the features present in the final averaged grating spectra were detected in each individual scan. Among the molecular species detected in Orion, CO appears to be the most widespread. At angular distances higher than $\sim 180''$ from the central region the CO lines weaken, and the spectra are dominated by the [O I] (146 μm) and [C II] (158 μm) fine–structure lines originated in the PDR region.

Four CO pure rotational lines have been detected in the LWS grating between 160–186μm. The flux values for external positions are listed in Table 1.

2. Physical Properties of the Cloud

We have estimated the temperatures necessary to account for the fluxes of the different rotational lines identified in the LWS bands (see Table 1) across the whole Orion-IRc2 cloud. CO line profiles in the millimeter and infrared range were calculated using a Large Velocity Gradient (LVG) code.

Our model is based on the simple following hypothesis:

- Temperature and density are uniform along the line of sight.
- The CO column density is N(CO)=10^{19} cm^{-2} (Blake et al 1987).
- The linewidth is $\Delta v = 10\,\mathrm{km\,s^{-1}}$ for positions close to the center ($\pm 90''$, $\pm 90''$). For distant points from IRc2 the adopted linewidth is $\Delta v = 5$ km s^{-1}.

which provides us with a full temperature description of the cloud at 80$''$ resolution.

Throughout the *whole* cloud we find $T \geq$80–100 K, which probably arises from two contributions : the hot gas of the PDR and the molecular gas.

Around the central position ($\pm 90''$, $\pm 90''$) we find evidence for hot gas with $T \geq 100$ K. Across the whole cloud the typical densities are of the order of $\sim 10^5$ cm^{-3}, with higher values at (0$''$,$-$90$''$) (coincident with a star forming core: S6, in Schmid–Burgk et al 1990).

The most severe limitation of these data comes from the contamination of the CO lines by strong adjacent features (e.g. OI at 145.6μm, OH at 163.2 μm , H_2O). In practice the flux for J$\geq$18 transitions could not be directly measured in a reliable way.

More accurate analysis estimating these contributions and adopting a more realistic density distribution are on going.

References

Blake, G.A., Sutton, E.C., Masson, C.R., Phillips, T.G., 1987, ApJ, 315, 621.
Cernicharo et al., 1998, in preparation
Leeks, S., et al.,1998, in preparation
Schmid–Burgk, J., Güsten, R., Mauersberger, R., et al., 1990, ApJ, 362, L25

DISTANCE SCALES AFTER HIPPARCOS

M. W. FEAST
Astronomy Department, University of Cape Town,
Rondebosch, 7700, South Africa

Abstract. A general review is given of the use of Hipparcos data to calibrate basic distance indicators.

1. Introduction

The Hipparcos mission changed the whole face of stellar astrometry. Not only did it provide parallaxes and proper motions for a large number of stars of very high precision but these were based on a system which was uniform over the sky and referred to the International Celestial Reference System (ICRS). Thus the parallaxes and proper motions can be considered to be on a quasi-inertial system and the uncertainties of correcting from relative to absolute values are avoided.

The present review explores those areas in which Hipparcos has made important contributions to the calibration of distance indicators which are important for both galactic and extragalactic astronomy. Whilst the impact of Hipparcos on distance scales has been felt in many areas of astronomy, it has perhaps been most notable in its relevance to the calibration of the extragalactic distance scale and to the distances and ages of globular clusters. In this connection it is worth recalling that estimates of the ages of globular clusters are very sensitive to the cluster distances adopted. A change of 10 percent in the distance results in a change of about 20 percent in the estimated age.

In discussions of distance scales it is useful to compare them using the distance that each of them implies for the Large Magellanic Cloud (LMC). Once one calibrates an extragalactic distance indicator such as the Cepheids, this implies a distance to the LMC. In addition, the LMC contains metal-poor globular clusters some of which have been shown by recent HST work to be as old as the oldest galactic globular clusters (see

Astrophysics and Space Science is the original source of publication of his article. It is recommended that this article is cited as: *Astrophysics and Space Science* **263**: 209–214, 1999.

papers presented at IAU Symposium 190, Victoria, Canada, July 1998, in press). With the LMC distance known, the absolute magnitude of the RR Lyrae variables in these clusters is also known. This can then be applied to galactic globular clusters with good cm diagrams and this together with evolutionary tracks then fixes their age. It is interesting to notice that increasing the adopted distance to the LMC reduces the inferred age of the globular clusters (i.e. of the oldest known stars) whilst at the same time increasing the expansion age of the Universe as determined from the Hubble constant (H_0). Thus these two age estimates can only be brought into compatibility for one adopted distance of the LMC.

The following sections discuss various important distance indicators and the way Hipparcos data can be used to calibrate them. Throughout it will be convenient to compare results by giving the implied distance modulus of the LMC in each case.

2. Cepheids

The most direct way to calibrate the Cepheid period-luminosity (PL) relation is to use parallaxes of the Cepheids themselves. This is now possible for the first time using parallaxes from Hipparcos. In this way Feast and Catchpole (1997) obtained a PL calibration which leads to an LMC modulus of 18.70 ± 0.10. Because the Cepheid parallaxes are small it is essential in the analysis to use a method which avoids statistical bias problems (see Feast and Catchpole 1997, Feast, Pont and Whitelock 1998, Feast 1998 and, Koen and Laney 1998). This new LMC modulus is larger than that adopted in the HST extragalactic Cepheid programme (18.5). The implied 10 percent increase in distance scale already has interesting implications. Previously, expansion ages of the Universe in the range 9 to 12 Gyr have been widely quoted; whilst ages of about 15 Gyr have been quoted for globular clusters. The fact that the former are smaller than the latter has given rise to much debate. A 10 percent increase in distance scale revises the above figures to between 10 and 13 Gyr for the expansion age and about 12 Gyr for the oldest globular clusters (based on the absolute magnitudes of RR Lyraes in LMC clusters). In fact recent, revised, evolutionary models (Chaboyer et al. 1998) together with a 10 percent increase in the distance scale, reduce the implied globular cluster ages even further, to about 10 Gyr.

The proper motions of Cepheids from the Hipparcos catalogue can be used to check this result. It is possible to analyse the proper motions in a number of ways. However the proper motions themselves, and also the radial velocities, show very clearly the effects of differential galactic rotation. The safest way, therefore, of calibrating the Cepheid PL relation using the proper motions is to compare the value of the Oort constant, A, derived from radial velocities (which depends inversely on the distance scale adopted) and that derived from proper motions (which is nearly independent of the distance scale) (Feast and Whitelock 1997, Feast, Pont and Whitelock, 1998). The PL zero-point obtained in this way implies an LMC modulus of 18.74 ± 0.13, very close to that derived from the parallaxes. Feast and Whitelock (1997) also give revised galactic constants based on this distance scale including a revised kinematic distance to the galactic centre (R_0) of 8.5 ± 0.5kpc.

In deriving the LMC modulus from Cepheids, corrections were made for the fact that the LMC Cepheids have a lower metallicity than those in the solar neighbourhood. There has been considerable discussion in recent times on the effects of metallicity on the Cepheid scale. Such discussions do not always make clear that there are two distinct effects involved. One of these is the effect on the Cepheid colours which affect the reddening corrections deduced. The other effect is on the PL relation itself. The magnitude of the combined metallicity effect obviously depends on the colours used for the reddening determinations and the wavelength at which the PL relation is used. in the work of Feast and Catchpole (1997) the reddening corrections in the LMC were based on BVI work (Caldwell and Coulson 1995, Caldwell and Laney 1991). This has the advantage that the effects of metallicity in the colour- colour diagram are due simply to changes in photospheric blanketing and this can be reliably computed. Reddening corrections using a period - colour relation can be less certainly corrected for metallicity since both differential blanketing and a change of photospheric temperature at a given period are involved. Metallicity corrections to the PL in (Johnson) V (=PL(V)) were taken by Feast and Catchpole (1997) from Laney and Stobie (1994). These (small) corrections are based on the assumption that the PL relation in M_{bol} is independent of metallicity. Whilst early theoretical work (see e.g. Iben and Renzini 1984) suggested that this was closely true, the matter has recently come under debate again. Laney (1999) has now shown empirically that, if anything, a metal poor Cepheid is likely to be slightly brighter in M_{bol} than a metal-normal one of the same period, though the effect is likely to be very small. His result follows from a discussion of the temperatures and radii of Cepheids in our Galaxy, the LMC and the SMC. It is known (Laney and Stobie 1986) from infrared photometry of SMC and LMC Cepheids, that at a given period the more metal-poor a Cepheid is, the hotter it is. Laney has now drawn attention to the fact that SMC and LMC Cepheids follow closely the same period - radius relation as the galactic ones, the radii being derived from Baade-Wesselink type analyses. If anything the metal-poor Cepheids are slightly larger at a given period than metal-normal ones, though the effect is marginal. This result shows that the LMC Cepheids will be marginally hotter and, if anything, slightly larger than galactic ones of the same period. This means that using the Hipparcos galactic calibrations discussed above will lead to a slightly underestimated LMC modulus though the effect is likely to be very small. In their work Feast and Catchpole adopt an LMC Cepheid metal deficiency of a factor of 1.4. Recent spectroscopic work suggests that the deficiency is somewhat greater (a factor of about 2 rather than 1.4) (see e.g. Smith 1999). In that case the LMC modulus given above will be a slight underestimate. A change from 1.4 to 2.0 in metal deficiency would increase the LMC modulus from 18.70 to 18.76.

3. Mira Variables

Miras provide an important distance indicator independent of the Cepheids. The Miras in the LMC show an excellent PL relation in $M_{\rm bol}$ and at K (2.2 microns)(Feast et al. 1989). Following initial work by van Leeuwen et al. (1997), the PL(K) relation has been calibrated by Whitelock (1998) using Hipparcos parallaxes and extensive infrared photometry of Miras. This leads to an LMC modulus of 18.64 ± 0.13. However it is known that the period - infrared colour relations of galactic and LMC Miras are slightly different. If this is due to a metallicity effect, as seems likely, this will affect the derived modulus. An approximate estimate of the abundance corrected modulus is ~ 18.8 (see Feast and Whitelock 1998).

Hipparcos data also provide an alternative way of calibrating the Mira PL relation. Reid (1998) has used Hipparcos parallaxes of subdwarfs to determine the distance of the globular cluster 47 Tuc. This cluster contains three Mira variables with periods near 200 day and these thus provide a calibration point for the PL relation. This leads to an estimate of the LMC modulus of 18.79 ± 0.17 (Feast and Whitelock 1998).

4. RR Lyrae Variables

Koen and Laney (1998) have recently analysed the Hipparcos parallaxes of field RR Lyrae variables. Applying their results to RR Lyraes in LMC globular clusters leads to an LMC modulus of 18.68 ± 0.22 (Feast 1999a). There is some uncertainty whether or not RR Lyraes in the galactic field are identical to those of the same overall metallicity in globular clusters (see e.g. Reid 1998). It is therefore valuable to avoid this possible problem by comparing RR Lyraes in LMC clusters directly with those in globular clusters of known distance in our own Galaxy. Such a comparison is now possible since Hipparcos provides a significant number of parallaxes of metal-poor subdwarfs which can be used to determine the distances of galactic globular clusters by main-sequence fitting methods. Subdwarf distances to M92, M15, and M68 lead in this way to an LMC modulus of

18.65±0.1 (Reid 1997). A somewhat different discussion of the subdwarf fit to M92 by Pont et al. (1998) leads to an LMC modulus in the range 18.50 to 18.61 depending on how the subdwarf data is treated. A subdwarf distance to the galactic globular NGC6397 (Reid 1998) leads to an LMC modulus of 18.71 ± 0.15. This latter result has the advantage that it compares galactic and LMC clusters of about the same metallicity ([Fe/H] ~ -1.8) so that no account need be taken of the variation of RR Lyrae (or horizontal branch) absolute magnitudes with metallicity which must otherwise be corrected for (see Feast 1999b).

Considerable interest has been aroused by statistical parallax determinations of the absolute magnitude of field RR Lyrae variables using Hipparcos proper motions (Fernley et al. 1998, Tsujimoto, Miyamoto and Yoshii 1998). The results of Fernley et al. when applied to the RR Lyraes in LMC clusters imply an LMC modulus of 18.25 ± 0.17, much lower than those discussed above. The Tsujimoto et al. result also leads to a low LMC modulus (18.37 ± 0.10) The reason for these low values is not yet fully understood. One possibility is that the galactic halo model assumed by the statistical parallax workers is not realistic enough. They assume a simple, classical, model of the halo. However it is now known that the halo is quite complex, containing streams which are probably the debris from infalling satellite galaxies. Such complexity may well affect the results of the statistical parallax work.

5. The Distance Modulus of the LMC: Summary

The results discussed above suggest that the modulus of the LMC probably lies in the range 18.60 to 18.80. This result depends on the use of Hipparcos data. The LMC modulus which is inferred from the ring around SN1987A is 18.58 ± 0.05 (Panagia 1999) and this is not inconsistent with the the result deduced from the Hipparcos data.

6. The Future

The distance scale based primarily on Hipparcos data would now seem to be reasonably well established at the $\sim$ 0.1mag level and provides a rather firm basis for extentions outwards to cosmologically interesting distances as well as for other problems such as the ages of globular clusters and the dynamics of our own Galaxy. However a number of apparent discrepancies remain. Some possible reasons for these have been given above but they will no doubt give rise to more extensive studies.

To significantly improve the reliability of the basic distance scale in the near future will probably be quite difficult. It will for instance require a very careful discussion of the effects of metallicity on the various indicators.

Such effects may be due not only to changes in overall metallicity but also to relative abundance changes among the elements.

A comparison of Cepheid distances to galaxies and those obtained using the luminosity of the red giant branch tip (RGB tip) as a distance indicator (Salaris and Cassisi 1998, Feast 1999b) suggests that the RGB tip can be a rather reliable indicator and no doubt this will be extensively used in the future. On the other hand the luminosity of red giant clump would seem to require large corrections depending on the age and metallicity of the sample involved (see Feast 1999b) in order for it to provide satisfactory distances. It may in fact be more useful as an indicator of stellar populations in systems whose distance can be determined in other ways.

It is clear that whilst much work remains to be done, the Hipparcos mission has already had a major impact on distance scale problems.

References

Caldwell, J.A.R., Coulson, I.M., 1985, MNRAS, 212, 879
Caldwell, J.A.R., Laney, C.D., 1991, in Haynes, R. & Milne, D., eds., IAU Symp. 148, The Magellanic Clouds, Kluwer, Dordrecht, p.249
Chaboyer, B., Demarque, P., Kernan, P.J., Lawrence, M.K., 1998, ApJ, 494, 96
Feast, M.W., 1998, Mem. Soc. It., in press
Feast, M.W., 1999a, in IAU Symp. 190, New Views of the Magellanic Clouds, Astr. Soc. Pacif., San Francisco, in press
Feast, M.W., 1999b in Whitelock, P.A., & Cannon, R., eds., IAU Symp. 192, The Stellar Content of Local Group Galaxies, Astr. Soc. Pacif., San Francisco, in press
Feast, M.W., Catchpole, R.M., 1997, MNRAS, 286, L1
Feast, M.W., Whitelock, P.A., 1997, MNRAS, 291, 683
Feast, M.W., Whitelock, P.A., 1998, in Caputo, F. & Heck, A., eds., Post-Hipparcos Cosmic Candles, Kluwer, Dordrecht, in press
Feast, M.W., Glass, I.S., Whitelock, P.A., Catchpole, R.M., 1989, MNRAS, 241, 375
Feast, M.W., Pont, F., Whitelock, P.A.,1998, MNRAS, 298, L43
Fernley, J., Barnes, T.G., Skillen, I., Hawley, S.L., Hanley, C.J., Evans, D.W., Solano, E., Garrido, R., 1998, A&A, 330, 515
Iben, I., Renzini, A., 1984, Phys. Rep., 105, 329
Koen, C., Laney, C.D., 1998, MNRAS, in press
Laney, C.D. 1999 in Whitelock, P.A. & Cannon, R., eds, IAU Symp. 192, The Stellar Content of Local Group Galaxies, Astr. Soc. Pacif., San Francisco, in press
Laney, C.D., Stobie, R.S., 1986, MNRAS, 222, 449
Laney, C.D., Stobie, R.S., 1994, MNRAS, 266, 441
Panagia, N., 1999, in IAU Symp. 190., New Views of the Magellanic Clouds, Astr. Soc. Pacif., San Francisco, in press
Reid, I.N., 1997, AJ, 114, 161
Reid, I.N., 1998, AJ, 115, 204
Salaris, M., Cassisi, S., 1998, MNRAS, in press
Smith, V.V., 1999, in IAU Symp. 190, New Views of the Magellanic Clouds, Astr. Soc. Pacif., San Francisco, in press
Tsujimoto, T., Miyamoto, M., Yoshii, Y., 1998, ApJ, 492, L79
van Leeuwen, F., Feast, M.W., Whitelock, P.A., Yudin, B., 1997, MNRAS, 287, 955
Whitelock, P.A., 1999, in preparation

CALIBRATION OF THE CLASSICAL CEPHEID PERIOD-LUMINOSITY RELATION FROM HIPPARCOS DATA

X. LURI, J. TORRA AND F. FIGUERAS
Universitat de Barcelona
Dept. d'Astronomia i Meteorologia, Diagonal 647,
08028 Barcelona

AND

A.E. GÓMEZ, M.J. GOUPIL AND J.P. BEAULIEU
Observatoire de Paris-Meudon
D.A.S.G.A.L., URA CNRS 335, F92195 Meudon CEDEX,
France

Abstract. The calibration of the classical Cepheid Period-Luminosity (PL) relation is a fundamental step for the determination of the cosmic distance scale. In this paper a calibration obtained from Hipparcos data is presented and used to estimate the LMC distance modulus. A value of $18.35^m \pm 0.13$ is obtained.

1. The LM method

To calibrate the PL relation we have applied the LM method. This is a method using Maximum-Likelihood estimation specifically designed for luminosity calibrations. It is described in Luri et al. (1996) and its main advantages are:

- All the available information for each star can be used: apparent magnitude, position, trigonometric parallax, proper motions and radial velocities as well as other relevant astrophysical parameters (color index, period, etc.).
- It has been designed to fully exploit Hipparcos data for luminosity calibrations.

Astrophysics and Space Science is the original source of publication of his article. It is recommended that this article is cited as: *Astrophysics and Space Science* **263**: 215–218, 1999.

- The observational selection criteria used to construct the sample of stars are taken into account. This is essential to obtain unbiased results (Brown et al., 1997).
- The effects of observational errors are taken into account.

– It gives at the same time the kinematical properties and spatial distribution of the sample.
– It is able to handle inhomogeneous samples composed of groups of stars with different luminosity, kinematical or spatial characteristics.
– It provides an individual distance estimate with optimal statistical properties.

2. The sample

All classical Cepheids from the Hipparcos catalogue (ESA, 1997) were included, even first overtone pulsators. The sample contains 238 stars.

All data (including periods) were taken from the Hipparcos catalogue, except the radial velocities that were taken from the *Hipparcos Input Catalogue* (Turon et al. 1992).

3. Modeling of the sample

To apply the LM method it is necessary to model the physics of the sample as well as the censorship applied for its construction. Each physical distribution (luminosity, velocity, spatial and error distribution) and the observational selection is described by a p.d.f. (probability density function) depending on a set of parameters. The values of these parameters are determined by applying the LM method to the sample.

3.1. ABSOLUTE MAGNITUDE DISTRIBUTION

– Linear mean PL relation: $M_v = \alpha + \beta \log(P)$
– Gaussian dispersion (σ_M) of the individual absolute magnitude around the mean relation [1]
– Interstellar absorption from the 3D model by Arenou et al. (1992)

3.2. VELOCITY DISTRIBUTION

– Velocity ellipsoid: $\Phi_K(U,V,W) = e^{-\frac{1}{2}(\frac{U-U_\odot}{\sigma_U})^2} e^{-\frac{1}{2}(\frac{V-V_\odot}{\sigma_V})^2} e^{-\frac{1}{2}(\frac{W-W_\odot}{\sigma_W})^2}$

[1]Actually, the real intrinsic distribution has a Top-Hat shape, but combined with the errors in the estimation of the interstellar absorption it is reasonably well approximated by a Gaussian

– Galactic rotation: Oort-Lindblad model at first order with $A = 14.4$km/s, $B = -12.8$km/s and $R_\odot = 8.5$kpc (Kerr & Lynden-Bell, 1986).

3.3. SPATIAL DISTRIBUTION

– Exponential disk: $\Phi_S(x, y, z) = e^{-|\frac{z}{Z_0}|}$

3.4. ERROR DISTRIBUTION

Approximation of Gaussian independent errors for $(\pi_t, \mu_\alpha, \mu_\delta)$, valid for large angular scales ($> 1°$), as in our sample. We also assumed Gaussian errors for v_r but we considered negligible the errors in apparent magnitude, galactic coordinates and period (m, l, b, P).

3.5. OBSERVATIONAL SELECTION

As we have included all the classical Cepheids contained in the Hipparcos Catalogue, the only observational selection is the one of the catalogue itself.

The main censorship of the catalogue is in apparent magnitude: the catalogue is complete up to the Hipparcos magnitude $H_p \simeq 7.9^m$ (the Survey) and its magnitude limit is $H_p \simeq 12.5^m$. We have modeled this censorship with a selection function giving completeness up to a certain apparent magnitude m_c (estimated by the method together with the rest of parameters) and a linear decrease of the completeness up to the limiting magnitude. Finally, we have considered the censorship on $(\pi_t, l, b, \mu_\alpha, \mu_\delta, v_r, P)$ not significant on average.

3.6. GROUPS

Our sample is a mixture of two types of Cepheids: fundamental pulsators and overtone pulsators. The stars have been classified a priori in these two types using the Fourier coefficients obtained from the Hipparcos light curves: 204 fundamental pulsators and 34 overtone pulsators. We have also made two supplementary hypothesis:

1. The slopes of the PL relations for both groups have been fixed to the values for the SMC obtained by the EROS team (Sasselov et al., 1997): $\beta_F = -2.72 \pm 0.07$ (fundamental) and $\beta_{OV} = -3.46 \pm 0.14$ (overtones). Only the zero points of the PL relations have been determined, α_F (fundamental) and α_{OV} (overtones).
2. We have assumed that both groups have the same kinematic and spatial distributions.

4. Results

We have applied the LM method with the modeling described above and we have obtained an estimation of the zero points of the PL relations:

Cepheids pulsating on the fundamental mode: $\alpha_F = -1.01 \pm 0.13$
Cepheids pulsating on the 1st overtone: $\alpha_{OV} = -1.26 \pm 0.20$

On the other hand, the values obtained for the kinematical and spatial distribution parameters are $(U_0 = -9.0 \pm 1.5, V_0 = -10.3 \pm 1.2, W_0 = -7.5 \pm 0.6)$ $km\ s^{-1}$, $(\sigma_U = 13.0 \pm 1.2, \sigma_V = 12.7 \pm 1.3, \sigma_W = 6.2 \pm 1.2)$ $km\ s^{-1}$ and $Z_0 = 97 \pm 7\ pc.$

5. LMC distance modulus

We have estimated the LMC distance modulus from the difference of zero points between our PL relation and the EROS apparent magnitude PL relation for the LMC (Sasselov et al. 1997). Using this difference and assuming:

- $E(B-V) = 0.1^m$ for the LMC (from Freedman et al., 1994)
- $R = \frac{A_v}{E(B-V)} = 3.311$
- A correction of 0.043^m to take into account the effects on the zero points of the metallicity difference between the LMC and our Galaxy (Laney & Strobie, 1994)

we have obtained:

$$(m-M)_0 = 18.35^m \pm 0.13$$

This result is in agreement with the "short" distance scale obtained from (for instance) RR-Lyrae (E. Solano in this volume and Luri et al. 1998) but not compatible with the "long" distance scale from other Hipparcos results for Cepheids (M. Feast in this volume).

References

Arenou, F. , Grenon, M. , Gómez, A.E., 1992, A&A, 258, 104
Brown, A., Arenou, F., Leeuwen F., Lindegren, L., Luri, X. 1992, ESA SP-402, 63
ESA, The Hipparcos Catalogue, ESA SP-1200
Freedman, W.L. et al. 1994, ApJ, 427, 628
Laney, Strobie 1994, *MNRAS*, 266, 441
Kerr, F.J. & Lynden-Bell, D. 1986, MNRAS, 221, 1023
Luri, X., Mennessier, M.O, Torra, J., Figueras, F. 1996, A&ASS, 117, 405
Luri, X., Gómez, A., Torra, J., Figueras, F., Mennessier, M.O. 1998, A&A, 335, L85
Sasselov, et al., 1997, A&A, 324, 471
Turon, C. et al., 1992, *The Input Catalogue*, ESA SP-1136

THE ABSOLUTE MAGNITUDE OF RR LYRAES: FROM HIPPARCOS PARALLAXES AND PROPER MOTIONS

E. SOLANO

INSA
P.O. Box 50727. 28080 Madrid (Spain)

Abstract. We have used HIPPARCOS proper motions and the method of Statistical Parallax to estimate the absolute magnitude of RR Lyrae stars. In addition, we have used the HIPPARCOS parallax of RR Lyr itself to determine its absolute magnitude. These two results are in excellent agreement with each other and give a zero-point for the RR Lyrae M_v,[Fe/H] relation of 0.77 ± 0.15 at [Fe/H]$= -1.53$. This zero-point is in good agreement with that obtained recently by several groups using Baade-Wesselink methods which, averaged over the results from the different groups, gives $M_v = 0.73 \pm 0.14$ at [Fe/H]$= -1.53$. Taking the HIPPARCOS based zero-point and assuming a value of 0.18 ± 0.03 for the slope we find the distance modulus of the LMC is 18.26 ± 0.15. This value is compared with recent estimates based on other methods. Potential problems that may affect the results are outlined.

1. Introduction

RR Lyraes are one of the primary distance indicators. Within the Galaxy they are used to determine the distances to the globular clusters and hence their ages by matching their colour-magnitude diagrams with theoretical isochrones. Since the globular glusters are amongst the oldest objects in the Galaxy, this sets a lower limit to the age of the Universe. They are also used to determine distances to other Local Group galaxies where they serve as a check on brighter distance indicators (e.g. cepheids).

Astrophysics and Space Science is the original source of publication of his article. It is recommended that this article is cited as: *Astrophysics and Space Science* **263**: 219–222, 1999.

2. Determination of the distance to LMC

Ideally, the best way of calculating the distance to LMC would be to use accurate values of trigonometric parallaxes. Unfortunately, the LMC is too far for the HIPPARCOS accuracy. Moreover, field RR Lyraes are even too far for HIPPARCOS: 65 out of 173 RR Lyraes measured with HIPPARCOS have relative errors greater than 100% and 57 have negative parallaxes. Only one star (RR Lyr itself) shows an accurate trigonometric parallax ($\sigma_\pi/\pi = 0.13$). This forces to use indirect methods (i.e. Baade-Wesselink and Statistical Parallax) to determine the absolute magnitudes of these stars.

2.1. STATISTICAL PARALLAX ANALYSIS

The algorithm used for the Statistical Parallax analysis is described in Hawley et al. (1986). The input data used in the analysis are given in Fernley et al. (1998a). Using a kinematically homogeneous sample ([Fe/H] < -1.3; Layden et al. 1996)) an absolute magnitude $M_v = 0.77 \pm 0.17$ at [Fe/H]$= -1.66$ is obtained. Combining this result with that obtained using the trigonometric parallax of RR Lyr which gives $M_v = 0.78 \pm 0.29$ at [Fe/H]$= -1.39$ we get $M_v = 0.77 \pm 0.15$ at [Fe/H]$= -1.53$. These results give no information about the slope of the M_v,[Fe/H] relation, however this has been recently discussed by Fernley et al. (1998b) who estimate a value of 0.18 ± 0.03. Thus we obtain

$$M_v = 0.18 \pm 0.03([\mathrm{Fe/H}] + 1.53) + 0.77 \pm 0.15 \qquad (1)$$

Using observations of RR Lyraes in 5 LMC clusters a mean dereddened magnitude m_v of 18.98 and a mean [Fe/H] of -1.8 have been calculated. With these data, a distance modulus $(m - M)$ of 18.26 ± 0.15 is obtained using Eq. (1).

2.2. COMPARISON WITH OTHER DISTANCE INDICATORS

In Table 1 we compare the distance modulus given above with those obtained using other distance indicators. We can see how the most discrepant results are those obtained using RR Lyraes and cepheids. In this section some potential problems that may affect the results obtained using these two indicators are briefly discussed.

2.2.1. *RR Lyraes*

- *Intrinsic differences between field and cluster RR Lyraes:* Some authors (e.g. Sweigart 1997) have pointed out the existence of anticorrelations Na-O, Mg-Al in globular cluster red giants that are not present in field stars. The origin of these anticorrelations seems to be related to the internal rotation which would produce a bluer and brighter morphology for the RR Lyraes in globular clusters. Chaboyer et al. (1998) also claimed that cluster RR Lyraes could be typically more evolved and, hence, brighter than field RR Lyraes. On the contrary, some other authors (e.g. Catelan 1998) have shown that field and cluster RR Lyraes have the same distribution in the period-temperature diagram suggesting very similar luminosities. Moreover, Baade-Wesselink analysis found no differences for cluster and field RR Lyraes irrespective of [Fe/H].

– *Depth effects in the LMC:* Unlikely to be significant in the present case since the mean m_v for the RR Lyraes is based on observations in 5 clusters which are spread evenly across the face of the LMC.

2.2.2. *Cepheids*

– *Strong dependence with the method:* In Table 1 it is shown that, unlike RR Lyraes, the distance modulus obtained using cepheids strongly depends on the method used.
– *Binarity:* Szabados (1997) argued that the sample of cepheids used by Feast & Catchpole (1997) contains a large number of binaries which would lead to a systematic underestimation of the parallaxes (and hence an overestimation of the LMC distance modulus).
– *Metallicity effects on the period-luminosity relation:* A dependence of the period-luminosity relation (metal-poor cepheids being fainter and, therefore, the distance modulus to LMC smaller) has been suggested by a growing number of authors (e.g. Sasselov et al. 1997).

TABLE 1. Large Magellanic Cloud distance moduli

Indicator	Distance Modulus	Reference
RR Lyraes	18.26 ± 0.15	Fernley et al. (1998a)
RR Lyraes (B/W analysis)	18.31 ± 0.14	Fernley (1994)
RR Lyraes (LM method)	18.37 ± 0.23	Luri et al. (1998)
Cepheids	18.70 ± 0.10	Feast & Catchpole (1997)
Cepheids (Cluster stars)	18.33 ± 0.10	Feast (1995)
Cepheids (B/W analysis)	18.49 ± 0.05	Gieren et al. (1998)
Cepheids (LM method)	18.29 ± 0.17	Luri et al. (1998)
Miras	18.54 ± 0.18	van Leewuen et al. (1997)
SN 1987A	18.37 ± 0.04	Gould & Uza (1998)
SN 1987A	18.43 ± 0.10	Sonneborn et al. (1997)
SN 1987A	18.58 ± 0.03	Panagia et al. (1998)
SN 1987A	18.67 ± 0.08	Lundqvist & Sonneborn (1998)
Main Sequence fitting	18.60 ± 0.07	Gratton et al. (1997)
Tip of the RGB	18.60 ± 0.14	Salaris & Cassisi (1999)
Eclipsing binaries	18.42 ± 0.07	Guinan et al. (1998)

References

Catelan, M., 1998, ApJ, 495, 81
Chaboyer, B., Demarque, P., Kernan, P.J., Krauss, L.M., 1998, ApJ, 494, 96
Chaboyer, B., Demarque, P., Sarajedini, A., 1996, ApJ, 459, 558
Feast, M.W., 1995, ASP Conf. Ser., 83, 209
Feast, M.W., Catchpole, R.M., 1997, MNRAS, 286, L1
Fernley, J., 1994, A&A 284, L16
Fernley, J., Barnes, T.G., Skillen, I., Hawley, S.L., Hanley, C.J., Evans, D.W., Solano, E., Garrido, R., 1998a, A&A, 330, 515
Fernley, J., Carney, B., Skillen, I., Cacciari, C., Janes K., 1998b, MNRAS, 293, L61
Gieren, et al., 1998, ASP Conf. Ser., in press
Gould, Uza, 1998, ApJ 494, 118
Gratton, et al., 1997, ApJ 491, 749
Guinan, et al., 1998, ApJ Lett., in press
Hawley, S.L., Jefferys, W.H., Barnes, T.G., Wan, L., 1986, ApJ, 302, 626
Layden, A.C., Hanson, R.B., Hawley, S.L., Klemola, A.R., Hanley, C.J., 1996, AJ, 112, 2110
Lundqvist, Sonneborn, 1998, ASP Conf. Ser., in press
Luri, X., et al., 1998, A&A 335, L81
Panagia, et al., 1998, ASP Conf. Ser., in press
Salaris, Cassisi, 1999, MNRAS, in press
Sasselov, D.D., et al., 1998, A&A, 324,471
Sonneborn, et al., 1997, ApJ 477, 848
Sweigart, A.V., 1997, ApJ, 474, L23
Szabados, L., 1997, ESA-SP, 402, 657
van Leewuen, et al., 1997, MNRAS, 287, 955

RADIATIVE TRANSFER UNDER INHOMOGENEOUS CONFIGURATIONS

P. FABIANI BENDICHO

Instituto de Astrofísica de Canarias.
38200-La Laguna. Tenerife (Spain)

Abstract. We present, for the first time, three dimensional (3D) radiative transfer (RT) results with realistic atomic models (*multilevel*) and without using the *local thermodinamical equilibrium approximation* (*non-LTE*). We have developed a new code based on efficient iterative methods (Trujillo Bueno, and Fabiani Bendicho 1995; Fabiani Bendicho, Trujillo Bueno and Auer 1997) characterized by a very high convergence rate. With this 3D multilevel code and using a schematic atmospheric model we are able to demonstrate that one may need self-consistent multidimensional RT calculations in order to interpret high spatial resolution solar spectroscopic observations.

1. Introduction

In order to solve the radiative transfer problem under realistic conditions, the use of complex atomic models is required but also the presence of spatial inhomogeneities in the plasmas should be taken into account. This requires the self-consistent solution of the kinetic and RT equations, and the resulting problem is non-local and non-linear.

The improvement of the numerical solution methods of such a complex problem has been spectacular during the last years. We have now the opportunity to investigate the horizontal radiative transfer effects and to conclude about the validity of the plane-parallel approximation (1D) in a realistic diagnostic of stellar and solar plasmas.

Astrophysics and Space Science is the original source of publication of his article. It is recommended that this article is cited as: *Astrophysics and Space Science* **263**: 223–226, 1999.

2. Description of the code: the solution method.

To solve simultaneously the RT and the statistical equilibrium (SE) equations we need a fast iterative method effective even under extreme non-LTE conditions in very fine grids. The Gauss-Seidel thechniques have shown to be very efficients to solve this problem (Trujillo Bueno and Fabiani Bendicho 1995). However, for all operator-splitting methods, the convergence rate deteriorates when a finer spatial grid is considered. To avoid it we combine non-linear multigrid iterations with an efficient multilevel RT scheme based on Gauss Seidel iteration (Fabiani Bendicho, Trujillo Bueno and Auer, 1997).

Each iteration requires the solution of the RT equation for source function and opacity obtained from the previous iteration (i.e. the formal solution of the RT eq.). Our 3D formal solution method (Fabiani Bendicho and Trujillo Bueno, 1999) is based in the *short characteristics* method (Kunasz and Auer, 1988). We use periodic boundary conditions in the horizontal directions and we impose *monotonicity* from the parabolic interpolations (Auer and Paletou, 1994) in order to guarantee positivity for the intensity.

This method is optimal in multidimensional calculations: it has parabolic precision (i.e. the numerical error quadratically decreases when increasing the spatial resolution of the grid); the CPU time is linear with the number of spatial grid points and only two interpolations at each point are required.

3. The diagnostic of inhomogeneous plasmas

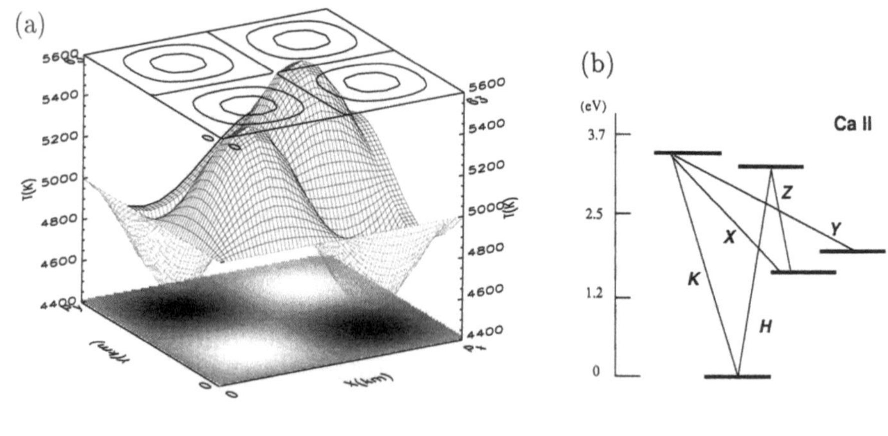

Figure 1. (a) Horizontal variation of the atmospheric model temperature.
(b)Energy level diagram for a Ca II atomic model with 5 bound levels. The line transitions (solid lines) are labelled with their usual symbols.

The inhomogeneous scenario assumed is a vertically isothermal atmosphere where we introduce horizontal fluctuations of the temperature (see fig. 1a).

$$T(x,y,z) = T_c + \Delta T \sin\left(\frac{2\pi x}{P_x}\right) \sin\left(\frac{2\pi y}{P_y}\right)$$

with $T_c = 5000\,\mathrm{K}$, $\Delta T = 500\,\mathrm{K}$ and P_x, P_y the horizontal periods. By changing the horizontal periodicity of the thermal fluctuation we can increase the degree of horizontal inhomogeneity. We assume the total hydrogen number density to be exponentially stratified in depth (horizontally homogeneous) with height scale $H = 100\,\mathrm{km}$.

With respect to the spectral lines simulated we have chosen a Ca II atomic model with the first five levels (fig 1b). For more details about the atomic model see Auer, Fabiani Bendicho and Trujillo Bueno (1995).

The schematic atmospheric model described will be enough to show us the horizontal transfer effects which occur when coupled atomic transitions are treated simultaneously. By using the 3D code described in the previous section, we compute the spatial dependent excitation state of the Ca II atoms consistent with the radiative field. We can then simulate the non-LTE line profiles emerging from the atmosphere.

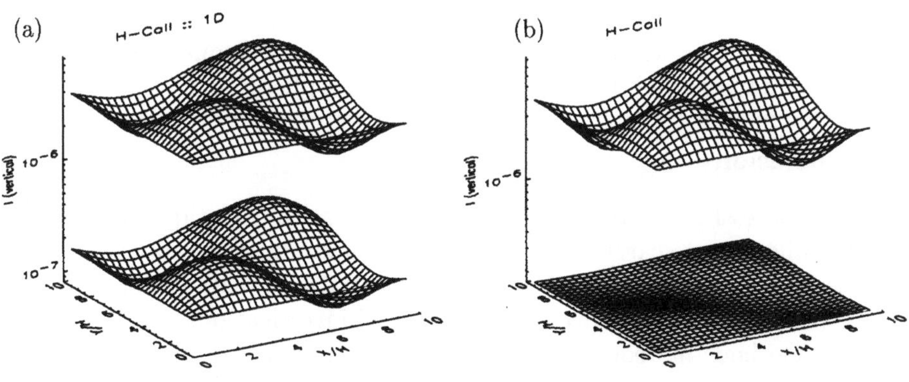

Figure 2. The vertically emergent intensity at the core (lower I) and continuum (higher I) frequency in the K line: (a) 1D and (b) 3D calculations with $P_x = P_y$ =1000 km.

In Fig. 2 we show the vertically emergent intensity at the line core frequency (upper part) and at the continuum freq. (lower part) of the CaII H line. When using the 1D approximation at each horizontal point, the atmospherical thermal fluctuation can be seen at all the line frequencies (Fig.a), but in the more realistic 3D calculations (Fig.b) the fluctuation disappears at the center of the line. If we try to explain this 3D emergent profiles

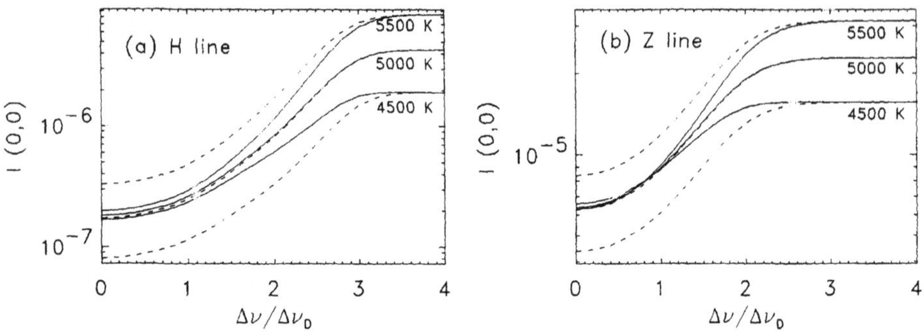

Figure 3. Vertically emergent profiles at the hottest, coolest and medium points. Dashed lines correspond to 1D calculations and solid lines to 3D calculations.

with plane-parallel calculations, we could wrongly conclude that the thermal fluctuation was only present in the lower layers of the atmosphere, but the top temperature were horizontally homogeneous. The results could be correctly interpreted only at the light of multidimensional radiative transfer effects.

The same behaviour will be present in the rest of the CaII lines and the *smoothing* in the core of the line will be stronger if we decrease the horizontal period of the inhomogeneity. For the infared triplet (see fig. 3) we can even obtain an intensity inversion at the core frequencies, so that we will see darker the hottest points of the atmosphere.

4. Conclusions

We have successfully combined powerful mathematical techniques in a manner that allows us to carry out radiative transfer studies for complex atomic models in inhomogeneous plasmas in all the spatial directions (3D). With this new code we can perform a wide range of investigations.

This method, moreover, has been already extended to the simulation of the four Stokes parameters (Fabiani Bendicho and Trujillo Bueno, 1998) and it will open our multidimensional investigations to the diagnostic of magnetic fields.

References

Trujillo Bueno, J., Fabiani Bendicho, P., 1995, ApJ, 455, 646
Fabiani Bendicho, P., Trujillo Bueno, J., Auer, L. H., 1997), A&A, 324, 161
Fabiani Bendicho, P., Trujillo Bueno, J., 1999, (in preparation)
Auer, L. H., Fabiani Bendicho, P., Trujillo Bueno, J., 1994, A&A, 292, 599
Kunasz, P., Auer, L./,H., 1988, Journal of Quant. Spectrosc. and Rad. Transfer, 39, 67
Auer, L. H., Paletou, F., 1994, A&A, 285, 675

WIND AND DISK DIAGNOSTICS IN PMS STARS. A CRITICAL ANALYSIS

E. ORTIZ AND C. EIROA
Dpto. Física Teórica, Facultad de Ciencias,
Universidad Autónoma de Madrid, 28049-Madrid, Spain

Abstract.

The relationships among traditional wind and disk diagnostics – Hα and [OI]λ6300 lines and IR luminosity excesses, respectively – and star parameters are critically analysed. The total sample includes 109 PMS stars – 20 Weak-line T Tauri (WTTS), 45 Classical T Tauri (CTTS) and 44 HAeBe stars–. Our results suggest that Hα is neither a wind nor an accretion tracer. Hα and [OI] emissions seem to correlate very well with the photospheric luminosity and not with $\Delta L_{IR}/L_{ph}$, a parameter related to the origin of the IR excesses.

1. Introduction

During the process of star formation mass accretion and ejection coexist. The origin of the powerful winds in the pre-main sequence phase of stars is one of the most controversial issues. Two main kind of models have been proposed: one suggesting that the wind is driven by the disk through the accretion process and a second one proposing that the direct wind source is the star. The observational supports of each models depend strongly on the goodness of the parameters used to trace each process. However, this aspect has not been rigorously analysed.

IR luminosity excess (ΔL_{IR}) is one of the most commonly disk diagnostic used (Cabrit et al. 1990; Corcoran & Ray 1998). Circumstellar disks can emit by two different processes: reprocessing and accretion. The IR luminosity produced by reprocessing is limited up to 50% of the stellar luminosity (Kenyon & Hartmann, 1987). When the IR luminosity excess is greater extra processes are required, like accretion (e.g. Kenyon & Hart-

Astrophysics and Space Science is the original source of publication of his article. It is recommended that this article is cited as: *Astrophysics and Space Science* **263**: 227–230, 1999.

mann 1987) or spherical envelopes (e.g. Adams et al. 1987). Nevertheless, it is not possible to isolate each contribution to the observed excesses.

The Hα line has been used as wind tracer for years and several models have been proposed to explain the observed profiles (e.g. Calvet et al. 1992; Mitskevich et al. 1993). Recently, models have appeared which try to explain this emission in a context of accretion (e.g. Hartmann et al. 1994). Current wind models can only reproduce normal P Cygni profiles and double-peaked ones with the fainter peak bluewards to primary one – i.e. *type B* (Reipurth et al. 1996) –. By contrast, accretion models only reproduce inverse P Cygni profiles and double-peaked ones with the fainter peak redwards (*type R*). This is an essential difference between both kinds of models. Another wind tracer often employed is the forbidden [OI]λ6300 line (Cabrit et al. 1990; Corcoran & Ray 1998).

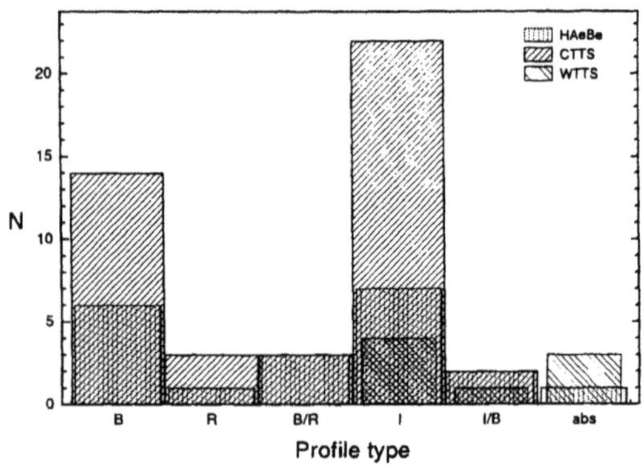

Figure 1. Histogram of Hα profile types for each group of PMS stars, according to the position of the secondary peak (Reipurth et al. 1996). *I* refers to single symmetric profiles, *B* to normal P Cygni profiles or double-peaked ones with the fainter peak bluewards to primary one, *R* to inverse P Cygni profiles or double-peaked ones with the fainter peak redwards and *abs* to absorption profile. Stars exhibiting profile variability are indicated as follow: I/B if they changed from type I to type B, B/R if changed from type B to R.

2. Hα emission

Figure 1 shows the distribution of Hα profile types in the 70 PMS stars of our sample exhibiting this line in emission (see Fernández et al., 1995 for whole details of the observations). The profile classification was carried out according to Reipurth et al. (1996).

Simple symmetric profile (type I) – not accurately reproduced by either wind nor accretion models – is the most common in all groups of PMS stars. This result does not seem to be seriously influenced by the resolution after the analysis of other observational studies (Finkenzeller & Mundt 1984; Reipurth et al. 1996). Remarkably, 29% of the stars have shown a type B profile, characteristic of wind models, while only 6% presented a type R profile, only reproduced by accretion models. Furthermore, 9% of the stars have exhibited variability from type B to R, which might be explained if Hα is formed by both mechanisms – wind and accretion – and the strength of each one is variable.

From these results our conclusion is that Hα does not seem to be an unambiguous diagnostic for wind or accretion.

3. Relationships between lines, IR excesses and L_{ph}

Photospheric luminosity (L_{ph}), IR luminosity excesses and line luminosities ($L_{H\alpha}$, $L_{[OI]}$) were estimated from photometric observations carried out in optical and IR ranges and those available in the literature. Whole details of the method can be found in Ortiz (1998).

A correlation between $L_{H\alpha}$ and IR excesses was observed either in near and mid IR (ΔL_{mIR}) and in far IR wavelength ranges (ΔL_{fIR}). This correlation had previously been reported by other authors (Cabrit et al. 1990 for T Tauri stars, and Corcoran & Ray 1998 for HAeBe stars). Nevertheless, due to the different processes involved in the origin of all these parameters – mainly accretion in ΔL_{mIR}, reprocessing in ΔL_{fIR} and wind and accretion in Hα –, it cannot be stated from this result alone that the line is related to accretion. Furthermore, in figure 2a a clear correlation between $L_{H\alpha}$ and L_{ph} is seen, extending from CTTS to HAeBe stars. WTTS appear segregated, probably reflecting the chromospheric origin of Hα in these stars, different from CTTS and HAeBe. The existence of this relationship contradicts previous studies (Cabrit et al. 1990 for T Tauri, and Corcoran & Ray 1998 for HAeBe stars), but it is in agreement with results from Böhm & Catala (1994) and Nisini et al. (1995) for HAeBe. Our results suggest that, regardless the origin of Hα, the total line emission is directly related to the star. In fact, L_{IR} excesses also scale with L_{ph}. This might be caused by an obvious relation of the reprocessing contribution to stellar luminosity, but also by a likely dependence of accretion to this latter (Ortiz, 1998).

In order to avoid the relation of $L_{H\alpha}$ and L_{IR} excesses with L_{ph} and to analyse whether any direct dependence between $L_{H\alpha}$ and IR luminosity excess exists, the parameter $\Delta L_{IR}/L_{ph}$ can be employed. As noted above, when $\Delta L_{IR}/L_{ph} > 0.5$, accretion or spherical envelopes are required to explain IR excesses. In figure 2b no tendency to exhibit larger $W_{H\alpha}$ when larger $\Delta L_{IR}/L_{ph}$ is observed. If such a relation between Hα and accretion actually exists, it might be masked by the different processes involved in the line formation and/or by the contribution of the spherical envelope to the IR excess, mainly in HAeBe.

Analogous results are obtained from the analysis of the [OI] line (Ortiz, 1998).

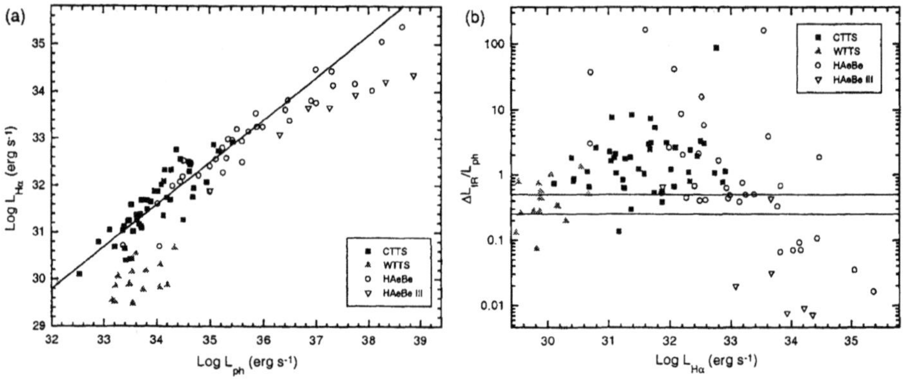

Figure 2. Plots of (a) $L_{H\alpha}$ versus L_{ph} and (b) $\Delta L_{IR}/L_{ph}$ parameter versus $W_{H\alpha}$ for stars in our sample, T Tauri stars from Cabrit et al. (1990) and HAeBe from Corcoran & Ray (1998). The straight line in (a) corresponds to least squares best fit. Symbols used for different groups of PMS stars are indicated in the key.

4. Conclusions

Observational supports to the disk driven model of winds are not so conclusive as it has been stated up to now. Wind and disk diagnostics normally employed are not appropriate. Hα line is not a good diagnostic parameter for neither wind nor accretion. On the other hand, Hα and [OI] emissions are directly related to the star in CTTS and HAeBe. Therefore, the correlations between L_{IR} excesses and lines might be due to a relation of all of these parameters with the star. This hypothesis is reinforced by the existence of a correlation between L_{IR} excesses and L_{ph} and by the fact that no dependence of emission lines to $\Delta L_{IR}/L_{ph}$ is observed. Nevertheless, a relationship between wind and accretion may not be dismissed, but proper diagnostics of both phenomena have to be searched.

References

Adams, F.C., Lada, C.J., Shu, F.H., 1990, ApJ, 312, 788
Böhm, T., Catala, C., 1994, A&A, 290, 167
Cabrit, S., Edwards, S., Strom, S.E., Strom, K.M., 1990, ApJ, 354, 687
Calvet, N., Hartmann, L., Hewett, R., 1992, ApJ, 386, 229
Corcoran, M., Ray, T.P., 1998, A&A, 331, 147
Fernández, M., Ortiz, E., Eiroa, C., Miranda, L.F., 1995, A&AS, 114, 439
Finkenzeller, U., Mundt, R., 1984, A&AS, 109, 141
Hartmann, L., Hewett, R., Calvet, N., 1994, ApJ, 426, 669
Kenyon, S.J., Hartmann, L., 1987, ApJ, 323, 714
Mitskevich, A.S., Natta, A., Grinin, V.P., 1993, ApJ, 404, 751
Nisini, B., Milillo, A., Saraceno, P., Vitali, F., 1995, A&A, 302, 169
Ortiz, E., 1998, Ph.D. Thesis, Universidad Autónoma de Madrid, Spain
Reipurth, B., Pedrosa, A., Lago, M.T.V.T., 1996, A&AS, 120, 229

CHROMOSPHERIC ACTIVITY OF WEAK-LINED T TAURI STARS

D. MONTES
Dpt. de Astrofísica. Facultad de C. Físicas.
Universidad Complutense de Madrid. E-28040 Madrid, Spain

Abstract. A summary of our ongoing high and medium resolution optical observational program, devoted to the study of chromospheric activity of pre-main sequence stars, is reported here. The study is centered in bona-fide weak-lined T Tauri stars (WTTS) in Taurus with known rotational period, and in WTTS recently discovered by the ROSAT All-Sky Survey. It is our purpose to quantify the phenomenology of the chromospheric activity of each star determining stellar surface fluxes in the more important chromospheric activity indicators (Ca II H & K, Hβ, Hα, Ca II IRT) as well as to obtain the Li I abundance, a better determination of the stellar parameters, spectral type, and possible binarity. By using these chromospheric excess emissions we will study the flux-flux and rotation-activity relationships in order to get insights into the mechanisms which drive solar-like stellar activity in this kind of low mass pre-main sequence stars.

1. Introduction

The weak-lined T Tauri stars (WTTS) are low mass pre-main sequence stars (PMS) with Hα equivalent widths ≤ 10 Å in which no signs of accretion are observed. The emission spectrum of these stars is not affected by the complications of star-disk interaction which often masks the underlying absorption lines as well as extincts the stellar light in classical T Tauri stars (CTTS). The WTTS are thus ideal targets to study the behavior of surface activity in the PMS stage of the stellar evolution. The chromospheric contribution to the different spectroscopic lines may be determined by means of the spectral subtraction technique (Montes et al. 1995a, b, c). While there are a large number of studies in UV, X-ray and radio wavelengths,

Astrophysics and Space Science is the original source of publication of his article. It is recommended that this article is cited as: *Astrophysics and Space Science* **263**: 231–234, 1999.

little research has been directed towards the study of the chromospheric activity using optical observations.

We have started a high and medium resolution optical observational program devoted to the study of chromospheric activity of late-type pre-main sequence stars. We include in our study both bona-fide WTTS in Taurus with known rotational period, and WTTS recently discovered by the ROSAT All-Sky Survey, which have been very little studied until now. The active-chromosphere contribution to the different spectroscopic features have been obtained by application of the spectral subtraction technique. The synthesized spectrum has been constructed with reference stars taken during the same observing run that the WTTS or from our libraries of late-type stars (Montes 1998).

2. Observations

High and medium optical spectroscopic observations of different kind of WTTS, covering most of the spectral lines widely used as chromospheric activity indicators, have been obtained during several observing runs using different telescopes and instrumental configurations: the 2.5 m Isaac Newton Telescope (INT) at the Observatorio del Roque de Los Muchachos (La Palma, Spain) using the Intermediate Dispersion Spectrograph (IDS); the 4.2 m William Herschel Telescope (WHT) (La Palma, Spain) using the Utrecht Echelle Spectrograph (UES); the 2.1 m telescope at McDonald Observatory (Texas, USA) using the Sandiford Cassegrain Echelle Spectrograph (SCES); and the 0.9-m and 2.1-m telescopes of the Kitt Peak National Observatory (KPNO) using the Penn State Fiber Optic Echelle (FOE). The spectral resolution achieved ranges from 0.09 to 1.0 Å.

3. Results

In the following we summarized our different observational programs and results obtained.

WTTS with known rotational period

Simultaneous Hα and Ca II H & K INT/IDS spectroscopic observations of a sample of 20 bona-fide WTTS in Taurus-Auriga molecular clouds, with photometric rotational periods well determined, have been taken. Our purpose is to quantify the phenomenology of activity in each star, study the flux-flux and rotation-activity relationships, and to compare these results with those obtained in active late-type stars and, in particular, in binary RS CVn systems (Montes et al. 1998b).

Preliminary results of these observations, and the particular case of V773 Tau and V410 Tau and LkCa7 were presented by Poncet et al. (1998).

The result for all the sample can be found in Montes & Miranda (1999). In this sample of twenty WTTS, eighteen stars present Hα emission above the continuum and two (HD283572 and HP Tau/G2) show a filled-in absorption line. LkCa15 and UX Tau A exhibit a complex Hα emission profile consisting of several absorption and emission components. All the stars in the sample, except for Tap 35, HD283572 and HP Tau/G2, present Ca II H & K emission lines above the continuum and some of them also show emission in higher Balmer lines (Hδ to Hθ). The subtracted Hα profile of the more active stars reveals the existence of broad wings suggesting the presence of a broad component, in addition to a narrow emission component. Similar broad wings have also been found in chromospherically active binary systems (Montes et al. 1998a) and could be interpreted as arising from microflares.

The double-lined spectroscopic binary V773 Tau

V773 Tau is a WTTS recently discovered by Welty (1995) to be a double-lined spectroscopic binary (K2V/K5V) with an orbital period of 51.075 days. We are analysing in detail (Montes, Ramsey & Welty 1999) the chromospheric activity of this system with spectra taken at different epochs, obtaining a very good orbital phase coverage. We have FOE spectra from 1987 to 1995; UES spectra in 1993 and 1994; and new spectra taken in 1998 with the SCES. A broad and variable Hα emission above the continuum is observed in the different epochs. Sometimes a double-picked feature is observed. The Gaussian fits to the observed profile indicate the presence of a broad component, that can be attributed to microflares, and two narrow components, that correspond to the chromospheric emission of the two components of the binary system. A flare in the 1993 observations is also detected as an increase in the chromospheric emissions and the presence of the He I D_3 line in emission.

ROSAT discovered WTTS

A large number of WTTS have been recently discovered by the ROSAT All-Sky Survey. Many of them have been found to be far away from the star formation clouds and whether these stars are really WTTS, or post TTS, or even young main sequence stars is a matter of ongoing debate. However, we have studied only the stars for which very recent studies clearly confirmed their PMS nature. The sample have been selected from the list of new WTTS stars in Taurus of Wichmann et al. (1996) and the lists of late-type stars south of Taurus of Neuhäuser et al. (1995) and Magazzù et al. (1997). From these lists we selected the stars that the spectroscopic studies of Magazzù et al. (1997) and Neuhäuser et al. (1997) clearly identified as WTTS from their greater Li abundance as compared with the Pleiades of the same spectral type.

Preliminary results of our SCES 1998 high resolution echelle spectroscopic observations of six of these WTTS, located in and around the Taurus-Auriga molecular clouds, can be found in Montes & Ramsey (1998). We have analysed, using the spectral subtraction technique, the Hα and Ca II IRT lines of RX J0312.8-0414NW, SE; RX J0333.1+1036; RX J0348.5+0832; RX J0512.0+1020; and RX J0444.9+2717. A broad and variable double-picked Hα emission is observed in RXJ0444.9+2717. Emission above the continuum in Hα and Ca II IRT lines is detected in RXJ0333.1+1036 and a filling-in of these lines is present in the rest of the stars. Our spectral type and Li I EW determinations confirm the PMS nature of these objects.

Additional information about WTTS can be found in the following web page I maintain:

`http://www.ucm.es/info/Astrof/wtts.html`

Acknowledgments We have made use of the La Palma ING data archive for retrieving some spectra. This work was supported by the Universidad Complutense de Madrid, the Spanish Dirección General de Enseñanza Superior e Investigación Científica (DGESIC) under grant PB97-0259, and by National Science Foundation (NSF) grant AST 92-18008.

References

Magazzù A., Martín E.L., Sterzik M.F., et al., 1997, A&AS 124, 449
Montes D., 1998, Ap&SS (this volume)
Montes D., Fernández-Figueroa M.J., De Castro E., Cornide M., 1995a, A&A 294, 165
Montes D., Fernández-Figueroa M.J., De Castro E., Cornide M., 1995b, A&AS 109, 135
Montes D., De Castro E., Fernández-Figueroa M.J., Cornide M., 1995c, A&AS 114, 287
Montes D., Fernández-Figueroa M.J., De Castro E., Cornide M., Poncet A., Sanz-Forcada J., 1998a, in *The Tenth Cambridge Workshop on Cool Stars, Stellar Systems, and the Sun*, eds. R.A. Donahue & J.A. Bookbinder, San Francisco, ASP Conf. Ser. 154, CD-1516
Montes D., Sanz-Forcada J., Fernández-Figueroa M.J., De Castro, Cornide M., Ramsey L.W., 1998b, Ap&SS (this volume)
Montes D., & Miranda L.F., 1999, A&AS (in preparation)
Montes D., Ramsey L.W., 1998, in *Solar and Stellar Activity: Similarities and Differences*, eds. C.J. Butler J.G. Doyle, ASP Conf. Ser., in press
Montes D., Ramsey L.W., Welty A.D., 1999, AJ, (in preparation)
Neuhäuser R., Sterzik M.F., Torres G., Martín E.L., 1995, A&A 299, L13
Neuhäuser R., Torres G., Sterzik M. F., Randich S., 1997, A&A 325, 647
Poncet A., Montes D., Fernández-Figueroa M.J., Miranda L.F., 1998, in *Cool Stars, Stellar Systems, and the Sun*, 10th Cambridge Workshop, eds. R.A. Donahue & J.A. Bookbinder, ASP Conf. Ser. 154, CD-1772
Welty A.D., 1995, AJ 110, 776
Wichmann R., Krautter J., Schmitt J.H.M.M., et al., 1996, A&A 312, 439

THE AGE OF BETA PIC TYPE STARS:

VEGA, FOMALHAUT, β PIC AND HR4796

D. BARRADO Y NAVASCUÉS
Max-Planck Institut für Astronomie
Königstuhl 17, D-69117 Heidelberg, Germany

Abstract.
We have established that several prominent β Pic stars are associated to moving groups. Based on this fact, we have estimated their ages. Our estimates are 200±100 Myr for Castor (which contains Vega and Fomalhaut), the same age for β Pic, and 8±2 Myr for HR4796 A.

1. The β Pic phenomenon

The β Pic type stars show IR excesses, which are associated the the presence of circumstellar dusty disks. The prototypes for this class of stars are β Pic, Vega and Fomalhaut. As a result of their proximity and brightness, these three stars have been studied in great detail, and estimates of the mass, radial distribution and structure, dust-grain properties of their disks have been made (e.g., Backman & Paresce 1993). It is generally believed that these circumstellar dusty disks are either the direct descendents of T Tauri disks or the secondary products of the planet formation process. Knowledge of the ages of β Pic stars is therefore one of the keys to understand the formation and evolution of their disks. However, there is only a handful of estimates of the ages of these systems, and no member of this class has yet been detected with certainty in an open cluster (although see Backman, Angelova & Stauffer 1998). All three prototypes are A stars, and estimates from their post-ZAMS evolution give approximate ages of 100, 200, and 400 Myr for β Pic, Fomalhaut and Vega, respectively (Backman and Paresce 1993). Recently, Barrado y Navascués et al. (1997a) estimated an age of 200±100 Myr for Fomalhaut, based on a number of properties (X-ray emission, rotation, lithium abundance, isochrones) of its physical companion GL879. Another relevant star belonging to this class, HR4796A, has been studied in a similar way by Stauffer et al. (1995). These authors

Astrophysics and Space Science is the original source of publication of his article. It is recommended that this article is cited as: *Astrophysics and Space Science* **263**: 235–238, 1999.

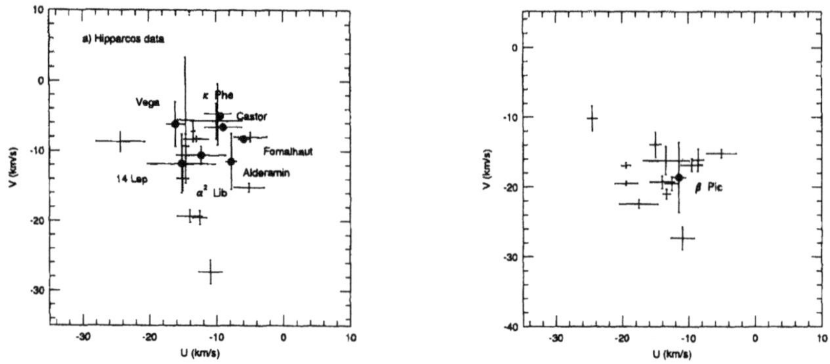

Figure 1. Components of the Galactic velocity of proposed components of the Castor (**a**) and β Pic (**b**) moving groups.

derived an age of 8±2 Myr. Lately, Holland et al. (1998) have imaged the thermal emission from the disks of the three prototypes at submillimeter wavelengths. They concluded that any Earth–like planet must have already formed. Moreover, the central holes found by Holland et al. (1998) in the orbit of Fomalhaut and by Jayawardhana et al. (1998) and Koerner et al. (1998) in the case of HR4796A, support the idea of the early formation of planets.

TABLE 1. Ages for different moving groups.

Group	Age	Stars of A & F spectral types
Castor	200±100	Castro, Vega, Fomalhaut, GL351A, GL255A, GL521.2A
"	"	κ Phe, 14 Lep, Alderamin, α^2 Lib, α^1 Lib
β Pic	200±100	β Pic
HR4796	8±2	HR4696 A

2. What is a kinematic group?

A groups of stars are considered associated if they fulfill the Soderblom & Mayor (1993) definition of a stellar kinematic group. The most restrictive requirement is that of having identical V component of the Galactic velocity, which is the projection of the velocity in the direction of the Galactic

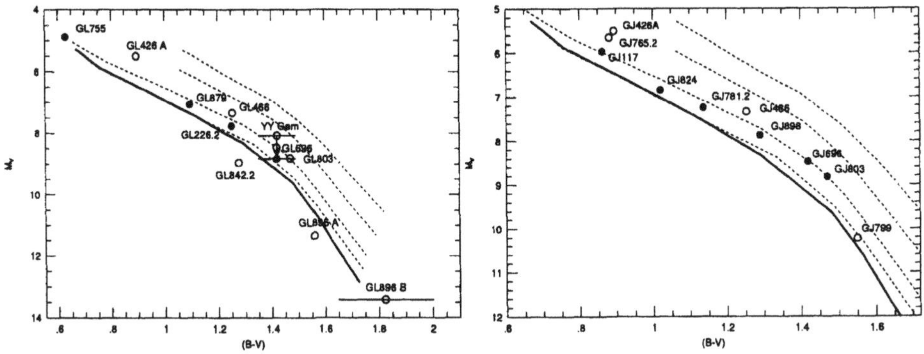

Figure 2. **a and b** Mv against (B–V) for proposed components of the Castor and β Pic moving groups. We show the late spectral type stars. Final members are shown as solid symbols, whereas the rejected members appear as empty circles. Isochrones from D'Antona & Mazzitelli (1994), with 3, 10, 35, 70 Myr and ZAMS, are included.

rotation. However, even if a group of stars satisfies these criteria, exists the possibility of having spurious members. However, when combined with additional data, such as lithium abundances, rotation, stellar activity, CMD, etc, they provide strong indications of common origin of, at least, the majority of the sample. However, the membership of individual stars cannot be known without doubt. In our case, both the Castor and the β Pic groups are likely to be real, as Figure 1 illustrates.

3. The ages of β Pic stars

We have recomputed the kinematic properties of several of dozens nearby stars, to try to estimate the ages of the prototypes. We have tried to verify if Castor (α Gem) and β Pic indeed have each of them a cohort of stars sharing the same space motion and age. We used kinematics, the location of the stars in Color-Magnitude Diagrams (Figure 2a,b), their activity (Figure 3a) and their lithium abundances (Figure 3b), to establish that these moving groups seem to be real and to reject several stars which were believed to be part of these groups. Additional details can be found in Barrado y Navascués (1998) and Barrado y Navascués et al. (1999).

The Castor moving group contains several A spectral type stars. Among them, Vega and Fomalhaut, two of the prototypes of this type stars. If these stars are coeval, their different levels of IR emission suggest that the time scale for the formation of planets is not universal. Furthermore, β Pic has a similar age. Due to large values of the estimated ages, these IR excesses would appear as a consequence of collisions and sublimation of larger bodies

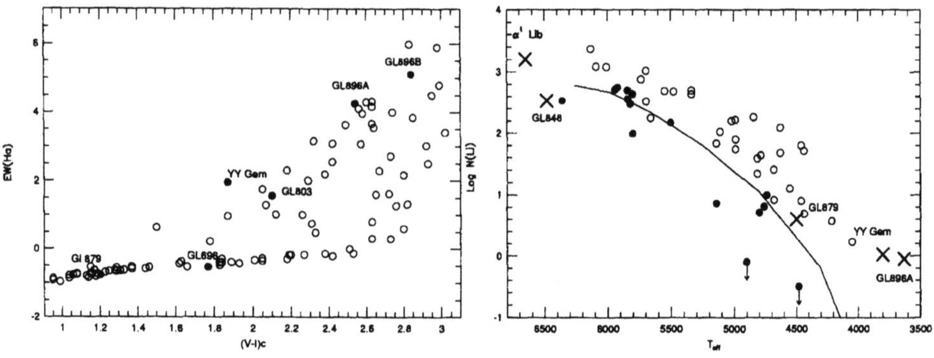

Figure 3. **a** Hα emission of proposed members of the Castor Moving Group (crosses) compared with those from Hyades members (open circles). **b** Lithium abundances against effective temperatures. Crosses represent the proposed members of the Castor Moving Group. M34 (200 Myr, open circles) and UMa Group (300 Myr, solid circles) stars are also shown.

and they would not arise from protoplanetary structures.

Since these associations include several late spectral type stars, we used their properties to estimate their ages. Table 1 summarizes the results. Our estimate for those ages are 200$\pm$100 Myr for the Castor and β Pic Moving groups and 8$\pm$2 Myr for HR4796.

References

Backman, D.E., Paresce, F., 1993, in *Protostars and Planets III*, University of Arizona Press, Tucson, USA. Eds. E.H. Levy and J. Lunine

Backman, D.E., Angelova, A.M., Stauffer, J.R., 1998, priv. comm.

Barrado y Navascués. D., 1998, A&A 339, 831

Barrado y Navascués. D., Stauffer, J.R., 1996, A&A 310, 879

Barrado y Navascués. D., Stauffer, J.R., Hartmann, L., Balachandran, S., 1997, ApJ 475, 313

Barrado y Navascués. D., et al., 1999, A&A, in prep.

D'Antona, F., Mazzitelli, I., 1994, ApJS 90, 467

Holland, W.S., et. al. 1998, Nature 392, 788

Jayawardhana, R., Fisher, S., Hartmann, L., Telesco, C., Pina, R., Fazio, G., 1998, ApJ 503, L79

Koerner, D.W., Ressler, M.E., Werner, M.W., Backman, D.E., 1998, ApJ 503, L83

Soderblom, D.R., Mayor, M., 1993, AJ 105, 226

Stauffer, J.R., Hartmann, L., Barrado y Navascués, D., 1995, ApJ 454, 910

BROWN DWARFS AND VERY LOW MASS STARS: TOWARDS A NEW AGE SCALE FOR YOUNG OPEN CLUSTERS

D. BARRADO Y NAVASCUÉS
Max-Planck Institut für Astronomie
Königstuhl 17, D-69117 Heidelberg, Germany

J.R. STAUFFER
Harvard-Smithsonian Center for Astrophysics
60 Garden St. Cambridge, MA02138, USA

AND

J. BOUVIER
Observatoire de Grenoble
F-38041 Grenoble, Cedex 9, France

1. Introduction

Open clusters are one of the main available laboratories in Astronomy. Among other things, the stellar evolutionary age scale relies primarily on them, so it is extremely important to estimate accurately the ages of different associations. Until now, the dating of clusters was based on either the Upper Main Sequence (MS), using stars evolving off it or already well beyond the turn–off point, or isochrone fitting. These techniques are model dependent and provide different ages for the same cluster. Very Low Mass Stars (VLM) and Brown Dwarfs (BDs) offer an alternative: It is well known that the lithium abundance is depleted with the age in stars (see the review by Balachandran 1994). This destruction depends also on mass, so dM stars deplete their lithium even before they arrive to the MS. However, the internal part of VLM/BDs is not hot enough to destroy lithium at this stage. In the lower part of the cluster MS, the border between members having no lithium and those showing it in their atmosphere is called the lithium depletion boundary (LDB). As a cluster gets older, the LDB moves toward less massive objects. Therefore, the detection of lithium in the brightest (e.g. bluer, hotter and more massive than others) BDs of a cluster puts a very strong limit on its age (D'Antona & Mazzitelli 1994; Martín & Montes

Astrophysics and Space Science is the original source of publication of his article. It is recommended that this article is cited as: *Astrophysics and Space Science* **263**: 239–242, 1999.

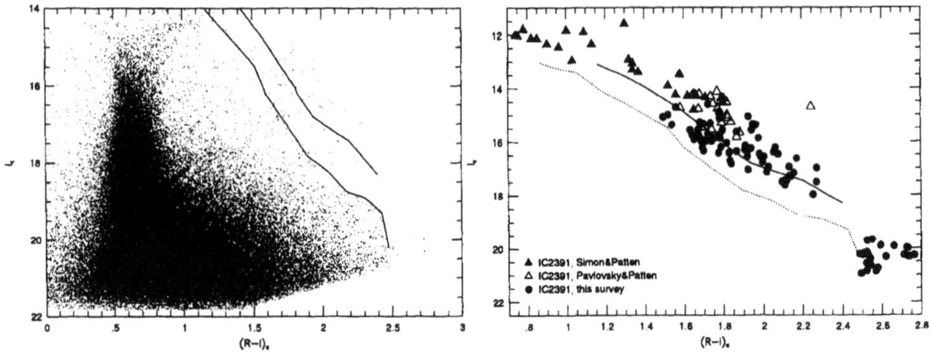

Figure 1. **a** Color-Magnitude Diagram for the area around IC 2391. **b** Our selection of possible members of IC 2391.

1997; Ventura et al. 1998). In this paper, we show how this technique works and apply it to several young open clusters.

2. Identification of candidate members: Optical and IR surveys

The initial step in order to identify new very low mass (VLM) stars and BDs has been to obtain optical photometry on the area around the cluster. In order to cover a significat fraction of a particular cluster in a reasonable amount of time, we have used in most of the cases detectors with several CCDs. Our targets have been:

i) The Pleiades.- We have used the CFHT (with the MOSAIC camera) and 48" telescope at MHO. We surveyed 2.5 and 1 sq. deg., respectively, discovering 17 and 6 BDs candidates in each survey (Bouvier et al. 1998; Stauffer et al. 1998a).

ii) α Per.- As a preliminary survey, we collected RI photometry with the 48" telescope at MHO. We discovered several candidates. Some of them were the targets of a spectroscopic campaign (next section).

iii) IC 2391.- The CTIO 4m telescope and the BTC camera allowed us to survey 2 sq. deg., detecting ~40 BDs candidates (Barrado y Navascués et al. 1999a). See Fig. 1a.

iv) M35.- CFHT+MOSAIC: This is a very well populated open cluster. Our study included several thousands candidates (stars) and few possible BD (Barrado y Navascués et al. 1999b).

v) NGC 2516.- CTIO 4m telescope + BTC. We covered 0.6 sq. deg., analysis in progress.

Since most of the lists of cluster candidates contain spurious members, it is necessary to select good candidates by obtaining near IR photometry.

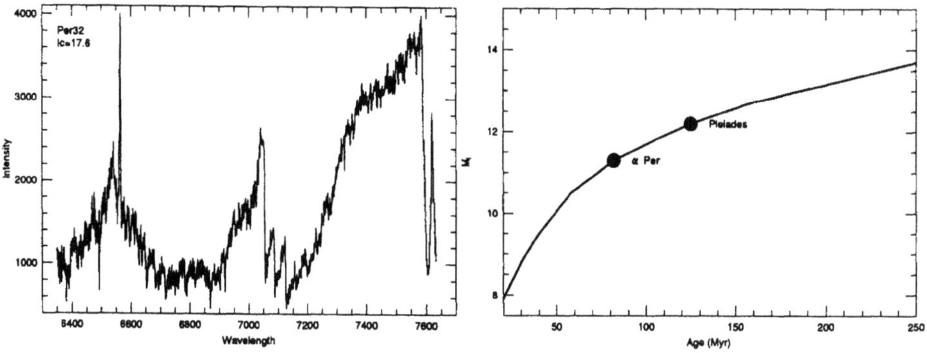

Figure 2. **a** Spectrum of a new member of α Per. **b** M_I versus age. The location of the lithium depletion boundary is shown for the Pleiades and α Per.

We have collected this information using the MHO 48" telescope and the NASA IRTF, in the case of the Pleiades. JHK photometry on the area around IC 2391 was obtained by the 2MASS project. All this information allowed us to construct robust lists of possible members. Figure 1b displays a Color-Magnitude diagram of IC 2391, and a comparison with previous surveys. A 30 Myr isochrone (solid line) and a Zero Age MS (dashed line) are included.

3. Spectroscopy: membership and the new age

Since these objects are too faint to have been detected with photographic plates in the past, most of them do not have measurements of their proper motions. Therefore, the final confirmation of membership comes from the spectroscopy. A intermediate resolution spectra (R=2000–5000) are good enough to obtain rough radial velocities (hence, establishing whether the target is a member). On the other hand, a detection of LiI 6708 Å reveals the BD nature of the candidate and also confirms the membership (Rebolo 1991). Using the Keck II telescope and the LRIS spectrograph we have collected spectra of VLM and BD candidates in the Pleiades and α Per open clusters. Figure 2a shows the spectrum of PER 32, a BD candidate discovered in the α Per region with the 48" MHO. The intense Hα line in emission and its radial velocity indicates that it is a real member of the cluster. However, we have not detected lithium, which indicates that this is a VLM star close to the LDB.

Figure 2b illustrates how ages are estimated. Following Stauffer et al. (1998b), the faintest member without lithium and the brightest member with it, bracket the magnitude of the LDB. We used the Chabrier & Baraffe

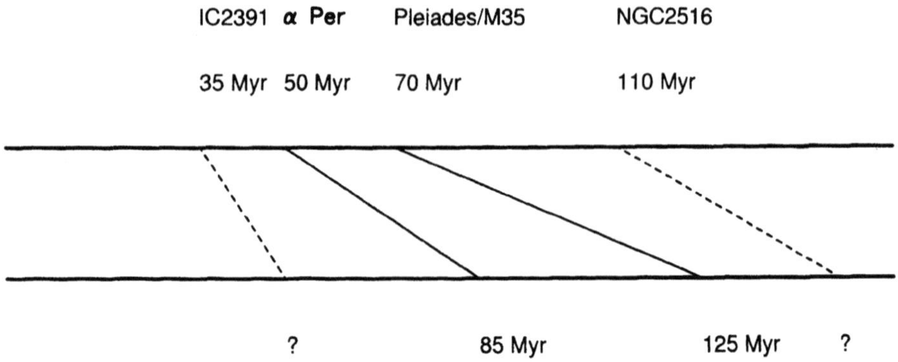

Figure 3. Comparison between the standard age scale (top) and the values estimated from the lithium depletion boundary for BD (bottom).

(1998) latest evolutionary model to covert these magnitudes, different for the Pleiades and α Per, into their corresponding ages. If the sampling is well done, the accuracy is better than 10%.

Figure 3 shows the old standard ages for the 5 clusters included in our study, together the new values for the Pleiades and α Per, which are significantly older than the values previously accepted. Our aim is to improve these values and to get new ones for the other clusters in the near future, in order to verify this interesting possibility. This would have a very important effect on several fields, including the time scale of properties of low mass stars (lithium abundance, rotation, stellar activity), pre-main sequence stars, and internal structure and evolutionary models.

References

Balachandran, S., 1994, in *8th Cambridge Workshop on Cool Stars, Stellar Systems, and the Sun*, ed. J.-P. Caillault, APS Conf. Series, 64, 234
Barrado y Navascués, D., et al. 1999a, ApJ, in preparation
Barrado y Navascués, D., et al. 1999b, this volumen
Bouvier, J., Stauffer, J.R., Martín, E.L., Barrado y Navascués, D., Wallace, B., Béjar, V.J.S., 1998, A&A, 336, 490
Chabrier, G., Baraffe, I. 1998, priv. comm.
D'Antona, F., Mazzitelli, I., 1994, ApJS 90, 467
Martín, E.L., Montes, D., 1997, A&A 318, 805
Rebolo, R, 1991, in IAU Symp. 145, *Evolution of stars: The Photospheric Abundance Connection*, eds G. Michaud & A. Tutukov, Dordrecht, Kluwer, 85
Stauffer, J.R., Schild, R.,Barrado y Navascués, D., Backman, D.E., Angelova, A., Kirckpatrick, D., Hambly, N., Vanzi, L., 1998a, ApJ 504, 805
Stauffer, J.R., Schultz, G., Kirkpatrick, J.D., 1998b, ApJ 499, L199
Ventura, P., Zeppieri, A., Mazzitelli, I., D'Antona, F., 1998, A&A 331, 1011

ON THE IONIZATION STRUCTURE OF CIRCUMSTELLAR DISKS AROUND BE TYPE STARS.

J.M. TORREJÓN AND G. BERNABEU

Depto. de Física, Ingeniería de Sistemas y Teoría de la Señal, Universidad de Alicante, Ap.99, E-03080, Alicante, Spain.

AND

J. FABREGAT

Departamento de Astronomia y Astrofísica, Universidad de Valencia, E-46100, Burjassot, Valencia, Spain.

Abstract. We present some insight on the ionization structure of the circumstellar disks around Be type stars based on the study of Balmer decrements for absorption-corrected emission lines. Owing to this analysis, three different ionization zones in the envelope are proposed: the inner one, completely ionized, of about $\sim 5R_{\star}$, an intermediate recombination zone extending up to $\sim [10-20]R_{\star}$, and an outer extended zone of neutral H.

1. Introduction.

Be objects are stars of spectral type B, and luminosity class III to V, whose spectrum show lines in emission and IR excess, originated in a circumstellar disk-like envelope ejected by the central star. They represent the last major class of variable stars that remain unexplained. It deals not with exotic objects since they can represent up to 30% of the B type stars. Yet, surprisingly, the phenomenon is far from being satisfactorily explained. The existing models are completely *ad hoc*: the mechanism of mass loss and angular momentum transfer needed to create the circumstellar envelope continues to be a mystery. This situation is due chiefly to the lack of observational constraints over the theoretical models.

Astrophysics and Space Science is the original source of publication of his article. It is recommended that this article is cited as: *Astrophysics and Space Science* **263**: 243–246, 1999.

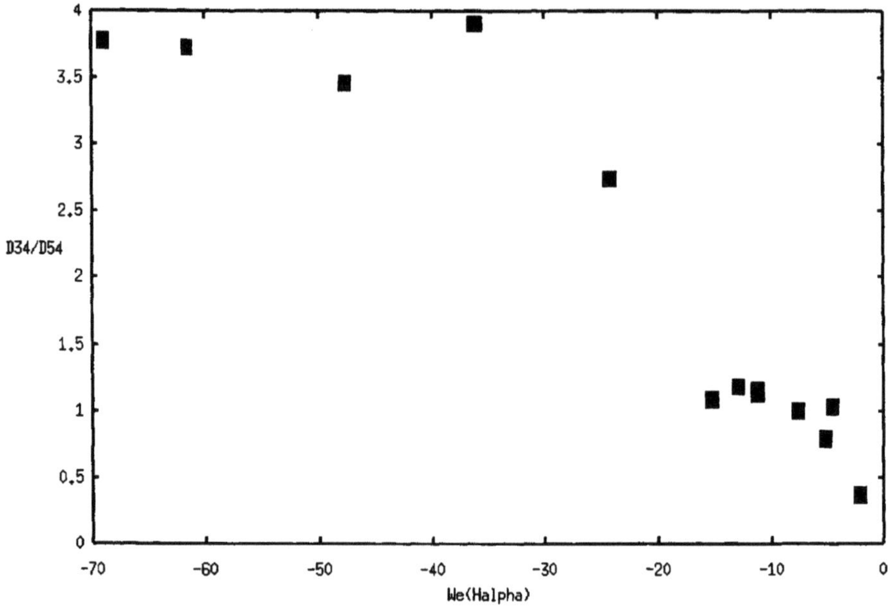

Figure 1. Decrement ratio vs. Hα equivalent width for programme Be stars.

2. The Observations.

The spectroscopy has been taken from Torrejón et al. (1997). In order to correct for the photospheric effects we have applied the calibration of Fabregat & Torrejón (1998). The results are plotted in Fig. 1. We plot the Balmer decrement ratio D_{34}/D_{54} against the size of the emitting envelope as measured by the equivalent width of the Hα line, corrected for photospheric absorption. The decrement ratio is used here as an estimator of the density of free electrons N_e, in such a way that N_e increases downwards (cf. Drake & Ulrich 1980) while the size of the emitting envelope increases leftward.

The observed values for the decrements in our sample are in the ranges $0.35 \leq D_{34} \leq 2.13$ and $0.40 \leq D_{54} \leq 1.00$. They are consistent with a H cloud at $T_e \sim 10^4$K, $N_e \sim 10^{12}$cm^{-3} and ionization rate $R_{1C} \sim 3$s^{-1} (Drake & Ulrich 1980).

As it can be clearly seen, the density of free electrons decreases with the size of the envelope, for relatively small envelopes, while for larger envelopes, i.e. those with $-W(H\alpha) \geq 35$ Å, the density reaches a stationary value.

3. Discussion.

If we assume that the density of free electrons scales with the temperature of the envelope, the behaviour observed in Fig. 1 can be explained if the temperature of the smaller envelopes is higher. This will be due to luminosity conservation $L_{\rm env} = 4\pi R^2_{\rm env}\sigma T^4_{\rm env}$, which can be satisfied by the smaller envelopes if their average temperature increases. Since the envelope gets its energy from the diluted stellar ionizing photons of the central star, the luminosity of the envelope will be a certain fraction of the star luminosity, i.e., $L_\star \sim L_{\rm env} \sim$ *constant*. Then, essentially, the luminosity of the central star will remain constant. This suggests that the major changes in the observables are produced by changes in the circumstellar envelope and not to changes in the physical properties of the central star itself.

By other hand, beyond $-W_e(H\alpha) \sim 35$ Å the density of free electrons reaches a constant value. This can be understood as the maximun reach of Lyman photons able to produce ionizations in the H atoms of the envelope. Beyond this *turnoff* point, the density remains essentially unchanged.

In order to estimate the *turnoff radius* we have applied the empirical relationship of Dachs et al. (1986):

$$W_e(H\alpha) = -1.5\text{Å}[(R_d/R_\star)^2 - 1]$$

with R_d being the radial distance inside the disk. Taking $-W_e(H\alpha) \simeq 35$ Å, the maximum reach of Lyman photons till the recombination zone will be of the order of $R_d \sim 5R_\star$. This is in agreement with the values deduced by Slettebak et al. (1992), namely, $[7{,}19]R_\star$ for Hα, $[5{,}12]R_\star$ for Hβ and $[5{,}11]R_\star$ for Hγ. If we take now $R_\star \simeq 7R_\odot$ as a typical stellar radius for a B type star, we have $R_d \simeq 35R_\odot$ or $R_d \simeq 2.4 \times 10^{12}$cm.

Now, the photospheric Lyman photons will have a reach inside the envelope of the order of:

$$l_{II} \sim (N_e\sigma_{II})^{-1} \sim 10^{24}N_e^{-1}\text{cm}$$

where $\sigma_{II} \sim (e^2/mc^2)^2 \sim 10^{-24}\text{cm}^2$ is the effective cross section for the scattering of photons by free electrons (Thomson scattering). If we now take N_e of the order of $\sim 10^{12}\text{cm}^{-3}$, as has been already deduced (this work, Slettebak et al. 1992, Dachs et al. 1990) then we can see how the inner radius of the disk can be explained as the outer boundary of the completely ionized region.

Beyond this point, things are harder to explain in this way. We would expect the photospheric photons moving freely a mean distance l_{II} from the central star and then become recombined in a layer of thickness:

$$l_I \sim (N_a \sigma_I)^{-1} \sim 10^{17} N_a^{-1} \mathrm{cm}$$

where $\sigma_I \sim r_a^2 \sim 10^{-17}\mathrm{cm}^{-2}$ is the effective radius of the atom and N_a the density of atoms (since in this region, electrons will be primarily bounded). Taking $N_a \sim 10^{12} (= N_e)$ we have $l_I \sim 10^5 << l_{II}$ which is much too small. Of course to assume a constant density through the path of the photon may not be quite realistic. However, trying with a density law of the form $N(r) = N(0)(r/R_\star)^{-n}$ and pushing the index n up to 5 (Dougherty et al. 1994), the deduced outer radius of the recombination zone is still too small to reconcile with the observations. It is certainly difficult to explain the observed size of the emitting region unless, for instance, a steep drop in density occurs beyond certain point. Anyway, ionization and recombination zones will have a maximum reach inside sufficiently large envelopes. Beyond this point, neutral H may exists but would be observable only under moderately high inclination angles, depending on disk geometry.

4. Conclusions.

From a study of Balmer decrements for photosphere-corrected emission lines we find that the inner radius of the circumstellar envelope can be explained as the outer boundary of a completely ionized region, in a Stromgren sphere-like mechanism. The observed radius of the emitting region is difficult to explain in this fashion unless a steep drop in density occurs inside the envelope. Our data also implies the average conservation of luminosity in Be type stars, which pools against drastic changes in the central star.

Homogeneous, wide wavelength coverage data is required to further constraint theoretical models.

References

Dachs, J., Rohe, D., Loose, A.S., 1990, A&A, 238, 227
Dougherty, S.M., Waters, L.B.F.M., Burki, G., et al., 1994, A&A, 290, 609
Drake, S.A, Ulrich, R., 1980, ApJS, 42, 351
Fabregat, J., Torrejón, J.M., 1998, A&A, 332, 643
Slettebak, A., Collins, G.W., Truax, R., 1992, ApJS, 81, 335
Torrejón, J.M., Fabregat, J., Bernabeu, G., Alba, S., 1996, A&AS, 122, 329

THE MOST LUMINOUS STARS IN OUR GALAXY

M.R. VILLAMARIZ AND A. HERRERO
Instituto de Astrofísica de Canarias
E-38200 La Laguna, Tenerife, Spain

Abstract. We present quantitative analyses of 7 galactic O stars of spectral types from O3 to O6. By means of classical hydrostatic and planeparallel models we find the so-called *He* and *Mass Discrepancies* for a number of stars of our sample. We also prove the need of *line-blocking* for the correct reproduction of these spectra. For stars with effective temperature around 50 000 K and above, there is no possibility of reproducing their spectra with these kind of models. Further analyses with unified models, including sphericity and mass loss, allow us to reproduce even the earliest spectra. They also show that the *Mass Discrepancy* is not only due to the neglect of sphericity and mass loss in the planeparallel analysis, and give us the mass loss rate of these stars. We find that HD15 570, with a present mass of 100 $M_{\odot}$, can be the most massive star known in the solar neighbourhood.

1. Introduction.

Quantitative analysis of spectra from early, very luminous and massive stars is a major topic in the understanding of several astrophysical environments. In particular, in the context of the evolution of these stars, the determination of accurate stellar parameters such as effective temperatures, luminosities, atomic abundances and mass loss rates, provides constraints to the different evolutive scenarios proposed for them.

An example of the importance of these spectral analyses is the work by Herrero et al. (1992), in which a large sample of O and early B stars was analized by means of classical planeparallel hydrostatic models. Among the results of this work are the so-called *He Discrepancy* and *Mass Discrepancy*, which show that a large number of the stars in the sample are He rich (compared to the predictions of standar stellar evolution) and that supergiants show masses well below the ones that correspond to their posi-

Astrophysics and Space Science is the original source of publication of his article. It is recommended that this article is cited as: *Astrophysics and Space Science* **263**: 247–250, 1999.

tion in the HR diagram. These discrepancies between the predictions from stellar evolution theory and from spectral analyses, based on the theory of stellar atmospheres, have impulsed recent improvements in both disciplines (Langer 1992, Denisenkov 1994, Maeder 1998, Herrero et al. 1995, Santolaya-Rey et al. 1997).

The stars of the Herrero et al. (1992) sample covered the region between O6 and B1.5 spectral types, and the analyses showed greater difficulties when going to earlier types. In particular, above 40 000 K, HeI triplet and singlet lines gave different temperatures. Also the *Mass Discrepancy* was more severe for stars closer to the Eddington-Limit, the ones with higher temperatures and luminosity classes, so the extension of the study to earlier supergiants was the natural next step to complete this work.

So observations of seven stars from O3 to O6, mainly supergiants, belonging to open clusters in the solar vecinity, were carried out in 1991 and 1992 with the Isaac Newton 2.5 telescope at the Observatory of El Roque de los Muchachos, in La Palma. We observed the spectral region between 4000 and 5000 Å, and the region around H_α, with a spectral resolution of 0.6 and 0.8 Å respectively.

2. Planeparallel Analysis.

In order to obtain the stellar parameters that characterize the atmospheres of our target stars, we fit H and He photospheric lines in the optical espectra with our model profiles, first with the classical planeparallel models used by Herrero et al. (1992).

TABLE 1. Parameters determined for the program stars with planeparallel, hydrostatic models with *line-blocking*. Note that the parameters of Cyg OB2 #7 and HD 15 570 are only indicative, and could not be determined with plane parallel models. M_s and M_{ev} are the present spectroscopic and evolutionary masses, respectively.

Star	Classif.	$V_r \sin i$ km/s	T_{eff} kK	$\log g$ cgs	ϵ $\frac{N(He)}{N(He)+N(H)}$	$\log(L/L_\odot)$	M_s $M/M_\odot$	M_{ev} $M/M_\odot$
Cyg OB2#7	O3 If	105	51.0	3.66	0.12	6.23	46.2	111.8
HD 15570	O4 If	105	50.0	3.51	0.15	6.44	51.1	139.1
HD 15629	O5 V((f))	90	48.0	3.81	0.09	5.89	37.5	69.9
HD 15558	O5 III	120	46.5	3.71	0.07	6.16	63.8	91.7
HD 14947	O5 If	140	45.0	3.53	0.15	5.91	26.5	65.8
HD 210839	O6 I	250	41.5	3.47	0.25	5.98	38.7	67.0
HD 5689	O6	250	40.0	3.57	0.25	5.13	7.8	30.1

We follow the common technique of using a fixed NLTE model atmosphere of H and He, in radiative and hydrostatic equilibrium and with plane-parallel geometry (ALI, Kunze 1995), to solve the statistical equilibrium and transfer equations and then perform the formal solution for the lines of H and He chosen (with DETAIL & SURFACE, Butler & Giddings 1985).

In Herrero et al. (1994) the use of *line-blocking* helped to improve the problematic fits at higher temperatures. Using *line-blocking* means that metal opacities are taken into account when solving for the statistical equilibrium and transfer equations, which makes our models more realistic. We have included *line-blocking* extensively in the present analyses, whose results are given in Table 1.

3. Unified Analysis.

The fact that spectra of the most luminous stars are not properly reproduced with planeparallel models (see Table 1) and that the *Mass discrepancy* grows when going to stars closer to the Eddington Limit, tells us that stellar winds can be responsible for that. That is why we start a first study with the last generation of model atmospheres for early stars, the Unified Models, called so because they treat the stellar photosphere and wind as a whole. We use the code by Santolaya-Rey et al. (1997).

TABLE 2. Parameters determined using the spherical models with mass loss. Two values of the mass loss rate are given for HD 14 947 and HD 15 570, coming from a discrepancy between the fits of H_γ and H_α found in these two stars.

Star	V_∞ km/s	β km/s	$R/R_\odot$	$T_{\rm eff}$ kK	$\log g$ cgs	ϵ	$\log(\dot{M}/M_\odot)$	$M_{\rm s}$ $M/M_\odot$	$M_{\rm ev}$ $M/M_\odot$
Cyg OB2#7	2900	0.80	16.7	51.0	3.81	0.12	-4.90	65.4	111.8
HD 15570	2600	1.00	22.0	50.0	3.76	0.15	-4.95	101.1	139.1
							-4.60		
HD 15629	3000	1.00	12.7	48.0	3.91	0.09	-6.13	47.6	69.9
HD 15558	2800	0.75	18.5	46.5	3.86	0.07	-5.40	90.0	91.7
HD 14947	2400	1.00	14.8	45.0	3.68	0.15	-5.40	38.1	65.8
							-5.13		
HD 210839	2250	0.90	18.9	41.5	3.57	0.25	-5.50	48.2	67.0
HD 5689	2500	1.00	7.7	40.0	3.57	0.25	-6.70	7.8	30.1

The main improvements that this code introduces are the spherical geometry instead of the planeparallel one, not suitable for these stars extended atmospheres, and the inclusion of the wind structure and mass loss.

We perform the analysis by adding to the parameters obtained in the previous section the ones corresponding to the star extension and wind, namely, stellar radius, terminal wind velocity, the β exponent of the wind velocity law and the mass loss rate. For this first approach to the Unified Analysis we allow only the surface gravity, the β exponent, and the mass loss rate as free parameters. Fitting H_γ and H_α we obtain the results given in Table 2.

4. Conclusions.

The planeparallel analysis shows that *line-blocking* is needed to reproduce well these spectra, specially the HeI lines, and that, for the two most luminous stars of the sample, no set of parameters could be found to reproduce their lines. We could only estimate their parameters. We conclude that these models are only suitable for stars cooler than 50 000 K at the gravities of O supergiants.

For all the stars of the sample the spectroscopic mass, obtained from the surface gravity and stellar radius from the spectral analysis, is well below the evolutionary one, obtained by comparing evolutionary tracks from Schaller et al. (1992), with the star position in the HR diagram. Four of the seven stars of the sample show He enrichment in their atmospheres.

The unified analysis is capable of reproducing even the most luminous spectra, but for two of the stars we cannot find a unique value of the mass loss rate in agreement with both H_γ and H_α. We think that the adopted velocity law can be representing not well enough the situation in the wind of these two stars, where H_α is formed.

The spectroscopic masses are now closer to the evolutionary ones, but the inclusion of sphericity and mass loss seems not to be enough to explain the discrepancy to all its extent.

The star HD15 570, with a present spectroscopic mass of 100 $M_\odot$, can be the most massive star known so far in the solar neighbourhood.

References

Butler, K., Giddings, J., 1985, Newsl. on Anal. of Astron. Spectra 9, Daresbury Lab
Denisenkov, P.A., 1994, A&A, 287, 113
Herrero, A., 1994, Sp. Sc. Rev., 66, 137
Herrero, A., Kudritzki, R.P., Vilchez, J.M., et al., 1992, A&A, 261, 209
Herrero, A., Kudritzki, R.P., Gabler, R., Vílchez, J.M., Gabler, A., 1995, A&A, 297, 556
Kunze, D., 1995, Ph.D. Thesis, Ludwig-Maximillian University, Munich
Langer, N., 1992, A&A, 265, 17L
Maeder, A., 1998, in *II Boulder–Munich Workshop on Properties of Hot, Luminous Stars*, ASP Conf. Ser., 131, 85
Santolaya–Rey, A.E., Puls, J., Herrero, A., 1997, A&A, 488, 512
Schaller, G., Schaerer, D., Meynet, G., Maeder, A., 1992, A&AS, 96, 269

THE PISTOL STAR

F. NAJARRO
Intituto de Estructura de la Materia, CSIC
Serrano 121, Madrid, Spain

AND

D.F. FIGER
Department of Physics and Astronomy, University of California, Los Angeles, CA 90095-1562, USA

Abstract.
Results of an spectroscopic investigation of the Pistol star are presented. The near-infrared spectra and photometry data are fit with stellar wind models to find that the star is extraordinarily luminous, $L = 10^{6.7\pm0.5}$ $L_{\odot}$, making it one of the most luminous stars known. Coupled with the relatively cool temperature, $T_{\rm eff} = 10^{4.17^{+0.19}_{-0.06}}$ K, the star is clearly in violation of the Humphreys-Davidson limit. The derived line of sight velocity of the star assures its membership in the Quintuplet cluster. This, along with the inferred extinction, places the star at the Galactic Center.

1. Introduction

The Pistol Star is a member of the "Quintuplet Cluster" (Glass et al. 1987, Nagata et al. 1990) which is located 30 pc, in projection, from the Galactic Center. Figer et al. (1995) identified this star as a Luminous Blue Variable candidate (LBVc) based upon its luminosity – they estimated $L_{*}{\approx}10^{6.3}L_{\odot}$– its LBV-like K-band spectrum and its spatial relationship with the Pistol nebula. Being one of the most luminous stars known and given its evolutionary stage, the Pistol Star may have been the most massive when it formed. Thus, the Pistol Star constitutes an ideal object to provide constraints on star formation and evolution theories of massive stars. In this paper (see also Figer et al. 1998), we present results of a spectroscopic investigation of

Astrophysics and Space Science is the original source of publication of his article. It is recommended that this article is cited as: *Astrophysics and Space Science* **263**: 251–254, 1999.

the Pistol Star and briefly compare its derived stellar properties with other luminous stars.

2. Spectroscopic Analysis

For the spectroscopic analysis we proceeded as described in Najarro et al. (1994,1997). We used the iterative, non-LTE method presented by Hillier (1987,1990) to solve the radiative transfer equation for the expanding atmospheres of early-type stars in spherical geometry, subject to the constraints of statistical and radiative equilibrium and the assumption of steady state. The atmosphere was considered to consist of H, He, and N, though the latter was used basically to better account for cooling processes. The standard model was then prescribed by the stellar radius, R_*, the stellar luminosity, L_*, the mass-loss rate $\dot{M}$, the elements abundances, and the velocity field, $v(r)$. We found (see Table 1) two temperature ranges ("H" high and "L" low) fitting, to, within the errors, the Spectral Energy Distribution (SED) and the H and He I line profiles. It is important to note that, though this dichotomy cannot be removed in terms of the observed H and He I lines, the presence of Fe II and Mg II lines can be explained only in terms of the "L" model (Najarro et al. in prep.). Figure 1 shows the model fits to the lines and SED for the "L" model (see Figer et al. 1998 for further discussion).

TABLE 1. Model Parameters

Model	$\log(L_*/L_\odot)$	$\log T_{\rm eff}$	R_* $R_\odot$	$\dot{M}$ $10^{-5} M_\odot\,{\rm yr}^{-1}$	v_∞ km s^{-1}	$N_{\rm Ly,c}$ s^{-1}	$n_{\rm He}/n_{\rm H}$
L	6.61	4.15	340	3.8	95	$<3\times10^{41}$	0.13
H	7.20	4.33	300	2.8	95	4.4×10^{48}	0.12

Figure 2 displays the position of the "L" and "H" models in the HR diagram together with other, hot luminous stars in the Milky Way and Magellanic Clouds in quiescence (see Figer et al. 1998, for further details on stellar data). The error boxes for the "L" and "H" models are dominated by uncertainties in luminosity (i.e., extinction) rather than temperature (from the models). We assumed A_K=3.2±0.5. From Fig. 2 we see that the Pistol Star strongly violates the Humphreys-Davidson limit and is certainly on of the most luminous objects known. Stellar evolution calculations (Figer et al. 1998), suggest an initial mass of 200–250 $M_\odot$ and a current age of 1.7–2–1 Myrs, which confirm its present unstable evolutionary phase (LBV). We believe that such a high initial mass could be achieved due to the very little angular momentum that the star obtained when it formed. This allowed the

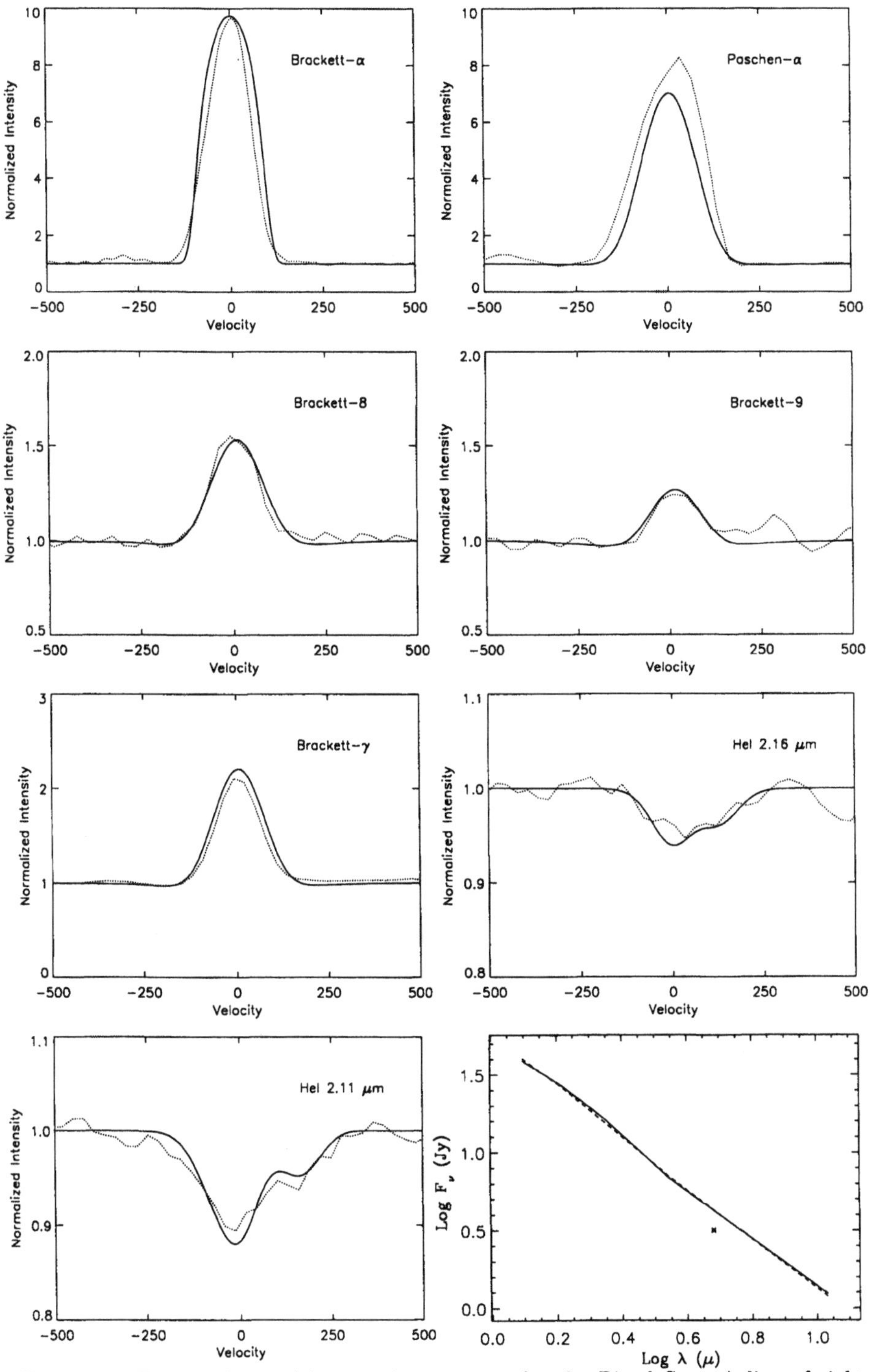

Figure 1. Fits to observed lines and continuum for the Pistol Star. A line of sight velocity of $\approx 130\,\mathrm{km\,s^{-1}}$ is derived

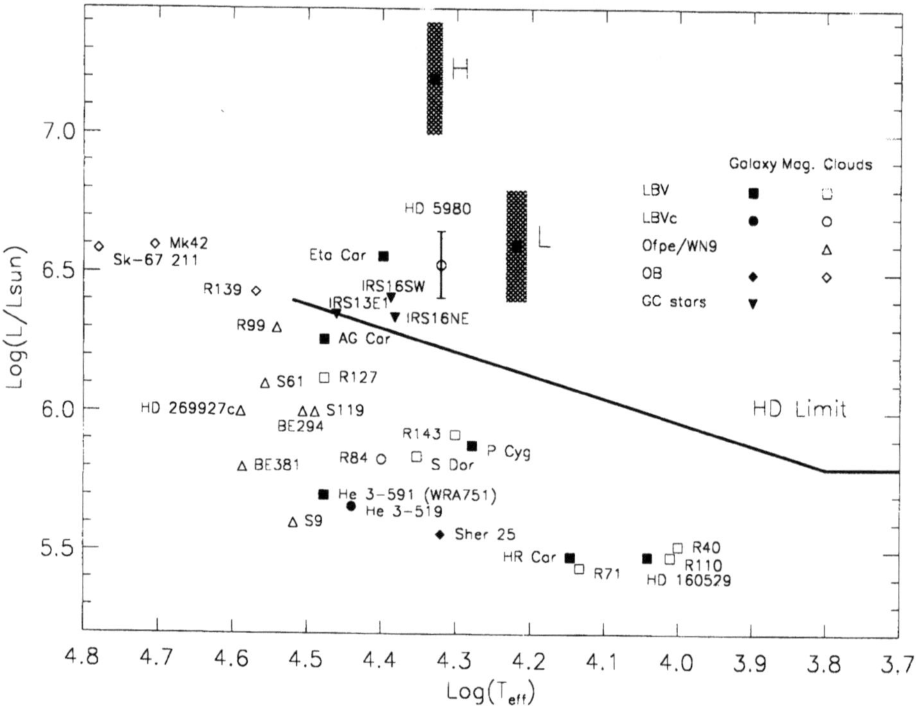

Figure 2. Position of the Pistol star in the HR diagram compared with other luminous early-type stars (see Figer et al. 1998).

star to evolve, with less than average mass loss, towards the cool side of the HR-diagram.

Acknowledgments

F. N. acknowledges DGYCIT grants PB96-0883 and ESP98-1351.

References

Figer, D.F., McLean, I.S., Morris, M., 1995, ApJ, 447, L29
Figer, D.F., Najarro, F., Morris, M., McLean, I.S., Geballe, T.R., Ghez, A.M., Langer, N., 1998, ApJ, 384, 404
Glass, I.S., Catchpole, R.M., Whitelock, P.A., 1987, MNRAS, 227, 373
Nagata, T., Woodward, C.E., Shure, M., Pipher, J.L., Okuda, H., 1990, ApJ, 351, 83
Hillier, D.J., 1987, ApJL, 63, 947
Hillier, D.J., 1990, A&A, 231, 116
Najarro, F., Hillier, D.J., Kudritzki, R.P., Krabbe, A., Genzel, R., Lutz, D., Drapatz, S., Geballe, T.R., 1994, A&A, 285, 573
Najarro, F., Krabbe, A., Genzel, R., Lutz, D., Kudritzki, R.P., Hillier, D.J., 1997, A&A, 325, 700

SPECTROSCOPIC ANALYSIS OF STARS IN CYG OB2

L.J. CORRAL, A. HERRERO AND R. VILLAMARIZ
Instituto de Astrofísica de Canarias, c/ Vía Lactea s/n, E-38200, La Laguna, España

AND

E. MARTIN
University of California, 601 Cambell Hall, Berkeley 9472, USA

1. Introduction

Herrero et al. (1992) found that from a sample of 25 galactic luminous OB stars, 11 showed helium abundances greater than those predicted by the stellar evolutionary models. This result is known as the "helium discrepancy". In order to study this problem in an homogeneus sample of young stars, we have observed and analyzed a sample of OB stars in Cyg OB2.

Cyg OB2 is a young stellar association that has been extensively studied by many authors (see Massey & Thompson, 1991, and references therein). One of the few O3 If stars known in the galaxy is a member of this association (Cyg OB2 No. 7 in the identification list of Schulte 1958). Accordingly to Massey & Thompson (1991), who studied Cyg OB2 with broad band photometry and spectroscopy, there are 108 stars earlier than B1.5 V. The average reddening deduced by these authors is high ($< E(B-V) > \sim 1.8$), as expected from the position of the association in the galactic plane ($l = 80^{\circ}\!.16$, $b = 1^{\circ}\!.3$). Using 67 stars with a known spectral type (from O3 If to B1.5 V), they found a distance module of 11.2±0.10 mag. Cyg OB2 is then a young association with a large number of early type stars and is an ideal place to study in a sistematic way the helium discrepancy problem.

2. Observations and analysis

The stars were observed at the Issac Newton 2.5 m. telescope with the Intermediate Dispersion Spectrograph during a run in 1995. Two gratings

Astrophysics and Space Science is the original source of publication of his article. It is recommended that this article is cited as: *Astrophysics and Space Science* **263**: 255–258, 1999.

TABLE 1. Parameters determined for the stars observed in Cyg OB2.

Star No.	Sp. Type	$T_{\rm eff}$	$\log g$	ϵ	$V_r \sin i$
22	O4 III(f)	47.0	3.61	0.12	125
516	O5.5 V((f))	44.0	3.61	0.15	135
8c	O5 If	48.0	3.77	0.09	145
9	O5 If	44.5	3.52	0.09	135
11	O5 If$^+$	43.0	3.42	0.09	120
29	O7 V	40.0	3.83	0.09	180
4	O7 III((f))	39.0	3.52	0.07	125
20	O9.5 V	35.0	4.00	0.09	25
19	O9.5 III	30.0	3.02	0.09	75
10	O9.5 I	31.0	3.11	0.09	85
21	B0.5 V	34.5	3.90	0.09	30

were used (H2400B and R1200B) which provided 0.6 and 1.3 Å resolutions respectively. We observed the spectral region from 4000 to 5000 Å. The data reduction was made with the IRAF package following standard procedures. The reduced spectra were rectified with a private procedure running under IDL.

To make the spectroscopic analysis we followed the procedures described by Herrero et al. (1992). We compared the observed profiles of H, He I and He II spectral lines with profiles calculated with plane parallel, line-blanketed atmospheric models in hydrostatic equilibrium. The projected rotational velocity ($V_r \sin i$) was calculated for each star using photospheric metallic lines. The models were calculated with different helium abundances. The spectral lines were broadened with the projected rotational velocity and compared to the observed profiles. The model parameters that best fit the observed profiles were chosen as the physical parameters of the star.

3. Results

We present in Figure 1 the HR diagram with the data obtained in the present analysis and the evolutionary tracks of Schaller et al. (1992) for non-rotating stars. In Table 1 we list the values of $T_{\rm eff}$, $\log g$ and ϵ (helium abundance by number) that best fit the observed profiles and the $V_r \sin i$ deduced for each star.

Herrero et al. (1992) pointed out that the majority of the stars that showed helium abundance are rapid rotators ($V_r \sin i > 200$ km s^{-1}). As

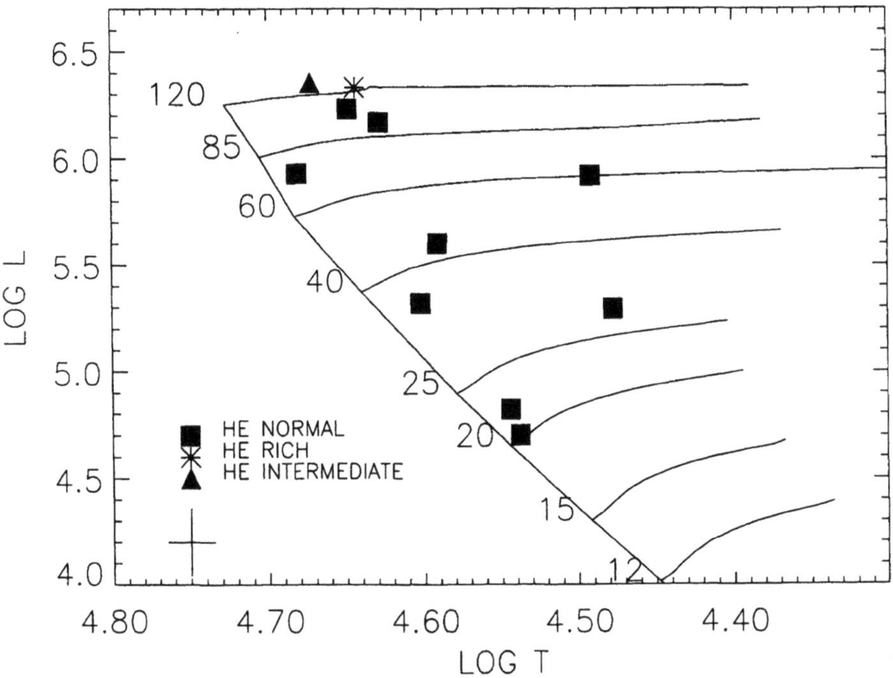

Figure 1. HR diagram of the stars of Cyg OB2. Tracks are from Schaller et al. (1992) for non-rotating stars. The cross at the bottom left corner represents the error bars.

we can see in Table 1, the projected rotational velocity of the stars of Cyg OB2 is moderate, six of them have rotational velocities between 120 and 145 km s^{-1} and only one reaches 180 km s^{-1}. The helium content found in all stars is normal because, in the only case that we found an enhanced helium content (star 516, $\epsilon = 0.15$), the evolutionary tracks of Schaller et al. (1992) give a surface helium abundance of $\epsilon = 0.14$.

4. Conclusions

The results presented here are in contrast with those of Herrero et al. (1992), where 44% of the stars analyzed showed an enhanced helium content. One possible cause is the age of the sample. According to Massey & Thompson (1995), Cyg OB2 is a young association with an age between 1 and 5 Myr. If there is a mixing process that enhances the helium content shown in the atmosphere, maybe it could not have time to act yet. Another possible cause is the rotation. It has been suggested that rotation can activate or accelerate the mixing processes (eg. Maeder 1997). The moderated rotational velocity found indicates that the mixing processes associated must be also moderated.

We need to observe and analyze more OB stars in associations with different physical constraints (age, rotational velocities, etc.) and known distances in order to evaluate the influence of age and rotation in the helium discrepancy problem.

References

Herrero, A., Kudritzki, J.M., Vilchez, J.M., Kunze, D., Butler, K., Hasser, S., 1992, A&A, 261, 209
Maeder, A., 1997, A&A, 321, 134
Massey, P., Thompson, A.B., 1991, AJ, 101, 1408
Schaller, G., Schaerer, D., Maeder, A., Meynet, G., 1992, A&AS, 96, 269
Schulte, D.H., 1958, ApJ, 128, 41

RADIAL VELOCITY MEASUREMENTS WITH FXCOR

JOSÉ J. RODES AND GUILLERMO BERNABEU
Departament de Física, Enginyeria de sistemes i Teoria del senyal
Universitat d'Alacant

AND

JUAN FABREGAT
Departament d'Astronomía i Astrofísica
Universitat de València

Abstract.
We present measured radial velocities of two peculiar early B type stars, to check their membership to the open cluster NGC 663. The radial velocities have been obtained with the cross correlation technique, using as a template stars from the list of Liu et al. (1989).

We have obtained accuracies around 3-4 km/s in our radial velocity measurements. From our results we conclude that both stars are probably non members of the cluster.

1. Introduction

An important problem in the determination of the radial velocities for young early-type stars is the usual high velocity of rotation. The spectral lines are very therefore to determine the center line is difficult to obtain accurately.

For these kinds of stars the use of cross-correlation techniques has provided more accurate radial velocity measurements. However, the use of this technique is difficulted by the lack of a set of well-established early-type velocity standards. The IAU standard system is defined only by late-type stars. We have used as cross-correlation templates a list of early-type stars with radial velocities measured in an homogeneous system, given by Liu et al. (1989). The internal accuracy of these measurements is 2 km/s.

Astrophysics and Space Science is the original source of publication of his article. It is recommended that this article is cited as: *Astrophysics and Space Science* **263**: 259–262, 1999.

TABLE 1. B type cross-correlation template stars (Liu et al. 1989)

star	spectral type	VR (km/s)
ζ Cas	B2 V	2.1 ± 1.1
HD 166182	B2 V	-14.1 ± 0.6
HD 184171	B3 IV	-20.4 ± 0.7

We are studying if the stars LS I +61° 235 and BD +60° 329 belong to the open cluster NGC 663. The former is the optical counterpart of the X-Ray source RX J0146.9+6121. The photometric study of Fabregat et al. (1996) concludes that there is a high probability of cluster membership. If confirmed, it would be the first Be/X-ray binary in a open cluster known so far. BD +60° 329 has a spectral type B0V, bluer than the cluster main sequence turnoff at B2 (Fabregat et al. 1996). The cluster membership would imply that it is a blue straggler, the first one identified in NGC 663.

2. Observations

Spectroscopic observations were carried out from the 1.0 m Jacobus Kapteyn Telescope, located at the Roque de los Muchachos Observatory in La Palma (Spain). Spectra were obtained using the Richardson Brealey Spectrograph with the 1200 grating and the TEK4 CCD chip. Two observing runs were performed: the first one in August 1995 and the second one in March 1996.

3. Results and discussion

We estimated the radial velocities for LS I +61° 235 and BD +60° 329. Our results are summarized in table 2 and table 3, respectively. In table 1 we list the early-type stars we have used as templates.

3.1. LS I +61° 235

The mean radial velocity and its standard error are: $\overline{V_h}$ = - 4.7 km/sσ= 3.8 km/s. The mean radial velocity of the cluster is -35 ± 5 km/s, so this star is probably non member of the cluster. However, as LS I +61° 235 is member of a X-ray binary system, we can attribute the difference between the radial velocities to orbital motion. This possibility is very unlike. On the one hand, the mass of a typical early B type star is around 15 $M_\odot$ and the mass of the compact object is around 1.4 $M_\odot$. An amplitude of 30 km/s in the radial velocity means an orbit period less than 10 days.

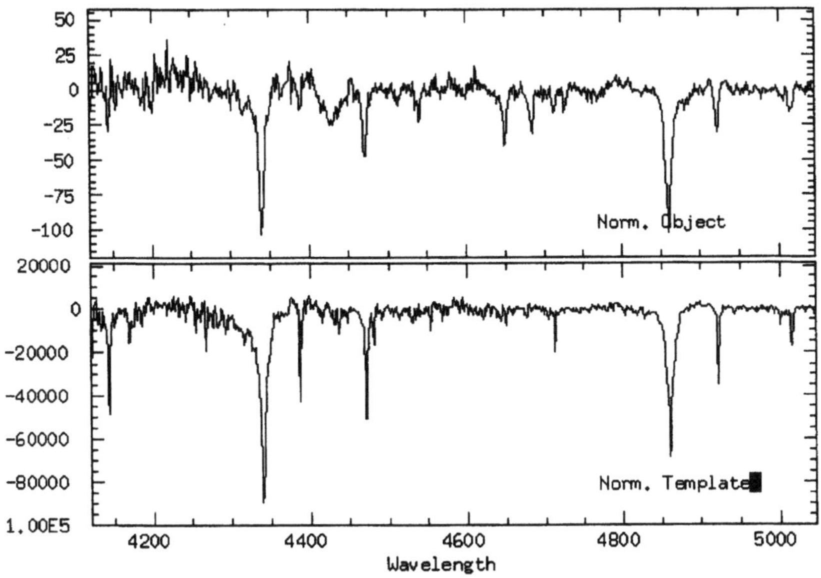

Figure 1. Normalized spectrum of the star BD +60° 329 (top) and normalized spectrum of the template star ζ Cas (bottom)

TABLE 2. Cross-correlation results of the object spectrum with three template spectra.

object spectra	template spectra	heliocentric velocities (km/s) relative	absolute
LS I +61° 235	ζ Cas	-7.3	-5.2
LS I +61° 235	HD 166182	5.9	-8.2
LS I +61° 235	HD 184171	19.0	-0.6

All known orbital periods in Be/X-ray binaries are larger than this value. Moreover, the relation between the spin and orbital periods for Be/X-ray binaries (Corbet 1986), predicts an orbital period larger than 400 days.

3.2. BD +60° 329

The mean radial velocity and its standard error are: $\overline{V_h}$= -23.0 km/sσ= 2.5 km/s. The mean radial velocity of the cluster and the mean radial velocity of the star are compatible at 2σ level, so the result in this case is uncertain.

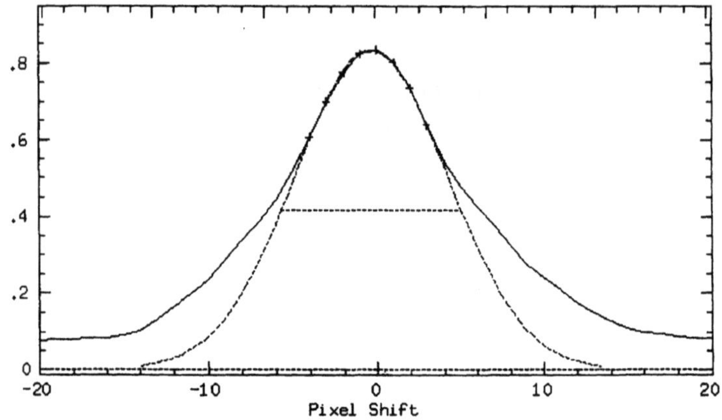

Figure 2. Cross-correlation function of the spectrum of BD +60o 329 against the spectrum of ζ Cas as a template. The derived heliocentric velocity for this cross- correlation is 21.2 km/s

TABLE 3. Cross-correlation results of the object spectrum with three template spectra.

object spectra	template spectra	heliocentric velocities (km/s) relative	absolute
BD +60° 329	ζ Cas	-23.3	-21.2
BD +60° 329	HD 166182	-11.8	-25.9
BD +60° 329	HD 184171	-1.5	-21.9

4. Conclusions

We have used as standards of radial velocities the stars proposed by Liu et al. (1989) and we have tested it with our own spectra. Then we have estimated the radial velocities of two early-type stars using cross-correlation techniques. Our errors are larger than twice the errors presented by Liu et al., mainly due to the lower signal to noise (around 30) of our target spectra.

We are planning to continue this work with new spectra with higher s/n in order to improve our radial velocities. This will lead to further improvements in the precision of the radial velocities presented in this poster.

References

Corbet, R.H.D., 1986, MNRAS, 220, 1047
Fabregat, Juan et al., 1996, A&ASS, 119, 1
Liu, T., Janes, K.A., Bania, T.M., 1989, AJ, 98, 626

STELLAR WINDS AND MASS LOSS IN A-TYPE SUPERGIANTS

E. VERDUGO
ISO Data Centre, P.O. Box 50727, E-28080-Madrid, Spain
A. TALAVERA
LAEFF/INTA, P.O. Box 50727, E-28080-Madrid, Spain
AND
A.I. GÓMEZ DE CASTO
Instituto de Astronomía y Geodesia (CSIC-UCM),
Fac. cc.Matemáticas, Universidad Complutense,
Av. Complutense s/n, E-28040-Madrid, Spain

Abstract. We examine the relationship between the surface escape velocities and wind terminal velocities of A-type supergiants according to the radiation-driven wind theory.

1. Introduction

The best indicators of mass-loss from hot stars are the ultraviolet resonance lines, whose profiles are very sensitive to the presence of a stellar wind and its characteristics. In the case of A-type supergiants, studies of the wind properties are still scarce. Although the radiation pressure seems to be the most plausible mechanism to drive the wind in these stars, no systematic studies have been carried out on a significant sample in order to test the correctness of this theory.

We have analyzed the UV spectrum of 35 A-type supergiants in order to study the wind of these stars.

2. The Sample

Our data set is based on UV spectra of luminous galactic A-type supergiants observed with IUE at high resolution. The 35 stars selected cover a range in spectral type from A0 to A5 and luminosity classes Ia, Iab and

Astrophysics and Space Science is the original source of publication of his article. It is recommended that this article is cited as: *Astrophysics and Space Science* **263**: 263–266, 1999.

Ib. A complete description of the sample, the IUE observations and data reduction is provided in Verdugo et al. (1998a).

3. Terminal Velocities

A-type supergiants show essentially two types of UV resonance line profiles characteristic of mass outflow: saturated profiles with extended violet absorption wings or profiles composed of discrete absorption components (DACs). Traditionally, the terminal velocity of a stellar wind has been observationally defined as the modulus of the largest negative velocity seen in absorption in the P Cygni profiles of UV resonance lines. For saturated UV profiles the maximum velocity at which zero residual intensity is recorded ($v_{\rm black}$) provides an accurate measure of the wind terminal velocity.

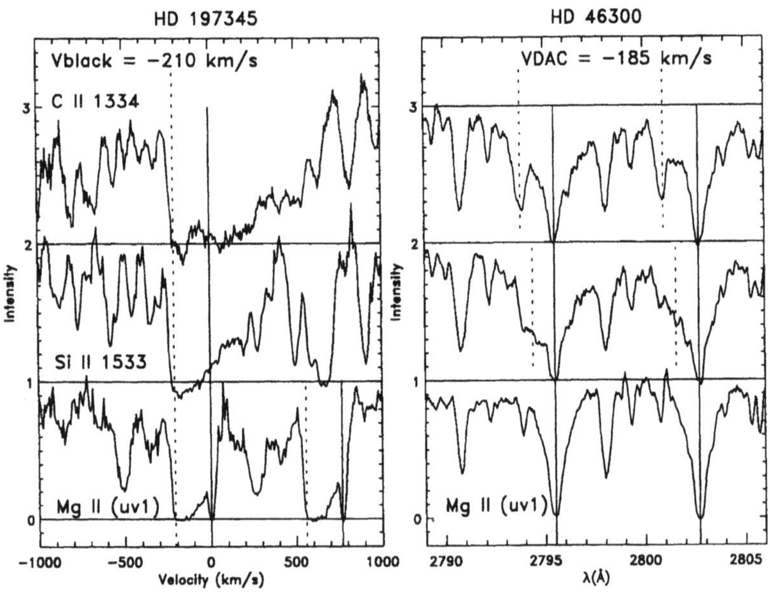

Figure 1. Examples of the two methods for determining v_∞: (left) from saturated profiles, (right) from discrete absorption components. Full and dashed vertical lines indicate the rest wavelength of the lines and $v_{\rm black}$ and $v_{\rm DAC}$ respectively.

We have measured v_∞ from the saturated profiles of Mg II (uv1), C II (1334 Å) and Si II (1533 Å). Unfortunately, several stars of our sample do not show this type of profiles. From unsaturated profiles we have estimated v_∞ from the final velocity reached by the DACs. The accuracy of this determination is lower than that of stars with saturated profiles and only provides a lower limit to v_∞. Examples of these measurements are shown in Figure 1.

4. The Relationship between v_∞ and v_{esc}

According to the radiation-driven wind theory, the terminal velocity scales with the escape velocity as $v_\infty/v_{esc} \sim [\alpha/(1 - \alpha)]^{1/2}$ (e.g. Abbot 1982). The "force multiplier parameter" α depends on the optical depth of the driving lines. If all the driving lines were optically thick, then α would be 1. If all the driving lines were optically thin, α would be 0.

We represent in Figure 2 our measured values of v_∞ from v_{black} and v_{DAC} *versus* the effective escape velocities, $v_{esc} = \sqrt{2GM(1-\Gamma)/R}$, where Γ is the ratio of stellar to Eddington luminosity, M the stellar mass and R the photospheric radius. The stellar parameters have been obtained from Verdugo et al. (1998b). The v_∞ measured from saturated profiles are marked with points and error bars. These uncertainties are not standard deviations but an estimation of the possible range of v_∞. The v_∞ measured from v_{DAC} are marked with crosses.

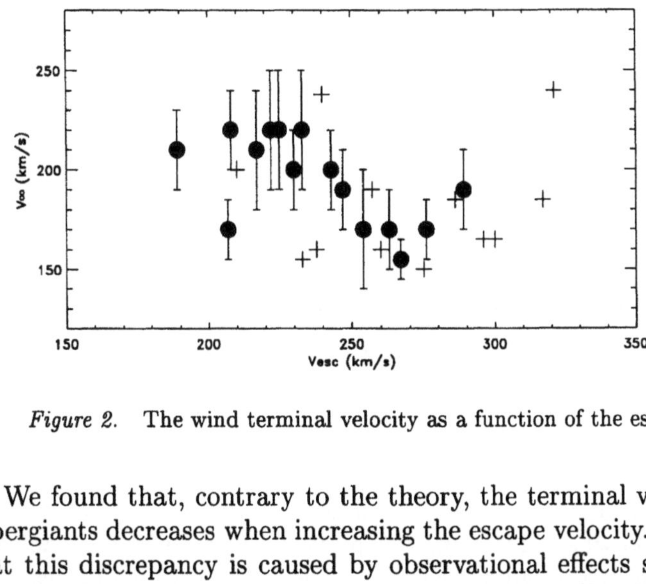

Figure 2. The wind terminal velocity as a function of the escape velocity.

We found that, contrary to the theory, the terminal velocity in A-type supergiants decreases when increasing the escape velocity. It can be argued that this discrepancy is caused by observational effects such as the small amount of accurate measurements or the errors in the measuring process. Moreover, we have also detected variations in v_{black} and v_{DAC}. This fact makes the observed slope sensitive to variations of the profiles.

5. Ratio v_∞/v_{esc}

In spite of the reasons given above, we found that the observed scatter in v_∞ towards higher escape velocities correlates with assorted stellar parameters. In Figure 3 the ratio v_∞/v_{esc} is represented against v_{esc} together with the theoretical predictions (solid curves) for different values of α. The negative slope observed in Fig. 2 could be related to the degree of ionization and the density of the wind. Higher values of α indicate higher degree of ionization in the wind, the radiation pressure produces higher terminal velocities and a relatively low density wind. When the temperature drops a little the wind will recombine to lower ionization stages which produce a smaller value of α and hence lower terminal velocities and a higher density wind.

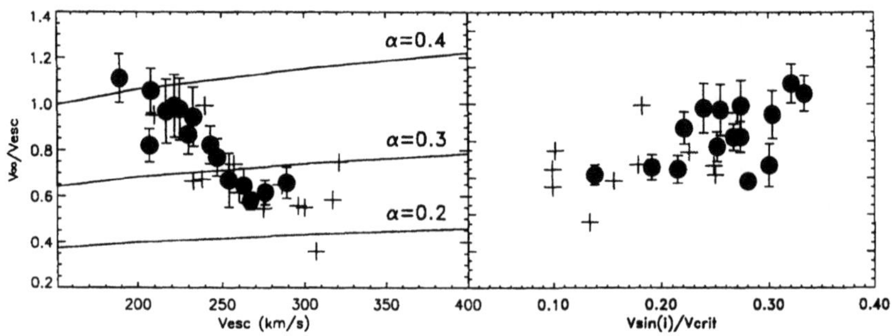

Figure 3. The ratio $v_\infty/v_{\rm esc}$ as a function of escape velocities (left) and projected rotational velocities (right). See the text for details.

However, we also found that the observed trend could be caused by the different rotational velocities of the stars. In fact, some dependence of $v_\infty/v_{\rm esc}$ on the ratio $v_{\rm rot}/v_{\rm crit}$ is predicted by the models (Friend & Abbot 1986; Pauldrach et al. 1986). The $v_{\rm crit}$ is the equatorial rotational velocity at which the centrifugal force matches the gravitational force: $v_{\rm crit} = v_{\rm esc}/\sqrt{2}$. We have, accordingly, examined our data set for evidence of this correlation. The results are represented in Fig. 3. Accurate rotational velocities are needed to confirm this possibility.

6. Conclusions

We have observed that the terminal velocity of the wind in A-type supergiants decreases when increasing escape velocity in direct contradiction with the radiation-driven wind theory. It could be caused by observational effects, by the different rotational velocities of the stars or could be a "real" scattering due to different degree of ionization and density of the winds.

References

Abbot, D.C., 1982, ApJ, 259, 282
Friend, D.B., Abbot, D.C., 1986, ApJ, 311, 701
Pauldrach, A., Puls, J., Kudritzki, R.P., 1986, A&A, 164, 86
Verdugo, E., Talavera, A., Gómez de Castro, A.I., 1998a, A&AS, submitted
Verdugo, E., Talavera, A., Gómez de Castro, A.I., 1998b, A&A, submitted

SPECTROSCOPIC SURVEY OF δ SCT STARS

E. SOLANO
INSA
P.O. Box 50727. 28080 Madrid (Spain)

AND

O.I. PINTADO
Instituto de Física
Universidad Nacional de Tucumán. Argentina

Abstract. The methods used to derive effective temperatures, projected rotational velocities and surface gravities of a sample of low and high-amplitude δ Sct stars are presented here. Effective temperatures were derived using Hα and Hβ line profiles whereas rotational velocities were calculated following the method described in Solano & Fernley (1997). Surface gravities were obtained using Hipparcos parallaxes and evolutionary models (Claret 1995). The results of this analysis will be published in a subsequent paper (Solano & Pintado 1999).

1. Introduction

This work is part of a series of papers devoted to characterize the δ Sct group by determining their fundamental physical parameters and relating them to the pulsational parameters, i.e., period and amplitude. δ Sct stars are A-F type stars with a range of luminosities that extends from the main sequence to two magnitudes above it. These stars are characterized by having periods shorter than $0^{d}.3$, amplitudes ranging from some thousandths to some tenths of magnitude and complicated light curves. The pulsation behaviour ranges from single–mode or multimode radial to mixed radial/non–radial oscillations.

Astrophysics and Space Science is the original source of publication of his article. It is recommended that this article is cited as: *Astrophysics and Space Science* **263**: 267–270, 1999.

2. Observations and Data Reduction

The observations were performed in February 1996 at the Complejo Astronómico el Leoncito (CASLEO) using the 2.15 m. telescope. A resolving power of 26 000 was achieved. The echelle spectra were reduced using the IRAF reduction package. Polynomial calibrations of wavelength as a function of pixel number were calculated for every night: a second order polynomial fit was found adequate. The ripple correction was performed for each night and order using the observed spectra of Procyon and the *Atlas of Procyon* (Griffin & Griffin 1979). Nevertheless, because of changes in curvature from spectrum to spectrum, the removal of the blaze function was not perfect and a third order polynomial was fitted to the spectral continuum.

3. Stellar parameters

3.1. ROTATIONAL VELOCITIES

The projected rotational velocities (v sin$i$) have been measured following the method described in Solano & Fernley (1997) (hereafter, Paper I). In short, the method uses the relation between v sin$i$ and the frequencies where the Fourier transform of the rotational profile reaches a relative minimum. A linear limb–darkening law with $\epsilon = 0.6$ has been assumed. We estimate less than 2% the difference in the total emergent flux between the linear and cuadratric laws. Moreover, the dependence of ϵ with wavelength is also negligible. According to Díaz-Cordovés et al. (1995) we obtain $0.55 \leq \epsilon \leq 0.75$ assuming an interval of T_{eff}=6500 K – 8500 K , $\log g = 3.5$ – 4.0 and spectral ranges of $\lambda\lambda$ 4170 – 6820 Å. In Paper I it was shown that a difference of $\Delta\epsilon = 0.15$ implies an increment in v sin$i$ less than 1 km s^{-1} .

A limiting factor in the calculation of v sin$i$ is the sampling frequency. Defining this frequency as $\sigma = 0.5/\Delta\lambda$, a minimum value of v sin$i$ $\approx$ 18–20 km s^{-1} can be achieved with the resolving power quoted in Sect. 2.

3.2. EFFECTIVE TEMPERATURES

It is well established that the Balmer lines are good temperature indicators for $T_{eff} \leq 8\,500$ K because of their small gravity and metallicity dependence (Paper I). In this work we tried to calculate effective temperatures using the Hα and Hβ line profiles. However, the spectral range of the orders where these lines appear was not wide enough to fully embrace the wings causing a continuum indetermination and giving rise to large errors in the effective temperatures. An alternative method based on the variations in intensity of the wings of the Balmer lines at two wavelengths (λ 6567.308 Å, λ 6578.808 Å for Hα, λ 4865.342 Å, λ 4879.342 Å for Hβ) was used (Fig. 1). These wavelengths were selected in such a way that they lie in the region of the Balmer line profile where the dependency with temperature is maximum. The effective temperatures were calculated comparing the observed intensity ratio with the intensity ratio measured in a grid of synthetic Balmer profiles (Kurucz 1993) convolved with a rotational profile of the corresponding $v \sin i$. Prior to this comparison, the observed spectra were shifted in wavelength to make coincide with the synthetic spectra. The lack of sensitivity to the gravity permits to fix the gravity of the grid of models at $\log g = 4.0$. An error of $\Delta T_{\rm eff} = \pm$ 200 K has been assumed.

It is known that the treatment of the convective flux in the ATLAS 9 version used to compute the 1992 models stored in CD-ROM 13 generates discontinuities in the grids of the Balmer lines around 7300 K. To check how these discontinuities affect the values of $T_{\rm eff}$ we have compared the effective temperatures in the range 7200 K – 7500 K calculated using the Balmer lines and the Moon & Dworetsky (1985) calibration (hereafter, MD85) which is not affected by this problem. The average difference is -17 K $\pm$ 116 K (Balmer - MD85) which is negligible with respect to other sources of error in our analysis.

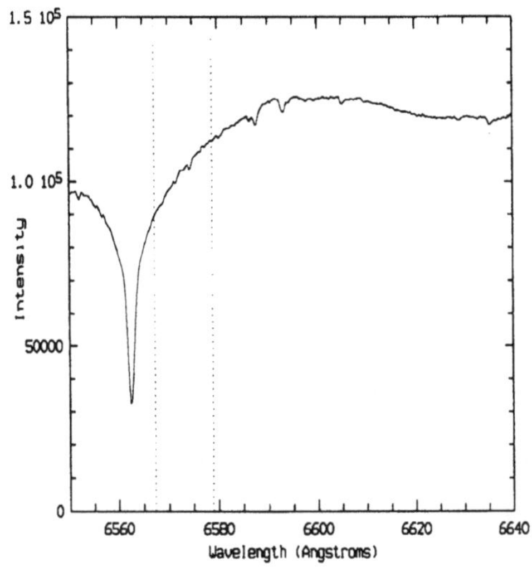

Figure 1. Location of Hα in the spectral order. The wavelengths used in the determination of the effective temperature are shown.

3.3. SURFACE GRAVITIES

Surface gravities were derived from the parallaxes given in the Hipparcos Catalogue (1997). In short, the method relies on the basic relations

$$M_{\rm bol} = m_v + 5 \log \pi + 5 + {\rm BC} \quad (1)$$
$$L = L_\odot \times 10^{0.4(M_{\rm bol,\odot} - M_{\rm bol})} \quad (2)$$

where $M_{\rm bol}$ is the bolometric absolute magnitude, m_v is the apparent visual magnitude, π is the trigonometrical parallax determined by Hipparcos, BC is the bolometric correction and L is the luminosity. $L_\odot$ and $M_{\rm bol,\odot}$ refer to the solar luminosity and absolute bolometric magnitude respectively. The bolometric correction was taken from Alonso et al. (1995) for $T_{\rm eff} \leq 7250$. For higher temperatures, the values given by Lang (1974) were adopted. The mass of the star was calculated from its position in the $\log L - \log T_{\rm eff}$ diagram by interpolating in the isochrones by Claret (1995). The derived mass was used as input value in the $\log T_{\rm eff} - \log g$ diagram to obtain the surface gravity. A chemical composition X = 0.70, Z = 0.02 was adopted. The mixing length and core overshooting parameters were fixed to 1.52 and 2.0 respectively. Only stars with $\sigma_\pi/\pi \leq 0.1$ and amplitude less than $0^{\rm m}.1$ were considered. For comparison, surface gravities using MD85 were also calculated. Surface gravities calculated using Hipparcos are only slightly higher than those derived using MD85 (Hipp − MD85 = 0.05 ± 0.12). A conservative value of $\Delta \log g = 0.15$ dex has been estimated based on errors in the bolometric correction and in the calculated mass.

References

Alonso, A., Arribas, S., Martínez-Roger, C., 1995, A&A, 297, 197
Claret, A., 1995, A&AS, 109, 441
Díaz-Cordovés, J., Claret, A., Giménez, A., 1995, A&AS, 110, 329
ESA, 1997, *The Hipparcos and Tycho Catalogues*, ESA SP-1200
Griffin, R., Griffin, R., 1979, *A photometric atlas of the spectrum of Procyon λλ3140-7470 Å*. The Observatory, Cambridge.
Kurucz, R.L., 1993, CD-ROM 1-23, Smithsonian Astrophysical Observatory
Lang, K.R., 1974, *Astrophysical Formulae*, Springer-Verlag.
Moon, T.T., Dworetsky, M.M., 1985, MNRAS, 217, 305 (MD85)
Solano, E., Fernley, J., 1997, A&AS, 122, 131 (Paper I)
Solano, E., Pintado, O.I., 1999, in preparation

IDENTIFICATION OF λ BOOTIS STARS USING IUE SPECTRA

E. SOLANO
INSA
P.O. Box 50727. 28080 Madrid (Spain)

AND

E. PAUNZEN
Institut für Astronomie der Universität Wien,
Türkenschanzstr. 17, A-1180 Wien (Austria)

Abstract. An analysis of the stars included in the catalogue of λ Bootis by Paunzen et al. (1997) and with IUE Low Resolution observations is presented here. Using line-ratios of carbon to heavier elements (Al and Ni) allows us to establish unambiguous membership criteria for the λ Bootis group. These criteria have been used to look for new λ Bootis candidates in the IUE Final Archive.

1. Introduction

Historically there has been a lack of agreement in the definition of the λ Bootis stars mainly due to the absence of features in the optical spectra of these objects suitable to be used as discriminating criteria. In an attempt to solve this situation a new definition of λ Bootis stars is proposed in Paunzen et al. (1997): λ Bootis stars are defined as "Population I, metal-poor (except of C,N,O and S) A to F-type stars". Based on this definition a catalogue of λ Bootis stars with properties firmly established has been built up.

Some authors (e.g. Faraggiana et al. 1990) have pointed out that the lack of discriminating criteria in the optical domain can be alleviated if the ultraviolet range is used. The main goal of this work is to identify features in the UV that can be used as indicators of the λ Bootis phenomenon. Two are the main reasons to perform this analysis. Firstly, the number of λ Bootis stars with UV observations has significantly increased since the

Astrophysics and Space Science is the original source of publication of his article. It is recommended that this article is cited as: *Astrophysics and Space Science* **263**: 271–274, 1999.

last compilation by Faraggiana et al. (1990) and secondly, the Archive of the IUE satellite has been improved with the creation of the IUE Final Archive: all IUE images are now uniformly processed and calibrated making possible a homogeneous analysis.

2. Analysis

λ Bootis stars included in Paunzen et al. (1997) with available IUE observations have been analyzed. The spectra of these stars have been compared to corresponding spectra of Population I A-F type stars with solar abundances ("standards" hereafter) . Spectra of field horizontal branch stars (hereafter, FHB) from the list by Jaschek et al. (1985) were also studied. Effective temperatures and surface gravities for the standard and λ Bootis stars were derived from Moon & Dworetsky (1985) whereas the most recent values in the literature were adopted for the FHB stars. The list of objects and the adopted physical parameters are given in Solano & Paunzen (1998).

The IUE spectra were normalized using a global pseudo-continuum through the highest points. Equivalent widths were measured performing a numeric integration between two continuum points. Because of the low spectral resolution (FWHM $\approx$ 6 Å) we cannot study individual lines but only blended features named according to the main contributor as identified from the computation of the synthetic spectrum using ATLAS 9 (Kurucz 1993). The following absorption features have been selected:

- Carbon: C I 1657 Å; C I 1931 Å
- Aluminium: Al II 1670 Å
- Nickel: Ni II 1741 Å (blend of Ni II 1741 Å;Ni II 1748 Å; Ni II 1751 Å; Ni II 1754 Å)

Equivalent widths were measured only for those lines not affected by any external error source (saturation, cosmic rays, bright spots, reseau marks). As expected from the definition of λ Bootis stars, metallic lines appear weaker than in standard stars. Moreover, the two carbon features have equivalent widths similar to those of the standard stars confirming the results by Venn & Lambert (1990) and Stürenburg (1993) in the optical domain.

2.1. RATIO BETWEEN ABSORPTION LINES

The differences between λ Bootis and standard as well as FHB stars can be enhanced by using line-ratios of elements with solar abundance (e.g. Carbon) to the heavier elements (e.g. Aluminium, Nickel). Figure 1 shows the computed line-ratios. Since the intrinsic equivalent widths for CI 1657 are stronger than those of CI 1931 we have only used the former to establish

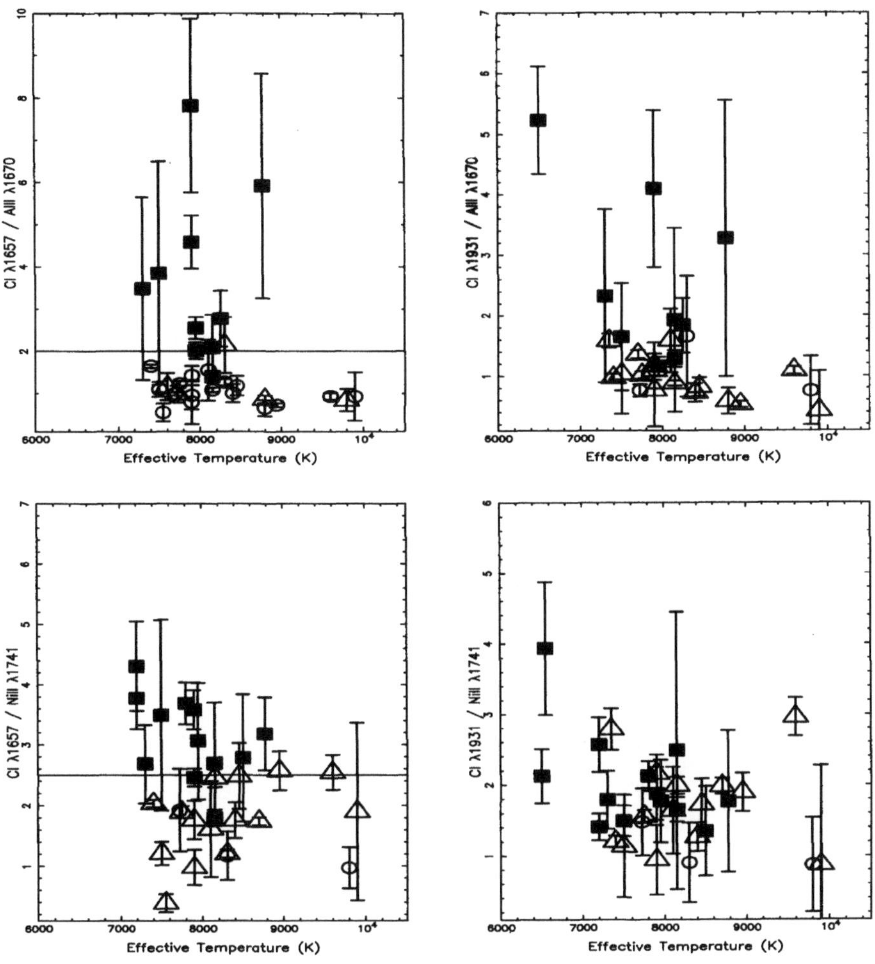

Figure 1. Line-ratios of the absorption lines described in Section 2 for λ Bootis (squares), field horizontal branch (circles) and standard stars (triangles).

limits for the membership to the λ Bootis group. We consider a star to be member of the λ Bootis group if:

- C I 1657/Al II 1670 > 2
- C I 1657/Ni II 1741 > 2.5

These limits have been set in such a way that they are valid for the entire range of effective temperatures. Small uncertainties due to variations in log g and [M/H] have been taken into account making, therefore, these limits independent of physical parameters. Moreover, they are also independent of evolutionary status as a clear distinction between λ Bootis and FHB stars can be made.

3. Search of new candidates

Based on the criteria given above a search in the IUE Archive of new λ Bootis candidates has been performed. A total of 5143 SWP spectra corresponding to late B, A and F-type have been studied. No LWP spectra were examined since no conclusive criteria were defined for this wavelength range. 22 new λ Bootis candidates have been identified (Solano & Paunzen 1999).

References

Faraggiana, R., Gerbaldi, M., Boehm, C., 1990, A&A, 235, 311
Jaschek, M., Baschek, B., Jaschek, C., Heck, A., 1985, A&A, 152, 439
Kurucz, R.L.,1993, CD-ROM 1-23, Smithsonian Astrophysical Observatory
Moon, T.T., Dworetsky, M.M., 1985, MNRAS, 217, 305
Paunzen, E., Weiss, W.W., Heiter, U., North, P., 1997, A&AS, 123,93
Solano, E., Paunzen, E., 1998, A&A, 331, 633
Solano, E., Paunzen, E., 1999, in preparation
Stürenburg, S., 1993, A&A, 277, 139
Venn, K.A., Lambert, D.L., 1990, ApJ, 363, 234

LIBRARIES OF HIGH AND MID-RESOLUTION SPECTRA OF F, G, K, AND M FIELD STARS

D. MONTES

Dpt. de Astrofísica. Facultad de C. Físicas.
Universidad Complutense de Madrid. E-28040 Madrid, Spain

Abstract. I have compiled here the three libraries of high and mid-resolution optical spectra of late-type stars I have recently published. The libraries include F, G, K and M field stars, from dwarfs to giants. The spectral coverage is from 3800 to 10000 Å, with spectral resolution ranging from 0.09 to 3.0 Å. These spectra include many of the spectral lines most widely used as optical and near-infrared indicators of chromospheric activity. The spectra have been obtained with the aim of providing a library of high and mid-resolution spectra to be used in the study of active chromosphere stars by applying a spectral subtraction technique. However, the data set presented here can also be utilized in a wide variety of ways. A digital version of all the fully reduced spectra is available via FTP and the World Wide Web (WWW) in FITS format.

1. Introduction

Spectral libraries of late-type stars with medium to high resolution and large spectral coverage are an essential tool for the study of the chromospheric activity in multiwavelength optical observations using the spectral subtraction technique (Barden 1985; Montes et al. 1995, 1999). Furthermore, these libraries are also very useful in many areas of astrophysics such as the stellar spectral classification, determination of atmospheric parameters (T_{eff}, $\log g$, [Fe/H]), modeling stellar atmospheres, radial velocity templates, study of variable stars, spectral synthesis applied to composite systems, and spectral synthesis of the stellar population of galaxies.

Astrophysics and Space Science is the original source of publication of his article. It is recommended that this article is cited as: *Astrophysics and Space Science* **263**: 275–278, 1999.

2. Observations

The spectra included in these libraries have been taken during several observing runs using different telescopes and instrumental configurations: the 2.5 m Isaac Newton Telescope (INT) at the Observatorio del Roque de Los Muchachos (La Palma, Spain) using the Intermediate Dispersion Spectrograph (IDS); the 4.2 m William Herschel Telescope (WHT) (La Palma, Spain) with the Utrecht Echelle Spectrograph (UES); the 2.2 m telescope at the German Spanish Observatory (CAHA) in Calar Alto (Almería, Spain), using a Coudé spectrograph; and the 0.9-m and 2.1-m telescopes of the Kitt Peak National Observatory (KPNO) (Arizona, USA) using the Penn State Fiber Optic Echelle (FOE) spectrograph. The spectral resolution achieved ranges from 0.09 to 3.0 Å. We have obtained a nearly complete optical region coverage from 3900 to 10000 Å. The signal to noise ratio is larger than 100.

3. The libraries

I have summarized here the three libraries of high and mid-resolution optical spectra of late-type stars I have recently published.

3.1. HIGH AND MID-RESOLUTION SPECTRA IN THE CA II H & K, Hα, Hβ, NA I D_1, D_2, AND HE I D_3 LINE REGIONS

In Paper I (Montes et al. 1997) a library of high and mid-resolution (0.2 to 3 Å) spectra in the Ca II H & K, Hα, Hβ, Na I D_1, D_2, and He I D_3 line regions of F, G, K and M field stars was presented.

The library includes a total of 170 spectra of a sample of 116 stars of which 71 are dwarfs (V) and 45 subgiant and giants (IV, III). The distribution in spectral type is as follows: 25 F type stars, 44 G, 31 K and 16 M. The spectra are available in FITS format at:

`http://www.ucm.es/info/Astrof/fgkmsl/fgkmsl.html`

3.2. HIGH-RESOLUTION UES ECHELLE SPECTRA

In Paper II (Montes & Martín 1998) a library of UES echelle spectra of a sample of F, G, K and M field dwarf stars was presented. This library is an extension of Paper I to higher spectral resolution (0.09 to 0.19 Å) covering a large spectral range (4800 to 10600 Å). These spectra include some of the spectral lines most widely used as optical and near-infrared indicators of chromospheric activity and a lot of lines of interest to spectral classification and calibration of temperatures purposes, as well as other lines normally used for the application of the Doppler imaging technique (Vogt & Penrod 1983).

The library includes 105 spectra of a sample of 83 dwarfs. The distribution in spectral type is as follows: 12 F type stars, 20 G, 25 K and 26 M. The spectra are available in FITS format at:

http://www.ucm.es/info/Astrof/fgkmsl/UESfgkmsl.html

3.3. MEDIUM-RESOLUTION FOE ECHELLE SPECTRA

The spectral library presented in Paper III (Montes, Ramsey, & Welty 1999) is an extension of Paper I and II. The library consists of echelle spectra of a sample of F, G, K and M field stars, mainly dwarfs (V), subgiants (IV), and giants (III) but also some supergiants (II, I). The spectral resolution is intermediate, nominally R = 12000 ($\approx$ 0.5 Å in Hα), but the spectra have a nearly complete optical region coverage (from 3900 to 9000 Å) including some of the spectral lines most widely used as optical and near-infrared indicators of chromospheric activity such as the Balmer lines (Hα to Hϵ), Ca II H & K, Mg I b triplet, Na I D_1, D_2, He I D_3, and Ca II IRT lines, as well as temperature sensitive photospheric features such as TiO bands.

A total of 130 stars (50 dwarfs, 64 subgiants and giants, and 16 supergiants) are included in this library. The distribution in spectral type is as follows: 22 F type stars, 40 G, 44 K and 24 M. Many of these stars have been observed in several observing runs, and in some cases in several nights during the same observing run, being the total number of spectra 345. Using these spectra as well as those of Papers I and II, a study of possible short and long term spectroscopic variability of some of the multiply observed stars is possible.

The spectra are available in FITS format at:

http://www.ucm.es/info/Astrof/fgkmsl/FOEfgkmsl.html

Additional information about these and other libraries of stellar spectra available in the literature can be found in the following web page I maintain:

http://www.ucm.es/info/Astrof/spectra.html

Acknowledgments We have made use of the La Palma ING data archive for retrieving some spectra. This work was supported by the Universidad

Complutense de Madrid, the Spanish Dirección General de Enseñanza Superior e Investigación Científica (DGESIC) under grant PB97-0259, and by the american National Science Foundation (NSF) grant AST 92-18008.

References

Barden, S.C., 1985, ApJ, 295, 162
Montes, D., Martín, E.L., 1998, A&AS, 128, 485 (Paper II)
Montes, D., Fernández-Figueroa, M.J., De Castro, E., Cornide, M., 1995, A&A, 294, 165
Montes, D., Martín, E.L., Fernández-Figueroa, M.J., Cornide M., De Castro, E., 1997, A&AS, 123, 473 (Paper I)
Montes, D., Ramsey, L.W., Welty, A.D., 1999, ApJS, in press (Paper III)
Montes, D., Sanz-Forcada, J., Fernández-Figueroa, M.J., De Castro, E., Cornide, M., Ramsey, L.W., 1999, Ap&SS (this volume)
Vogt, S.S., Penrod, G.D., 1983, PASP, 95, 565

MULTIWAVELENGTH OPTICAL OBSERVATIONS OF CHROMOSPHERICALLY ACTIVE BINARY SYSTEMS

D. MONTES[1,2], J. SANZ-FORCADA, M.J. FERNÁNDEZ-FIGUEROA, E. DE CASTRO AND M. CORNIDE
[1] *Dpt. de Astrofísica. Facultad de C. Físicas. Universidad Complutense de Madrid. E-28040 Madrid, Spain*

AND

L.W. RAMSEY[2]
[2] *The Pennsylvania State University, Dpt. of Astronomy and Astrophysics, 525 Davey Lab, University Park, PA 16802, USA*

Abstract. We summarize here our ongoing project of multiwavelength optical observations aimed at studying the chromosphere of active binary systems, using the information provided for several optical spectroscopic features, that are formed at different heights in the chromosphere.

1. Introduction

The chromospherically active binaries (CABS) are detached binary systems with cool components characterized by strong chromospheric, transition region, and coronal activity. The RS CVn systems have at least one cool evolved component whereas both components of the BY Dra binaries are main sequence stars. The simultaneous observations of different lines, that are formed at different heights in the chromosphere (from the minimum temperature region to the higher chromosphere), are of special interest for stellar activity studies, since they provide very useful information about this stellar region. The best way to obtain the active-chromosphere contribution to these spectral lines is to subtract the underlying photospheric contribution using the spectral subtraction technique (Montes et al. 1995a, c). The synthesized spectrum has been constructed with reference stars taken from our libraries (Montes 1998). With the simultaneous analysis of these chromospheric activity indicators it is possible to study in detail

Astrophysics and Space Science is the original source of publication of his article. It is recommended that this article is cited as: *Astrophysics and Space Science* **263**: 279–282, 1999.

the chromosphere, discriminating among the different structures as plages, flares, and microflares.

2. Observations

High and intermediate optical spectroscopic observations of a large sample of CABS have been obtained during several observing runs using different telescopes and instrumental configurations: the 2.5 m Isaac Newton Telescope (INT) at the Observatorio del Roque de Los Muchachos (La Palma, Spain) using the Intermediate Dispersion Spectrograph (IDS); the 4.2 m William Herschel Telescope (WHT) (La Palma, Spain) and the Utrecht Echelle Spectrograph (UES); the 2.56 m Nordic Optical Telescope (NOT) (La Palma, Spain) using the Soviet Finish High Resolution Echelle Spectrograph (SOFIN); the 2.2 m telescope at the German Spanish Observatory (CAHA) in Calar Alto (Almería, Spain), using a Coudé spectrograph; the 2.1 m telescope at McDonald Observatory (Texas, USA) using the Sandiford Cassegrain Echelle Spectrograph. The spectral resolution achieved range from 0.05 to 0.5 Å. We have obtained a nearly complete optical region coverage (from 3900 to 9000Å), including most of the spectral lines widely used as chromospheric activity indicators.

3. Results

In previous work (Fernández-Figueroa et al. 1994; Montes et al. 1995a, b) we have studied the Ca II H & K and Hα lines in a large sample of CABS. The activity-rotation and flux-flux relations for these systems have been analyzed (Montes 1995; Montes et al. 1995a, c, 1996a). Currently we are analyzing in detail several optical spectroscopic features using echelle spectroscopy in order to determine the effects of stellar activity on spectral lines originating at different heights in the chromosphere. Na I D_1, D_2, and Mg I b triplet (upper photosphere and lower chromosphere), Ca II IRT lines (lower chromosphere), Hα, Hβ, Ca II H & K (middle chromosphere), and He I D_3 (upper chromosphere).

A flare on UX Ari

In our IDS/INT observations of UX Ari (Montes et al. 1996b) we found a dramatic increase in the excess Hα emission equivalent width by a factor of 2 in an interval of 1 day, indicating the beginning of a strong flare in this system. The presence of the He I D_3 in emission in coincidence with the enhancement of the Hα emission confirms the detection of a flare. The application of the spectral subtraction technique reveals that the core of the Na I D_1 and D_2 lines are also filled-in by chromospheric emission.

Simultaneous Hα, Na I D_1, D_2, and He I D_3 observations

We analyzed, using the spectral subtraction technique, IDS/INT simultaneous Hα, Na I D_1, D_2, and He I D_3 spectroscopic observations of 18 CABS (Montes et al. 1997). The subtracted Hα profile of the more active stars of the sample (Hα in emission above the continuum) has very broad wings, and is well matched using a two-components Gaussian fit (narrow and broad). The broad component is primarily responsible for the observed variations of the profile, and its contribution to the total EW increases with the degree of activity. So, we have interpreted this broad component emission as arising from microflaring activity that take place in the chromosphere of these very active stars (see also Montes et al. 1998b). Prominence-like extended material have been detected in a near-eclipse Hα observation of the system AR Lac. The excess emission found in the Na I D_1 and D_2 lines by application of the spectral subtraction technique and the behaviour of the Hα line in the corresponding simultaneous observations indicate that the filling-in of the core of these lines is a chromospheric activity indicator. For giant stars of the sample the He I D_3 line has been detected in absorption in the subtracted spectra.

EZ Peg

We have studied in detail the system EZ Peg using IDS/INT and UES/WHT observations (Montes et al. 1998a). We have found that the hot component is the active star of the system, showing strong emission in the Hα, Ca II H & K, Hϵ, and Ca II IRT lines, and a strong filling-in of the Hβ line. The He I D_3 could present a total filling-in due to microflaring. The higher luminosity class of the hot component, as our spectra seem to indicate, could explain why the hot component is the active star of the system.

SOFIN observations

A sample of CABS have been also observed with the SOFIN/NOT spectrograph. These high resolution echelle spectra cover the range from the Ca II H & K lines to the Ca II IRT. The preliminary analysis of the 1996 observations of 5 CABS (UX Ari, σ Gem, BF Lyn, DK Dra and OU Gem) can be found in Sanz-Forcada et al. (1998a). A large sample has been observed in 1998 and the preliminary results for one of the systems, MS Ser, can be seen in Sanz-Forcada et al. (1998b).

X-ray/EUV discovered CABS

As an extension of our study of well known CABS we have started to observe also CABS recently discovered by their X-ray and EUV emission detected by the ROSAT and EUVE satellites. We are analyzing the Hα and Ca II IRT lines of some of these stars with observations taken in 1998 at the 2.1m telescope at McDonald Observatory using the Sandiford Cassegrain Echelle Spectrograph. Worth mentioning is the system 2RE J0743+224. We found that this star, previously classified as single-lined spectroscopic binary, is a double-lined spectroscopic binary (SB2) with a K1 III primary and an orbital period of 10 days. During the observations an unusual long-duration flare took place. A Li I λ6708 Å line enhancement is detected during the flare, which we suggest is produced by spallation reactions (Montes & Ramsey 1998a, b).

High temporal resolution observations

Finally, we want to extend our study to fainter, short period, and rapid rotators CABS not studied until now with 2 and 4m class telescopes due to the phase smearing during the exposition. We will take observations with the recently dedicated 11m telescope HET (Hobby-Eberly Telescope, McDonald Observatory) to study this kind of stars. This telescope allows us also to perform very high time resolution spectroscopic observations to study microflaring (time variation of the Hα broad component), as well as the detection and evolution of prominence clouds and plage-like regions.

Additional information about our research in chromospheric activity can be found at: `http://www.ucm.es/info/Astrof/actividad.html`

Acknowledgments This work was supported by the Universidad Complutense de Madrid, the Spanish Dirección General de Enseñanza Superior e Investigación Científica (DGESIC) under grant PB97-0259, and by National Science Foundation (NSF) grant AST 92-18008.

References

Fernández-Figueroa M.J., Montes D., De Castro E., Cornide M., 1994, ApJS 90, 433
Montes D., 1995, Ph. D. Universidad Complutense de Madrid
Montes D., 1998, Ap&SS (this volume)
Montes D., Fernández-Figueroa M.J., De Castro E., Cornide M., 1995a, A&A 294, 165
Montes D., Fernández-Figueroa M.J., De Castro E., Cornide M., 1995b, A&AS 109, 135
Montes D., De Castro E., Fernández-Figueroa M.J., Cornide M., 1995c, A&AS 114, 287
Montes D., Fernández-Figueroa M.J., Cornide M., De Castro E., 1996a, A&A 312, 221
Montes D., Sanz-Forcada J., Fernández-Figueroa M.J., Lorente R., 1996b, A&A 310, L29
Montes D., Fernández-Figueroa M.J., De Castro E., Sanz-Forcada J., 1997, A&AS 125, 263
Montes D., Ramsey L.W., 1998a, in *Solar and Stellar Activity: Similarities and Differences*, eds. C.J. Butler J.G. Doyle, ASP Conf. Ser., in press
Montes D., Ramsey L.W., 1998b, A&A, 340, L5
Montes D., Sanz-Forcada J., Fernández-Figueroa M.J., De Castro E., Poncet A., 1998a, A&A 330, 155
Montes D., Fernández-Figueroa M.J., De Castro E., Cornide M., Poncet A., Sanz-Forcada J., 1998b, in *The Tenth Cambridge Workshop on Cool Stars, Stellar Systems, and the Sun*, eds. R.A. Donahue & J.A. Bookbinder, ASP Conf. Ser. 155, CD-1516
Sanz-Forcada J., Montes D., Fernández-Figueroa M.J., De Castro E., Cornide M., 1998a, in *The Tenth Cambridge Workshop on Cool Stars, Stellar Systems, and the Sun*, eds. R.A. Donahue & J.A. Bookbinder, ASP Conf. Ser. 155, CD-1450
Sanz-Forcada J., Montes D., Fernández-Figueroa M.J., De Castro E., Cornide M., 1998b, in *Solar and Stellar Activity: Similarities and Differences*, C.J. Butler and J.G. Doyle, eds., ASP Conf. Ser., in press

OPTICAL CLASSIFICATION OF IRAS POST-AGB CANDIDAT

O. SUÁREZ, M. MANTEIGA AND A. ULLA
Departamento de Física Aplicada, Universidade de Vigo
E-36200 Vigo, Spain

P. GARCíA-LARIO
ISO Data Centre, Vilspa
Apdo. 50727, E-28080 Madrid, Spain

AND

A. MANCHADO TORRES
Instituto de Astrofísica de Canarias, E-38200 La Laguna, Spain

Abstract. We present the first results of a survey carried out on a sample of optically bright post-AGB candidates selected from the IRAS Point Source Catalogue, many of them previously unidentified in the literature. The selection criteria are based on their characteristic far infrared colours. Low resolution spectroscopy of more than 150 sources has been obtained during 6 years. Our preliminary analysis confirms that most of the candidates observed are true post-AGB stars, some of them still strongly reddened. Some young planetary nebulae, a few young stellar objects and active galactic nuclei are also present in the sample. The distribution of post-AGB stars with different spectral types in the IRAS two-colour diagram and its evolutionary implications are briefly discussed.

1. Introduction

Post-Asymptotic Giant Branch (Post-AGB, hereafter) stars are rapidly evolving low and intermediate mass stars in the transition phase from the AGB to the planetary nebula (PNe, hereafter) stage. AGB stars are very bright in the infrared, but heavily obscured in the optical by thick circumstellar envelopes formed as a consequence of the strong mass loss. Shortly after the end of the AGB, the mass loss suddenly stops, the star is no longer variable and the effective temperature of the central star increases. That leads to a change in the star spectral type, migrating from late to early

Astrophysics and Space Science is the original source of publication of his article. It is recommended that this article is cited as: *Astrophysics and Space Science* **263**: 283–286, 1999.

spectral types. In the time this change takes place, the circumstellar envelope vanishes and the star moves towards the right in the IRAS two-colour diagram (García-Lario 1992). Only with the advent of IRAS data it has been possible to efficiently recognize stars in this phase, since in the optical they appear identical to low gravity Population I supergiant stars.

2. Selection criteria

The selection criteria used to define our sample are those described in Manchado et al. (1989). A slight modification allows the inclusion of sources not detected in the 12μm IRAS band but satisfying the rest of criteria, assuming an upper limit of 0.25 Jy for the flux in this band.

According to previous results obtained by Pottasch et al. (1988), van der Veen et al. (1989) and García-Lario et al. (1997), this region of the IRAS [12]−[25] μm versus [25]−[60] μm two-colour diagram is mostly populated by PNe and stars in the previous AGB and post-AGB stage. Some overlap exists with regions where young stellar objects and Seyferts can be present.

In this preliminary study all objects satisfying our criteria and observable from the European Southern Observatory (ESO) were considered, excluding objects previously identified in the literature as young stellar objects or galaxies, and those without an optical counterpart in the ESO prints and a strong OH maser emission (probably still in the AGB stage, evolving as heavily obscured OH/IR stars).

3. Observations and data reduction

The observations here reported were conducted during 5 different runs spanning 6 years, from February 1990 to February 1995. All the observations were carried out at the 1.5m ESO telescope (La Silla, Chile). A total of 280 low resolution optical spectra, in most cases up to the near infrared range, were obtained for a sample of 151 objects satisfying our selection criteria. The spectral resolution ranges from 1.9 Å/pixel to 3.6 Å/pixel. The CCD images taken were reduced and calibrated with IRAF following standard procedures. As expected, we did not only find post-AGB stars, but also PNe, young stars, Seyfert galaxies and some objects without any optical counterpart, as shown in Figure 1.

4. Spectral classification

From the 97 spectra corresponding to evolved objects, a total of 56 were identified as post-AGB stars while the remaining 41 are PNe.

Spectra of supergiant stars in Silva et al. (1992) catalog were used as templates to classify the normalized spectra of objects in our sample. A

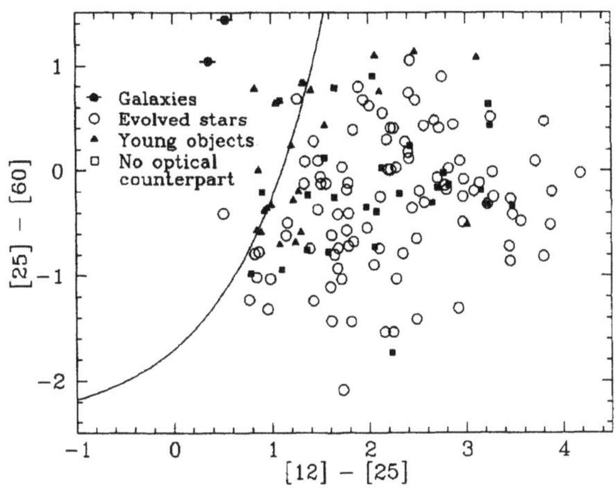

Figure 1. Distribution of the selected objects in the two-colour [12]-[25] μm versus [25]-[60] μm diagram

spectral region in the wavelength range from 3600 to 5050 Å was selected because it allows us to distinguish among classes from B to M with a good margin of confidence. Additional spectral regions in the red were used for late type stars to confirm their classification.

5. Results

Figure 2 shows the distribution of objects belonging to different spectral classes observed in the IRAS two-colour diagram together with different regions defined in order to facilitate the discussion below.

According to the model proposed (García-Lario 1992), stars on the AGB phase are located over the solid line in the two-colour diagram, that can be referred to as infrared main sequence (IRMS). The distribution of spectral types found in the IRAS two-colour diagram supports an evolutionary interpretation of the results obtained. Most of the objects in region II show early spectral types, while post-AGB stars with later spectral types appear concentrated close to the IRMS, in region I.

The distribution seen in region II can be explained as the consequence of the termination of the AGB phase and the dilution of the circumstellar envelope into the interstellar medium and the simultaneous increase of the effective temperature of the central star. Transition objects with nebular emission lines overimposed on a strong stellar continuum, proto-PNe, and stars with B or A spectral types are found in this region. Stars with more massive progenitors are expected to evolve faster and to show redder colours. These stars could be represented by the small group of objects

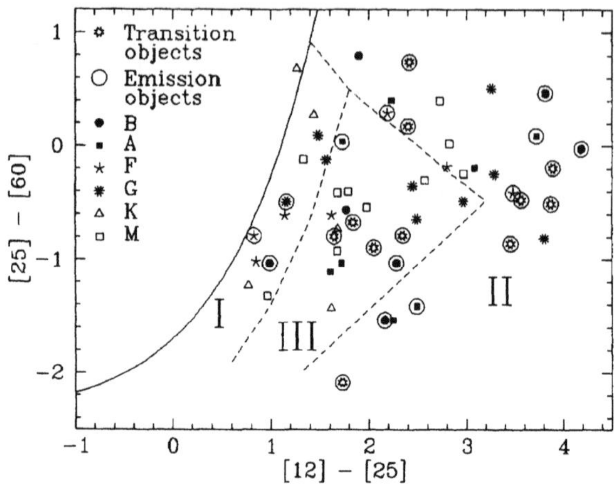

Figure 2. Distribution of post-AGB objects observed in the two-colour diagram [12]-[25] μm versus [25]-[60] μm

located in region II closer to the IRMS in the upper region of the IRAS two-colour diagram. In contrast, stars with low mass progenitors would evolve more slowly towards hotter effective temperatures, reaching early spectral types only when they have left the IRMS long time ago. They can be identified with the group of objects in region II in the lower regions of the diagram.

Stars in region I are mainly F, G, K or M type stars representing objects that have not yet evolved away from the IRMS, probably slowly evolving post-AGB stars with low mass progenitors.

The intermediate region, named III in figure 2, is populated by a mixture of objects mostly in their evolutionary way from the IRMS to region II.

Acknowledgements

This research was partially funded through Spanish DGICYT and DGESIC projects number PB94-1274 and PB97-1435-C02-02, respectively.

References

García-Lario, P., 1992, Ph. D. Thesis, University of La Laguna, Tenerife, Spain
García-Lario, P., Manchado, A., Pych, W., Pottasch, S.R., 1997, A&AS, 126, 479
Manchado, A., Pottasch, S.R., García-Lario, P., Esteban, C., Mampaso, A., 1989, A&A, 214, 139
Pottasch, S.R.,Bignell C., Olling R., Zijlstra A.A., 1988, A&A, 205, 248
Silva, D.R., Cornell, M.E., 1992, ApJS, 81, 865
van der Veen W.E.C.J., Habing H.J., Geballe, T.R., 1989, A&A, 226, 108

CONSEQUENCES OF THE COLLISION BETWEEN THE GAS EJECTED IN THE COLLAPSE OF A WHITE DWARF AND A LOW-MASS STAR

NURIA SERICHOL, DOMINGO GARCíA-SENZ AND EDUARDO BRAVO
Dpt. de Física i Enginyeria Nuclear (UPC) and Institut d'Estudis Espacials de Catalunya. Barcelona. Spain

1. The isolated millisecond pulsars enigma

The millisecond pulsars roughly amounted 10% of the known pulsar population. The most accepted explanation about their origin is that they are neutron stars, belonging to a compact binary system, whose spin has increased by the accretion of material from a nearby secondary. This explanation is supported by the existence of a large number of low-mass X-ray binaries, which are thought to be the progenitors of the binary pulsars. Nevertheless, a complication arises as a small subset of these millisecond pulsars are found isolated. In order to understand how the neutron star is left alone, once it has been spun up by accretion, it is necessary to invoke a mechanism to eliminate the secondary. In this regard, several mechanisms have been proposed such as: evaporation of the secondary caused by the radiation from the pulsar (Ruderman et al. 1989), ionization of the binary due to an encounter with a third star, unstable Roche lobe overflow leading to the transformation of the degenerate secondary in a heavy disc orbiting around the neutron star (Ruderman & Shaham 1985). There are other scenarios, although none of them free of objections (see for instance Bhattacharya, 1996, for a critical review).

In this work we have followed a different approach to explain the possible origin of the isolated millisecond pulsars, exploring an idea suggested long ago by Iben & Tutukov (1984). In the Iben & Tutukov proposal, a massive white dwarf ($M_{\rm wd} \simeq 1.4M_{\odot}$) in a compact binary system increases its rotation velocity by effect of the angular momentum transfer from the accretion disc. As a consequence of accretion, the compact star increases not only its rotation velocity but also its mass. At some point the white dwarf's

Astrophysics and Space Science is the original source of publication of his article. It is recommended that this article is cited as: *Astrophysics and Space Science* **263**: 287–290, 1999.

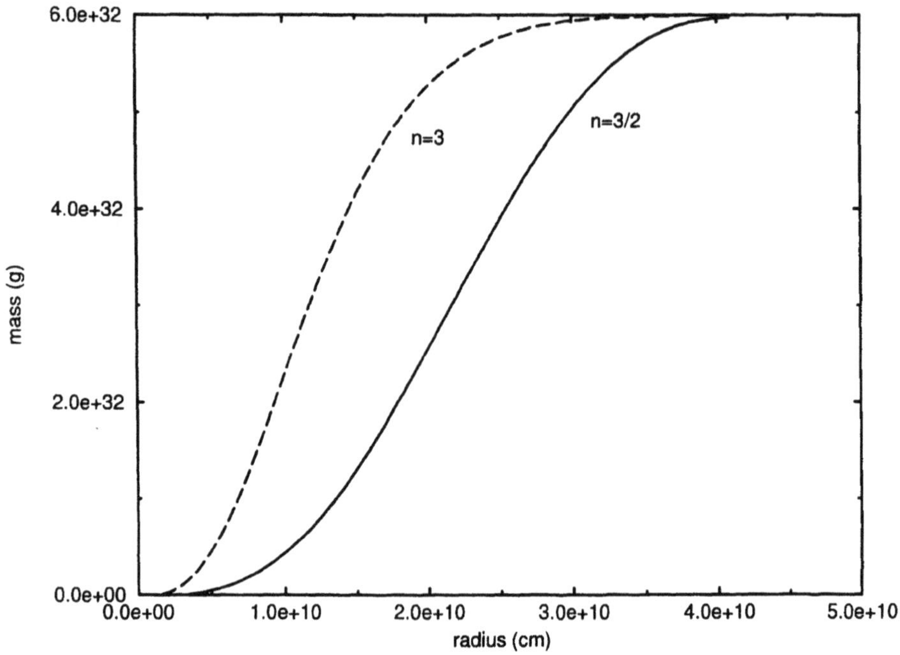

Figure 1. Mass distribution for polytropes of index n=3/2 and n=3 with total mass $0.3M_{\odot}$

mass becomes greater than the Chandrasekhar mass limit, and collapses to a neutron star (Canal, Isern & Labay 1990). During that accretion-induced-collapse (AIC), the rotation period shortens by a huge factor, while a small quantity of material located near the surface of the original white dwarf may be ejected at high velocity ($\simeq 0.1M_{\odot}$ at 10^4 km s^{-1}). Although in this scenario the explosion is less violent than in a conventional supernova event, the effect of the blast wave on the companion star could be important. Low-mass stars have convective interiors that can be approximately described by an index $n = 3/2$ polytrope, which is much less centrally condensed than a radiative type $n = 3$ polytrope. Thus, low-mass stars have a higher concentration of matter near the surface where ablation takes place (Figure 1), being easier to destroy than the more massive ones. The destruction of the secondary would leave an ultrarapid pulsar isolated.

2. Results of the hydrodynamic simulation of the collision

A smooth particle hydrodynamics code was used to study the collision between the high-velocity ejecta and a $0.3M_{\odot}$ main sequence star. The low-mass star was simulated by using 22400 particles. The spherical ejecta had a size of a few times the radius of the secondary and a thickness of a third of the orbital separation D. It was represented with 7600 particles of equal mass. The orbiting distance was chosen as to allow the transfer of matter from the low-mass star to the white dwarf via Roche-lobe overflow. The main parameters of the simulation are shown in Table 1. The meaning of the columns are as follows: M_s, R_s and D stand for the secondary mass and radius and the orbital separation, $v_{\rm ej}$ is the velocity of the ejecta, $M_{\rm in}$ and $p_{\rm in}$ are the incident mass and the incident momentum, and the last column is the ratio between the incident kinetic energy of the ejecta and the binding energy of the secondary.

TABLE 1. Input parameters for the initial model

$M_s(M_\odot)$	$R_s(R_\odot)$	$D(R_\odot)$	$v_{\rm ej}$ (km/s)	$M_{\rm in}$ $(M_\odot)$	$p_{\rm in}$ (gr cm/s)	E_k/E_B
0.3	0.594	2.3	10000	4.9×10^{-3}	9.7×10^{39}	38.8

As it was pointed out in previous calculations (Fryxell & Arnet 1981), the effect of the hypersonic collision is two-fold. The direct momentum transfer from the gas to the star strippes a portion of the envelope of the secondary, giving at the same time a radial velocity component of a few dozens of kilometers per second to the remnant. On the other hand, the energy deposition which takes place in such an inelastic collision heats the core. The subsequent thermal ablation of part of the core material leads to a new increase of the radial velocity owing to the rocket-effect. From the figures of Table 1 we expect that both mechanisms will be of considerable importance in order to decide the outcome of the secondary. The large E_k/E_B ratio (due primarily to the low binding energy of the secondary) promises a good deal of ablation.

The simulation started once the ejecta came into contact with the outer edge of the secondary. As time went on, the impact of the ejecta blew all the material of the outermost envelope while sending strong shock waves to the interior of the star. At $t \simeq 150$ s the gas of the ejecta had already passed through the star and the interaction ceased. Some particles belonging to the ejecta reversed their velocity during the collision. The shock wave, which was travelling right to the center of the star, compressed and heated the material, increasing its internal energy. A temperature peak of about $7-8 \times 10^7$ K and a density of several gr/cm^3 were observed near the center of the low-mass star, which would be enough to switch-on the combustion of hydrogen. However, nuclear combustion is not expected to be relevent in the time-scale of the interaction. After $t = 189$ s, the particles of the gas representing the ejecta were removed from the simulation. At $t = 298$ s the calculation was stopped. By then, the total energy distribution, $E_{\rm kin} + E_{\rm int} + E_{\rm grav}$, showed that the energy was only slightly negative in a narrow region close to the surface, which encompassed 15% of the mass of the star. In addition, the outermost 5% of the mass was already stripped at the beginning of the collision. On average, the innermost stellar material is drifting in the direction of the collision at a velocity of $\simeq 50$ km s^{-1}. Even though more models should be run to follow in detail the evolution of the remnant, the total destruction of the star is the most plausible outcome in view of the distribution of total energy in the last calculated model.

Our results suggest that the AIC of a white dwarf in presence of a low-mass secondary could lead to the isolation of the nascent pulsar. Work is in progress to study the outcome of collisions under a wider range of parameters of the initial configuration, such as the velocity of the ejecta, the orbital separation, the mass of the secondary and others. Much work should also be done to clarify the reliability of the scenario, especially the mechanism by which a compact binary with the required masses and orbital parameters has come into being.

References

Bhattacharya, D., 1996, IAU Symp. 165, 243
Canal, R., Isern, J., Labay, J., 1990, ARA&A, 28, 183
Fryxell, B., Arnett, D., 1981, ApJ, 243, 994
Iben, I., Tutukov, A., 1984, ApJS, 54, 354
Ruderman, M., Shaham, J., 1985, ApJ, 289, 244
Ruderman, M., Shaham, J., Tavani, M, 1989, ApJ, 336, 507

THERMAL EVOLUTION OF NEUTRON STARS

JUAN A. MIRALLES, JOSÉ A. PONS
Departament d'Astronomia i Astrofísica
Universitat de València, SPAIN

AND

SANJAY REDDY
Department of Physics and Astronomy
SUNY at Stony Brook, NY, USA

Abstract. We present results from simulations of protoneutron star thermal evolution using neutrino opacities that are consistently calculated with the equation of state. When hyperons are allowed to appear in the system, we obtain metastable configurations that after the deleptonization stage become unstable. Concerning the evolution of old neutron stars, we present the results of our investigation on the effect of the Joule heating due to magnetic field dissipation. We conclude that this mechanism can be efficient in maintaining the surface temperature of the star above $3 \times 10^4 - 10^5$ K during a very long time (≥ 100 Myr), comparable with the decay time of the magnetic field.

1. Introduction

At birth, neutron stars (NSs) are very hot and lepton rich. This heat and the excess of lepton number is radiated away by neutrinos during the first tens of seconds of its live (Burrows and Lattimer 1986; Pons et al. 1998). The detection of these neutrinos from the SN1987A has confirmed the standard scenario widely accepted on the formation of NSs. However, the undetected photon radiation (or its effect) coming from the surface of the NS and the absence of pulsed radiation from any stellar remnant have led some authors to suggest that the protoneutron star (PNS) formed in the aftermath of a successfuly supernova explosion can collapse after deleptonization into a black hole (Prakash et al., 1997). Some models proposed to explain this

Astrophysics and Space Science is the original source of publication of his article. It is recommended that this article is cited as: *Astrophysics and Space Science* **263**: 291–294, 1999.

mechanism are based on the presence of hyperons or any boson condesate (either pions or kaons). In this paper we present recent results on the evolution of PNSs employing an equation of state (EOS) which allows the inclusion of hyperons. The neutrino opacities are consistently calculated with the EOS.

Following this epoch, in which neutrinos are diffusing through the star, a stage of transparency comes (see Tsuruta 1998 for a comprehensive review). Neutrinos produced leave the NS almost without interaction. The neutrino luminosity has dropped below the terrestrial detectors thresshold and we cannot expect to get direct information from this radiation. Only the detection of photons emitted from the surface can be used to infer the temperature and properties of the NS. Although neutrino radiation at this epoch is undetectable, is, by far, more important in determinnig the thermal evolution of the star than the photon radiation. After about 10^5 years the temperature at the interior of the star is low enough to reduce the neutrino emission below the photon emission from the surface. The NS has arrived to the photon cooling phase, in which the photon emission and possible heating sources (Shibazaki and Lamb 1989; Umeda et al. 1993; Van Riper, Link and Epstein 1995; Miralles, Urpin and Konenkov 1998) determine the subsequent evolution.

2. Protoneutron star evolution

To study the first tens of seconds in the evolution of a PNS we have solved the transport equations within the frame of the so-called *diffusion approximation.* In this approximation, neutrinos are supposed to be in thermal equilibrium with the matter, although they are allowed to be out of chemical equilibrium. We assume quasi-static evolution. This means that velocity terms are neglected in the transport equations and the hydrostatic equilibrium equations are satisfied during the whole evolution. The EOS we use in our simulations is based on a finite temperature field-theoretical model where interactions between baryons are mediated by the exchange of σ, ω and ρ mesons. The baryons we consider in the system are the nucleons, n, p, and the hyperons Λ, Σ and Ξ. The neutrino diffusion coefficients we use in our simulations are consistently determined from the EOS, taking into account the effects of Pauli blocking, composition and the change in the baryon effective mass due to strong interaction (Reddy, Prakash and Lattimer, 1998).

With all the ingredients described above, we made calculations of PNS evolution for different initial models (temperature profiles, etc.), total masses and adding or not hyperons to the system (Pons et al., 1998). The main conclusions we can deduce from the results are that neutrino luminosities and mean energies do not depend much upon the details of the initial model, but are sensitive to the total mass. The presence of hyperons allows the existence of metastable configurations: those which are stable at birth but upon deleptonization they become unstable against gravitational collapse into black hole. The evolution time for those stars is longer as they are closer to the maximum mass supported by a cold, deleptonized star.

3. Joule heating and the thermal evolution of old neutron stars

To study the influence of the magnetic field dissipation on the thermal evolution of old NSs we consider the magnetic field confined in the crust of the star. The evolution of such a field is governed by the induction equation without the velocity term since the matter of the crust solidifies after formation. We restrict our considerations to a dipolar field which can be described by the vector potential $\mathbf{A} = (0, 0, A_\varphi)$, $A_\varphi = S(r,t)\sin\theta/r$, where r and θ are the spherical radius and polar angle, respectively (Sang and Chanmugam, 1987).

In terms of $S(r,t)$, the induction equation can be written as

$$\frac{\partial^2 S}{\partial r^2} - \frac{2S}{r^2} = \frac{4\pi\sigma}{c^2}\frac{\partial S}{\partial t}$$

with the boundary condition

$$\frac{\partial S}{\partial r} + \frac{S}{R} = 0$$

at the stellar surface $r = R$ and σ being the electrical conductivity. For a field confined to the crust, $S(r,t)$ should vanish in the deep layers.

The rate of heating per unit volume, after taking polar-average, can be written in terms of S as follows

$$\dot{q} = \frac{c^2}{24\pi^2 r^2 \sigma}\left(\frac{\partial^2 S}{\partial r^2} - \frac{2S}{r^2}\right)^2$$

The evolution of the magnetic field as well as the Joule heating is strongly sensitive to the electrical conductivity. In the crust, the electrical conductivity is determined by electron scattering with phonons and lattice impurities. In our calculations we use for phonon conductivity data from Itoh, Hayashi and Kohyama (1993) and for impurity conductivity we use analytical expressions from Yakovlev and Urpin (1980).

Our cooling calculations taking into account the additional heating caused by ohmic dissipation have been done within the *isothermal approximation* that follows from assuming constant temperature (corrected by the redshift factor) in the interior of the star ($\rho > 10^{10}$ g/cm^3) and a temperature drop at the surface given by the atmosphere model that relates the interior temperature with the effective temperature (outer boundary condition) and allows to calculate the photon luminosity of the star. In our calculations we use the atmosphere models from Van Riper (1988).

Calculations presented here are based on the standard cooling scenario which correspond to the standard neutrino emissivities. We consider 1.4 $M_\odot$ models of NSs constructed with the EOS of Friedman and Pandharipande (1981) and Pandharipande and Smith (1975).

The main conclusion we obtain from the calculations is that the decay of the crustal magnetic field can produce enough heat to maintain a sufficiently high surface temperature $\sim 3 \times 10^4 - 10^5$ K. Our calculations predict that Joule heating becomes important after a relatively short initial phase ($\sim$ 3–5 Myr depending on the model) when the NS cools down to a value of the surface temperature $\sim 3 \times 10^4 - 10^5$ K. The further thermal evolution slows down substantially: the characteristic cooling time becomes comparable to the decay time for the magnetic field, which turns to be very large (Urpin and Konenkov, 1997). Therefore, the NS can maintain a surface temperature practically unchangeed during extremely long time > 100 Myr.

Acknowledgements

This contribution presents the works carried out in collaboration with J. Lattimer, M. Prakash (PNS evolution) and V. Urpin and D. Konenkov (evolution of old NSs), to whom the authors are grateful. This work has been supported in part by the Spanish DGICYT (grant PB97-1432).

References

Burrows, A., Lattimer, J.M., 1986, ApJ, 307, 178
Friedman, B., Pandharipande, V.R., 1981, Nucl. Phys. A, 361, 502
Itoh, N., Hayashi, H., Kohyama, Y., 1993, ApJ, 418, 405
Miralles, J.A., Urpin, V., Konenkov, D., 1998, ApJ, 503, 368
Pandharipande, V.R., Smith, D., 1975, Nucl. Phys. A, 237, 507
Pons, J.A., Reddy, S., Prakash, M., Lattimer, J.M., Miralles, J.A., 1998, ApJ, in press
Prakash, M. Bombaci, I., Prakash, M., Ellis, P., Lattimer, J.M., Knorren, R., 1997, Phys. Rep., 280, 1
Reddy, S., Prakash, M., Lattimer, J.M., 1998, Phys. Rev. D, 58, 013009
Sang, Y., Chanmugam, G.. 1989, ApJ, 323, L61
Shibazaki, N., Lamb, F.K., 1989, ApJ, 346, 808
Tsuruta, S., 1998, Phys. Rep., 292, 1
Umeda, H., Shibazaki, N., Nomoto, K., Tsuruta, S., 1993, ApJ, 408, 186
Urpin, V., Konenkov, D., 1997, MNRAS, 297, 167
Van Riper, K.A., 1988, ApJ, 329, 339
Van Riper, K.A., Link, B., Epstein, R.I., 1995, ApJ, 448, 294
Yakovlev, D., Urpin, V., 1980, SvA, 24, 303

STANDARD SPECTRAL EVOLUTIONARY SEQUENCE FOR TYPE IA SUPERNOVAE

G. GÓMEZ
Instituto de Astrofísica de Canarias
Vía Láctea s/n, E-38200 La Laguna, Tenerife, Spain

AND

R. LÓPEZ
Dep. Astronomia i Meteorologia. U. de Barcelona
Avda. Diagonal 647, E-08028 Barcelona, Spain

Abstract. Spectroscopic monitoring of a wide range of supernovae of different types and at different evolutionary stages has been carried out. The observations were taken using the same instrumentation throughout, and the data have been gathered together in the CANARIAN SUPERNOVAE ARCHIVE (CSNA). The data have been used to carry out an extensive study on type Ia supernovae. As a result, we have been able to establish a template spectral sequence. The sequence is extremely useful in, for example, permitting the identification of intrinsically peculiar objects, and it allows more reliable dating in cases in which photometric information is lacking. The homogeneity of type Ia supernovae is unclear because variations of second order have been detected. However, the validity of the template spectral sequence remains unaffected.
In addition, a case-by-case study has been carried out on certain supernovae that seem especially appealing: (i) our discovery of the type Ia twin supernovae 1992R and 1992ac; (ii) analysis of the peculiar type Ia supernovae 1991F which supports the existence of a new supernova sub-group.

1. Introduction

The Supernova Monitoring Project (SMP) was carried out at the Observatorio del Roque de los Muchachos between 1990 and 1994. The first aim of the Project was to compile a database of supernova spectra. This database

Astrophysics and Space Science is the original source of publication of his article. It is recommended that this article is cited as: *Astrophysics and Space Science* **263**: 295–298, 1999.

presents some advantages relative to those previously published; namely: 1) the spectra were obtained with a homogeneus instrumental setup; 2) a wide spectral range coverage (from $\lambda \sim 3500$ Å to $\lambda \sim 9500$ Å), and 3) wide temporal monitoring, from photospheric to the old nebular phases (i.e., from the early postmaximum to ~ 1 year).

2. Observations

The spectra were obtained with the faint object spectrographs FOS1 and FOS2 and with the ISIS spectrograph, using the Isaac Newton and the William Herschel telescopes. The spectral resolution is typically ~ 15 Å and their S/N > 50. The observational strategy was to set up an *Eulerian grid*, which consists in scheduling the observing time at regular intervals in order to take spectra of the detectable supernovae at the date of each observation, thus including supernovae at different evolution phases (see Gómez, López & Sánchez 1996 and Gómez & López 1998 for details).

3. Results

The CSNA Database contains 59 type Ia supernova spectra which correspond to 27 objects.

It is widely accepted that the bulk of Type Ia supernovae (SNe Ia) is highly homogeneous throughout its whole evolution with time. Because the spectra compiled in the CSNA have the advantage of having been recorded with the same instrumental setup, with the consequent avoidance of any effects caused by the diversity of the data sources, our database has allowed us to confirm the "first-order" homogeneous spectroscopic behaviour.

The "first-order" spectral homogeneity of SNe Ia suggested the use of the spectra of the CSNA in order to perform a standard type Ia supernova spectral evolutionary sequence. This standard sequence could be useful in yielding clues leading to the detection of some "peculiar" type Ia supernova events. Furthermore, it could be useful in assigning a phase to a given object from its spectrum when no photometric data are available. The standard spectral sequence we have defined is shown in Figure 1.

Furthermore, our database has allowed us to detect some events whose spectra are clearly peculiar throughout their entire evolution, the most important being SNe 1991F and 1991bg. The peculiarity of the first one was fully established on the basis of our standard spectral sequence. All the properties of SN 1991F lead us to conclude that this supernova belongs to the recently defined subluminous SN 1991bg-like supernova subclass (see Gómez & López 1995 for details).

Our database has also been used to perform a finer analysis, comparing the spectra of different supernovae at similar evolutionary phases. The

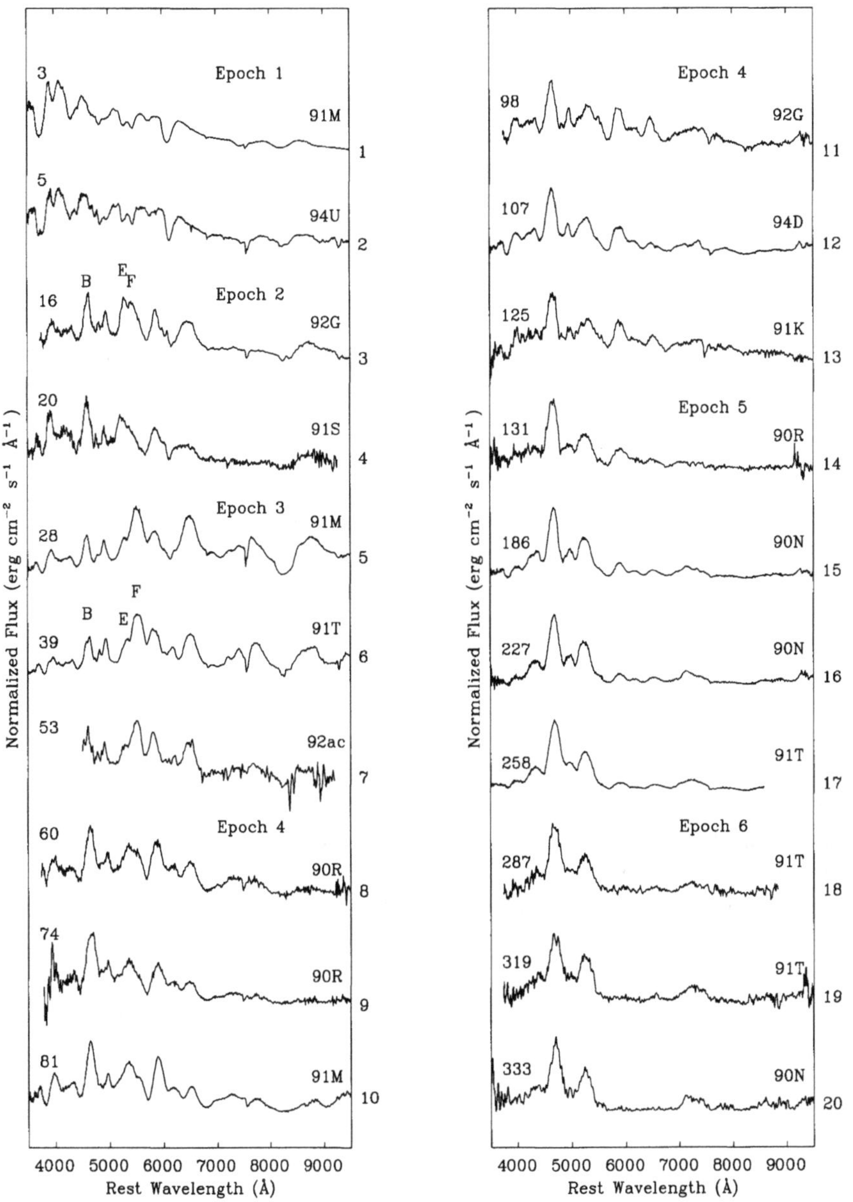

Figure 1. Standard Spectral Sequence for "normal" supernovae type Ia. The spectra were distributed into six "epochs" according with their respective phases. The spectra belonging to a given "epoch" show close similarities in the shapes and in the relative intensities of their spectral features, and differ through these characteristics when compared with the spectra of different "epochs". The spectra have beem ordered (indicated by a number on the right of each spectrum), from top to botom, by increasing phase.

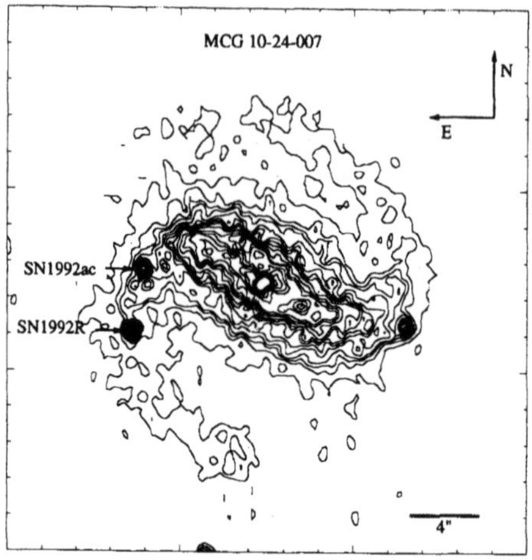

Figure 2. Central portion (28″x 28″) of the MCG 10-24-007 image in the R band taken on June 27 1992 at the 1 m JKT. The two supernovae are visible on the galaxy arm.

results point towards the existence of "second-order" differences related to strengths, velocities and evolution over time of some spectral features. However, these "second-order" variations do not invalidate the standard spectral sequence we have defined.

Finally, the standard spectral sequence was used to properly identify the phases of the "twin supernovae"; i.e., SNe 1992R and 1992ac (this second event, discovered within the SMP). These two type Ia supernovae were found in the same galaxy (MCG 10-24-007), nearly simultaneously (separated by less than two months). These two supernovae can be distinguished in the arm of the galaxy, separated by $\sim$ 4", in Figure 2.

This kind of work provides important tools for the full understanding of the properties of "close" type Ia supernovae. Knowledge of their behaviour is a prerequisite for their use in cosmological determinations of SNe Ia in highly redshifted galaxies.

References

Gómez, G., López, R., 1995, AJ, 109, 737
Gómez, G., López, R., Sánchez, F., 1996, AJ, 112, 2094
Gómez, G. & López, R., 1998, AJ, 115, 1096

CCD PHOTOMETRY OF VERY YOUNG OPEN CLUSTERS.

A. MARCO, G. BERNABEU AND J.M. TORREJÓN
Departamento de Física, Ingeniería de Sistemas y Teoría de la Señal, Universidad de Alicante, Ap.99, E-03080, Alicante, Spain.

AND

J. FABREGAT
Departamento de Astronomía y Astrofísica, Universidad de Valencia, E-046100, Burjassot, Valencia, Spain.

Abstract. We are developing a long term programme to determine abundances of peculiar stars, as a function of age, in open clusters with ages $\leq 5 \times 10^7$ yr. A fundamental part will be to deduce the ages and physical properties of these clusters with higher accuracy than in previous studies. As part of this effort we present here CCD *uvbyβ* photometry of two open clusters, NGC 884 and NGC 869.

1. Objectives

The long-term objective of this programme is the precise determination of abundances of peculiar stars, as a function of age, for very young open clusters. A chief part of this effort will be to determine the ages of the clusters by means of isochrone fitting to the observational HR diagram, as well as precise determination of reddening and distance. In order to do this, we present here standard *uvbyβ* CCD photometry for two open clusters, NCG 884 and NGC 869.

2. Introduction

The double cluster NGC 884 and NGC 869 is one of the richest young open clusters accessible to northern-hemisphere observers, and so is well documented in the literature. An extensive photographic study was carried

Astrophysics and Space Science is the original source of publication of his article. It is recommended that this article is cited as: *Astrophysics and Space Science* **263**: 299–302, 1999.

out by Oosterhoff (1937); MK spectral types for cluster members were determined by Bidelman (1943), Johnson & Morgan (1955), Schild (1965, 1966, 1967), and others; optical photometry in various systems was carried out by Johnson & Morgan (1955), Wildey (1964), Schild (1965), Crawford et al. (1970), Waelkens et al. (1990), and others; infrared photometry was obtained by Mendoza (1967) and Tapia et al. (1984). There is no general agreement on the distance moduli and the ages of both clusters. Crawford et al. (1970) conclude on the base of $uvby\beta$ photometry that both clusters have nearly the same age and distance, the distance modulus being 11.4±0.4 mags. Balona & Shobbrook (1984) have thought it necessary to correct this value for evolutionary effects and adopt a distance modulus of 11.16 for both clusters. On the other hand, Tapia et al. (1984) find confirmation of a previous suggestion by Schild (1967) that NGC 869 is younger and more nearby than NGC 884, with distance moduli of 11.7 and 12.0, respectively. In an attempt to improve these values our observations cover completely the faint end of the sequence until $M_V \sim 16$ mags. The determination of the interstellar absorption toward the clusters does not cause a major problem: from the most recent discussions, it appears that interstellar reddening is fairly uniform, with $E(b-y) = 0.41 \pm 0.02$ mags. Some stars present additional obscuration of circumstellar origin (Crawford et al. 1970; Tapia et al. 1984). A striking characteristic in the determination of abundances of peculiar stars in both clusters is the large frequency of Be stars that appears and the apparent absence of β Cephei stars. This contrasts with the small number of Be stars and the large number of β Cepheid stars found in the southern cluster NGC 3293, which is of similar age.

3. The observations

The data were obtained with the JKT telescope (La Palma) in December 1997, using the 1024 × 1024 TEK 4 chip CCD and $uvby\beta$ filters. Pixel size was 0.331 arcsec in such a way that the whole field covered by each frame was 5.6 × 5.6 arcmin. Three different observing times were scheduled for each filter and each cluster in order to reach the widest range of magnitudes. Standard stars were observed in the two clusters under study, and in NGC 6910, NGC 2169 and NGC 1039, using an intermediate exposure time.

3.1. REDUCTION PROCESS

The photometry has been obtained by PSF fitting using the DAOPHOT package (Stetson 1987) provided by IRAF, after bias and flat-field corrections. The atmospheric extinction correction were performed using the RANBO2 program, which implements the method described by Manfroid (1993).

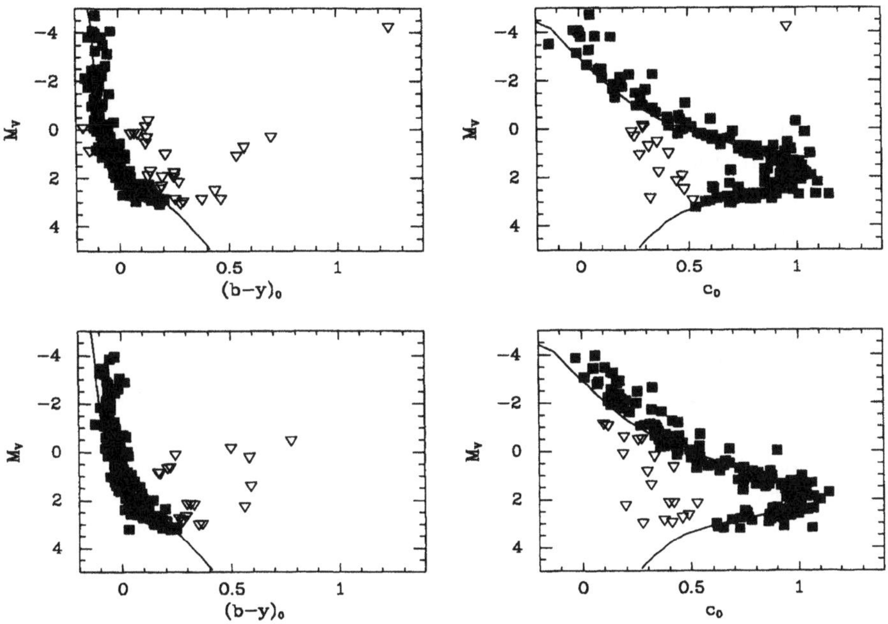

Figure 1. $M_V - (b-y)_0$ and $M_V - c_0$ for the open clusters NGC 884 (upper row) and NGC 869 (lower row). Triangles are stars classified as non-members.

Special care has been paid to the transformation to the standard system. A significative fraction of the observed stars in these clusters are highly reddened B type stars. Manfroid & Sterken (1987) and Crawford (1994) have shown that gross systematic errors may occur when the transformation process is applied to highly reddened stars, even if their colour range is fully bracketed by that of the standard stars. In order to avoid these effects we have elaborated a list of standards where only early-type stars (MK earlier than A3, where the maximum of Balmer lines appears) are included, covering a wide range of reddening values.

4. HR diagrams

In Fig. 1 we present the photometric diagrams of NGC 869 and NCG 884. The line represents the ZAMS given by Perry et al. (1987). Our data is consistent with the same uniform reddening, $E(b-y) = 0.41 \pm 0.02$ mags. given by Crawford (1970) for both clusters. The obtained distance modulus is 10.86, for both clusters. This value is significantly larger than that given by Balona & Shobbrook (1984), namely 11.16.

We have represented with triangles those stars that have been classified as non-members on the basis of their position in the diagram, out of the main sequence in both diagrams. This is an important step in order to achieve our objectives.

As it can be seen, the photometry reaches very faint magnitudes (~ 16 mags). This will allow us to fit with higher accuracy the faint end of the sequence to deduce their physical properties and ages.

5. Conclusions

We present *uvbyβ* photometry of $\sim$ 360 stars in two very young open clusters, NGC 869 and NGC 884, $\sim$ 80 of which are catalogued for the first time. The great number of stars will allow us to fit the faint end of the main sequence so that we will can obtain the age and distance with higher accuracy than in previous studies.

On the basis of their positions in the $M_V - c_0$ and $M_V - (b-y)_0$ diagrams, we catalogue for the first time a number of stars as non-members.

Our data is consistent with the uniform reddening for both clusters, $E(b-y) = 0.41 \pm 0.02$ mags., given by Crawford (1970) while the distance modulus for both clusters deduced from our data is 0.3 mags. larger than the value, 11.16, given by Balona & Shobbrook (1984).

References

Balona, L.A., Shobbrook, R.R, 1984, MNRAS, 211, 375
Bidelman, W.P., 1943, ApJ, 98, 61
Crawford, D.L., 1994, PASP, 106, 397
Crawford, D.L., et al., 1970, AJ, 75, 822
Johnson, H.L., Morgan, W.W., 1955, ApJ, 122, 429
Manfroid, J., 1993, A&A, 271, 714
Manfroid, J., Sterken, C., 1987, A&AS, 71, 539
Mendoza, E.E., 1967, Bol. Obs. Tonantzintla y Tacubaya, 4, 149
Oosterhoff, P.T., 1937, Ann. Sterrewacht Leiden, 17, 1
Perry, C.L., Olsen, E.H., Crawford, D.L., 1987, PASP, 1184
Schild, R.E., 1965, ApJ, 142, 979
Schild, R.E., 1966, ApJ, 146, 142
Schild, R.E., 1967, ApJ, 148, 449
Stetson, P.B., 1987, PASP, 99, 191
Tapia, M., Roth, M., Costero, R., Navarro, S., 1984, Rev. Mex.Astron. Astrofis., 9, 65
Waelkens, C., et al., 1990, A&AS, 83, 11
Wildey, R.L., 1964, ApJS, 8, 439

THE MASS FUNCTION AT THE END OF THE MAIN SEQUEN THE M35 OPEN CLUSTER

D. BARRADO Y NAVASCUÉS
Max-Planck Institut für Astronomie
Könisgtuhl 17, D-69117 Heidelberg, Germany

J.R. STAUFFER
Harvard-Smithsonian Center for Astrophysics
60 Garden St. Cambridge, MA02138

J. BOUVIER
Observatoire de Grenoble
F-38041 Grenoble, Cedex 9, France

AND

E.L. MARTíN
University of California at Berkeley
601 Cambell Hall, Berkeley, CA 94720, USA

1. Introduction

The Open Cluster M35 (NGC 2168) is moderately nearby, with $(m\text{-}M)_0$=9.7 (Vidal 1973). It is very well populated, with a total mass estimated between 1000 and 3000 $M_\odot$. Although the cluster is at relative low galactic latitude (l^{II}=186.58, b^{II}=2.19), M35 is rich enough so that it is easy to separate the field stars and the cluster population. On the other hand, its reddening is not too high –E(B–V)=0.17, and its age is very close to the Pleiades value ($\sim 125\times 10^6$ yr, Stauffer et al. 1998), since their turn-off point are in the same location (Vidal 1973). We have carried out photometric observations of the cluster using the Canada–France-Hawaii Telescope (December, 1996) and the new mosaic camera (8000×8000 pixels), covering a region of 30'×30', equivalent to about a fourth of the total area of the cluster. We collected data in the magnitude interval $17.5 \leq I_c \leq 23.5$, $0.5 \leq (R\text{-}I)_c \leq 2.5$. Errors can be estimated as 0.05 for each filter. The survey is complete down to I_c=21, $(R\text{-}I)_c$=2.0. Details can be found in Barrado y Navascués et al. (1999).

Astrophysics and Space Science is the original source of publication of his article. It is recommended that this article is cited as: *Astrophysics and Space Science* **263**: 303–306, 1999.

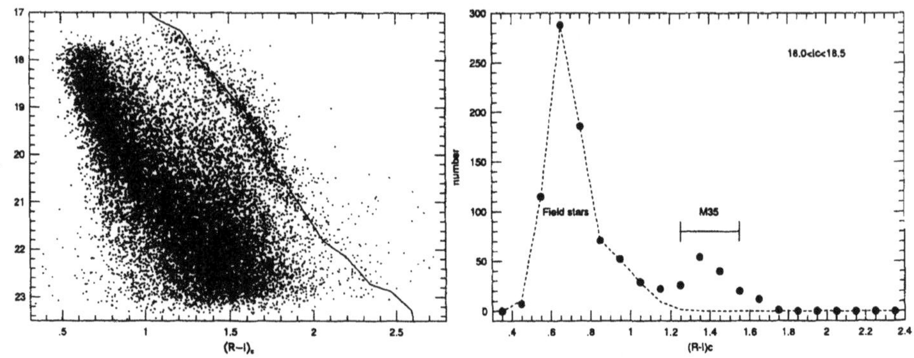

Figure 1. **a** Color-Magnitude Diagram for M35. **b** Histogram for one slice selected from the CMD. The number of stars in each (R-I) bin is shown as filled circles, whereas the fit for field stars appears as a dashed line.

2. The selection of candidates

We selected candidates to be members of the clusters based on the position of a Color-Magnitude diagram (CMD). Figure 1a shows I_c versus $(R–I)_c$. We have added an empirical ZAMS for field stars, derived from Leggett (1992), shifted with the reddening -$E(R\text{-}I)_c$=0.17,- and the distance module –$(m\text{-}M)_I$=10.05.– of M35. Note that the ZAMS accurately traces the relative concentration of stars in M35, as it does with Pleiades stars (Bouvier et al. 1998). Figure 1a shows clearly two accumulations of stars: Most of the detected stars lie between $0.5 < (R–I)_c < 1.7$, following a distribution parallel to the ZAMS, but 4 magnitudes fainter. A significant fraction of the stars are on the locus of M35. However, there is not a clear gap between the former and the latter groups. In any case, the diagram allows us to perform a tentative identification of candidates for membership to the cluster. They were selected based on the fact that the number of stars for a given bin in magnitude decreases as the color increases for field stars, except for the area close to the cluster, where a sudden increase appears. Therefore, we selected those stars in a band around the relative maximum, taking into account the errors and the effect of binarity.

The relatively smooth transition between the concentration of field stars, fainter and/or redder than M35, and this cluster indicates that our list of candidates is strongly poluted by field stars. In order to estimate this contamination, we have divided our CMD into strips parallel to the $(R–I)_c$ axis, 0.5^{mag} wide in I_c. Each strip was subdivided into intervals of $(R–I)_c$, 0.1^{mag} wide. Then, we counted the number of stars in each interval or box, N_*. We fitted 3 Gaussian curves to the pairs $[N_*,(R–I)_c]$ for a

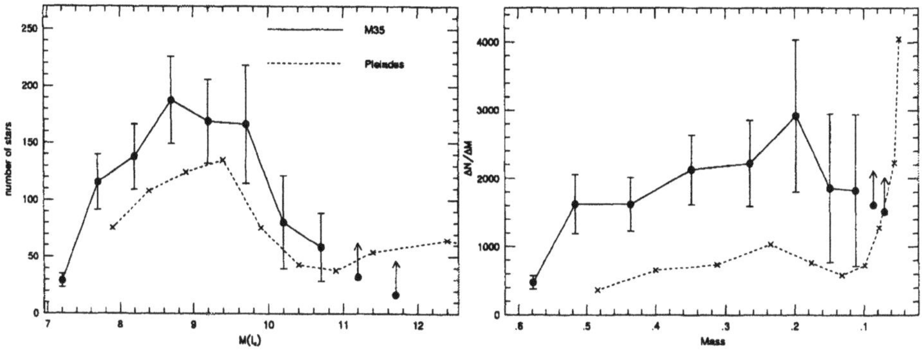

Figure 2. Comparison between the luminosity functions (**a**) and the mass function (**b**) of M35 and the Pleiades.

given I_c interval: one for the main population of field stars, another for the transition to M35, and the last one for the members of M35. The first two allowed us to estimate the contamination of field stars in the region where M35 lies (see Figure 1b). For stars fainter than I_c=22, we were not able to obtain reliable fits due to the lack of differentiation between field stars and the cluster, and, therefore, we cannot estimate the contamination by spurious members. Note, however, that the brown dwarf domain would start at $I_c \sim 22.2$, $(R-I)_c \sim 2.3$ (if M35 really is the same age as the Pleiades).

3. The Luminosity Function

We have subtracted the computed number of field stars for each box from the measured number, for the boxes corresponding to the location of M35. The final value is the number of cluster members and we have used it to compute the luminosity and initial mass functions. Note that we located the ZAMS in the diagram using the published values in the literature and appropriate conversions to the colors we used. We did not use this information when counting the number of stars in M35 and, therefore, our method is totally independent of distance modulus and reddening. However, we have obtained that the maximum density of stars, given a $(R-I)_c$ value, appears $\sim 0.1^{mag}$ below the ZAMS, as a visual inspection indicates in Figure 1a. However, this could be due to the assumed distance and reddening values.

Figure 2a shows the Luminosity Function (LF) of M35, once the contamination by field stars is removed (solid line). We have included the errors as vertical bars. The distribution peaks around I_c=19. After this point, the number of stars decreases with magnitude and a strong contamination by field stars appears. Our sample is complete until I_c=21, $(R-I)_c$=2.0. Therefore, the last two points are lower limits. The M35 LF is quite similar to the one characteristic of the Pleiades, shown as a dashed line in Figure 2a (Zapatero-Osorio, 1997). This distribution has been rescaled to match the values of M35, since both studies cover different angular areas (1 sq. deg. in the case of Pleiades against 0.25 sq. deg. for M35), and the latter cluster is more compact and is more populated (Leonard & Merritt, 1989). In any case, both distributions peak at similar value, and present a sudden decrease. Contrary to what happens in M35, the LF of the Pleiades increases below M_I=11. This difference is probably due to the fact that the Pleiades study is much deeper, complete in the range, whereas this is not true in our case.

4. The Mass Function

We estimated the masses using the I_c magnitudes. First, we corrected for the reddening and the distance modulus, obtaining M_I and $(R-I)_{c,o}$ for each box. Then, we convert the $(R-I)_{c,o}$ colors into $(V-I)_{c,o}$, using a polynomial expression between both colors which was obtained by fitting the single nearby dM stars listed in Leggett (1992). The $(V-I)_{c,o}$ allowed us to obtain bolometric corrections, using the conversion by Monet at al. (1992) and, with them and M_I, the bolometric magnitudes. Finally, masses were derived from M(bol) and the Baraffe (1997) isochrone, computed for an age of 120 Myr. Figure 2b displays the number of stars per mass interval against the mass. Data from M35 is shown as solid circles, whereas those from the Pleiades (Zapatero-Osorio 1997) appear, scaled for comparison purposes, as crosses. Errors bars are included for our data. The most important conclusion derived from this comparison is that both IMF are identical. If the results in the BD domain are extrapolated to M35, this last cluster should contain a significant population of BD. A more exhaustive analysis, considering the mass range 6-0.1 $M_\odot$, is presented in Barrado y Navascués et al. (1999).

References

Baraffe 1997, private comm.

Barrado y Navascués et al., 1999, A&A in preparation

Bouvier, J., Stauffer, R., Martín, E.L., Barrado y Navascués, D., Wallace, B., Béjar, V., 1998, A&A, 336, 490

Leggett, S.K., 1992, ApJS 82, 351

Leonard, P.J.T, Merritt, D., 1989, ApJ 339, 195

Monet et al. 1992, AJ 103, 638

Stauffer et al., 1998, ApJ 499, L199

Vidal, N.V, 1973, A&A Supp 11, 93

Zapatero-Osorio, M.R., 1997, PhD thesis

ASTROMETRY AND PHOTOMETRY OF OPEN CLUSTERS: NGC 1746, NGC 1750 AND NGC 1758

D. GALADÍ-ENRÍQUEZ AND C. JORDI
Dep. d'Astronomia i Meteorologia, Universitat de Barcelona
Martí i Franqués, 1. E-08028-Barcelona

AND

E. TRULLOLS
Dep. Matem. Apl. Telem., Universitat Politècnica de Catalunya
Av. Víctor Balaguer s.n., E-08800-Vilanova i la Geltrú

Abstract. We present an astrometric and photometric analysis of a 5 square degree region around NGC 1746, NGC 1750 and NGC 1758, close to the Galactic anti-center ($l = 179°, b = -10°$), in Taurus. We describe the acquisition, reduction and analysis of the observational data, including the development of innovative data analysis tools. The data comprise CCD *UBVRI* photometry, photographic *BVR* photometry and proper motions.

The clusters NGC 1750 and 1758 are poor and loose. We determine several physical parameters (distance, age, colour excess, positions, density profiles). The mutual separation, their relative velocity and their age difference allows us to conclude that these two clusters are physically independent. No real object in the area can be identified as NGC 1746.

1. The Area of NGC 1750

The existence and relationship of the possible clusters NGC 1746, NGC 1750 and NGC 1758 has been a matter of discussion during the last decades. Dreyer (1888), in his *New General Catalogue*, quoted these three entries. This star field is in the anti-center direction in Taurus ($l = 180°, b = -10°$).

The main previous studies of this zone were those by Cuffey (1937), Li (1954) and Straižys *et al.* (1992; hereafter referred to as SČM). These last authors deduced a distance to the Taurus dark clouds in this direction of 175 pc. They did not find any indication of the existence of NGC 1746, and they suggested that NGC 1750 and NGC 1758 are two real objects.

Astrophysics and Space Science is the original source of publication of his article. It is recommended that this article is cited as: *Astrophysics and Space Science* **263**: 307–310, 1999.

We refer to Galadí-Enríquez *et al.* (1998a) for an extensive review of the observational history and bibliography on the studied area.

2. Observations and reduction

2.1. *UBVRI* CCD PHOTOMETRY

We obtained Johnson-Cousins *UBVRI* CCD photometry for 3224 stars, $V_{\rm lim} = 20$. 1313 of them (down to $V_{\rm lim} \approx 15.5$) have complete photometry. The instruments were the 1.23 m telescope of CAHA and the 1.52 m telescope of OAN, both of them at Calar Alto Observatory. As a whole, more than 400 CCD images were processed.

2.2. PHOTOGRAPHIC MATERIAL

Our photographic analysis is based on astrograph and Schmidt plates from several origins. Six modern epoch Schmidt plates were obtained specifically for this work. The other plates were kindly supplied by different institutions. Some plates were used to derive photographic photometry for the whole area. Most of the plates were used also for astrometric purposes. 14 plates were measured at Paris (Centre d'Analyse des Images) with MAMA. The total number of plates is 30. The maximum time difference among them is 103 yr. The epoch difference among deep plates is 39 yr.

We were specially careful when dealing with magnitude effects on the star positions used to derive the astrometry. We applied an approach inspired in that proposed by Kozhurina-Platais *et al.* (1995) and Dinescu *et al.* (1996). Modern and old plates received a different treatment (Galadí-Enríquez *et al.*, 1998b). The key point of the process to correct magnitude effects is the comparison of the photographic star positions with the CCD measurements (free of magnitude effects). A sub-product of the comparison process is the production of a CCD-pseudo-plate, that was used for astrometric purposes just as any other plate.

2.3. PHOTOGRAPHIC ASTROMETRY

We applied an specific implementation of the iterative central overlap algorithm to compute plate-to plate transformations and to derive proper motions. Our approach is similar to that by Jones & Walker (1988). Proper motions are computed from different positions at different epochs, and they are fed-back to improve the geometric transformation among plates. After a few iterations, convergence is achieved and the procedure stops.

The resulting photographic astrometric and photometric catalogue contains data for 51846 stars. 39762 of them have photographic photometry, and 45036 have proper motions. The mean error in proper motion for bright stars is 1 mas yr^{-1}.

3. Cluster-field Segregation: Kernel Functions

We adopted a non-standard, fully non-parametric approach to solve the segregation problem. Due to the sparseness of the clusters found in the area, the classic parametric method (Vasilevskis *et al.* 1965, Sanders 1971) is not suitable. The advantages of the non-parametric method are discussed by Cabrera-Caño & Alfaro (1990). Details on our implementation of the empirical determination of frequency functions using kernel functions can be found in Galadí-Enríquez *et al.* (1998c). In synthesis, the main features of our method are:

- empirical determination of the probability density functions;
- use of normal circular kernels;
- application to three observational planes: kinematic (μ_x, μ_y), spatial (x, y) and photometric $(B-R, R)$;
- the sky area is divided into zones: inner circle for the clusters, outer corona for the field.

The empirical determination of the probability density functions (pdf) allowed us to discriminate two open clusters with specific photometric and, mainly, kinematic characteristics. The results indicated the existence of 79 members of NGC 1750 and 57 members of NGC 1758, down to $R \leq 16$ mag. No indication on the existence of NGC 1746 is found, and we conclude that this object does not exist.

The members were selected according to their membership probabilities in the three observational planes. The analysis of the resulting lists allowed us to determine the main physical parameters of the clusters.

4. Physical Parameters of NGC 1750 and NGC 1758

Among the main physical parameters of the clusters, we recall those quoted in Table 1. Colour excess was determined from additional Strömgren measurements and from the Vilnius data by SČM. The distances arise from ZAMS fit to the colour-magnitude diagrams. Ages are estimated by fitting Schaller *et al.* (1992) isochrones to the colour-magnitude diagrams. The physical parameters of the clusters indicate that they do not constitute a gravitationally bounded, binary system.

TABLE 1. Main characteristics of NGC 1750 and NGC 1758

Parameter	NGC 1750	NGC 1758
members identified ($R \leq 15$ mag)	79	57
center coordinates	$\alpha = 5^h4.3^m, \delta = +23°44'$	$\alpha = 5^h4.7^m, \delta = +23$
$\mu_\alpha \cos\delta$ (ICRS) [mas yr^{-1}]	-3.0	+0.7
μ_δ (ICRS) [mas yr^{-1}]	-6.0	-8.1
$E(B-V)$ [mag]	0.34 ± 0.07	0.34 ± 0.07
distance [pc]	630	760
age [Myr]	200 ± 50	400 ± 100
half-sample radius [arcmin]	10.2	6.6
core radius (King's profile) [arcmin]	9.1	2.6
integrated V [mag]	6.8	8.5
morphology	poor and loose	poor, central concentration
observed mass [$M_\odot$]	145	90

Acknowledgements

Many people have helped us in this investigation. Now we only can list their names: C. Alard, Ó. Alonso, K. Birkle, B. Chen, R. Chesnel, J. Fabregat, T. Girard, J. Guibert, J.C. Mermilliod, I. Platais, I. Ribas, N. Robichon, S. Röser, E. Schilbach, V. Straižys, K.P. Tian, S.E. Urban, W. van Altena, J. Zamorano, J.L. Zhao. Many thanks to the staffs of Centre d'Analyse des Images and Department of Astronomy of Yale University, specially to the Astrometry Group. This work has been supported by CYCIT (ESP95-0180, ESP97-1803). D.G.-E. acknowledges support from MEC (AP92 30526274), CIRiT (1996BEAI200346) and Universitat de Barcelona.

References

Cabrera-Caño, J., Alfaro, E.J., 1990, A&A, 235, 94
Cuffey, J., 1937, Annals Harvard Obs., 105, 403
Dinescu, D., Girard, T., Van Altena, W., Yang, T., Lee, Y., 1996, AJ, 111, 1205
Dreyer, J.L.E. (1888), *New General Catalogue of Nebulae and Clusters of Stars*, in Mem. Roy. Astron. Soc., 49
Galadí-Enríquez, D., Jordi, C., Trullols, E., Ribas, I., 1998a, A&A, 333, 471
Galadí-Enríquez, D., Jordi, C., Trullols, E., et al., 1998b, A&AS, 131, 239
Galadí-Enríquez, D., Jordi, C., Trullols, E., 1998b, A&A, 337, 125
Jones, B.F., Walker, M.F., 1988, AJ, 95, 1755
Kozhurina-Platais, V., Girard, T., Platais, I., et al., 1995, AJ, 109, 672
Li Hen, 1954, Annales de l'Observatoire Astronomique de Zǒ Sè, XXIII, 51
Sanders, W.L., 1971, A&A, 15, 173
Schaller, G., Schaerer, D., Meinet, G., Maeder, A., 1992, A&AS, 96, 269
Straižys, V., Černis, K., Meištas, E., 1992), Baltic Astronomy, 1, 125 (SČM)
Vasilevskis, S., Sanders, W.L., Balz, A.G.A., 1965, AJ, 70, 797

SUBMILLIARCSECOND ASTROMETRY OVER LARGE SKY SEPARATIONS

M.A. PÉREZ-TORRES, J.M. MARCAIDE AND J.C. GUIRADO
Dept. d'Astronomia i Astrofísica, U. de València, Spain

AND

E. ROS
Max-Planck-Institut für Radioastronomie, Bonn, Germany

Abstract. We observed the pair of radio sources 1150+812 and 1803+784 in November 1993 with an intercontinental array of radio telescopes, simultaneously recording at 8.4 and 2.3 GHz, using the very-long-baseline interferometric technique. We determined the angular separation between the two sources with submilliarcsecond accuracy by using differential techniques. This result validates the feasibility of high precision differential astrometry for radio sources separated in the sky by almost 15°, and opens the avenue to its application to larger samples of radio sources.

1. Introduction

Very Long Baseline Interferometry (VLBI) is a powerful astrometric technique. Centimeter-wavelength VLBI group-delay astrometry of extragalactic radio sources routinely provides precisions at the milliarcsecond (mas) level, thus allowing a celestial reference frame to be built with milliarcsecond accuracy (Ma et al. 1990). One step further in high precision astrometry is phase-delay differential astrometry, in which the differences between the phase delays of two radio sources are used to determine their relative separation. For close source pairs, this technique yields accuracies of a few microarcseconds (μas), i.e. the case of the double quasar 1038+528 A and B (Marcaide and Shapiro, 1983), whose components are separated by less than an arcminute. For sources separated up to a few degrees, accuracies

Astrophysics and Space Science is the original source of publication of his article. It is recommended that this article is cited as: *Astrophysics and Space Science* **263**: 311–314, 1999.

of about 0.1–0.3 mas are usually achieved (Guirado et al. 1995, Lara et al. 1996, Ros et al. 1999). The basic idea underlying differential astrometry is that the differenced phase-delays from the source pair will be largely free of unmodelled systematic effects. If the sources are very close, cancellation of systematic effects is almost complete, whereas if the sources lie at increasingly angular distances, the cancellation is less perfect, and the determination of their angular separation is correspondingly less accurate.

In this work, we address the question of the applicability of the differential astrometric technique to a pair of strong radio sources, 1150+812 and 1803+784, separated by almost 15°. We demonstrate the feasibility of the technique for this pair of radio sources, and measure their relative separation with submilliarcsecond accuracy.

2. Observations and data reduction

We observed the radio sources 1150+812 (QSO) and 1803+784 (BL Lac) on November 1993 during 12 hours, with an intercontinental interferometric array of radio telescopes, simultaneously recording at 8.4 GHz and 2.3 GHz. The data were correlated at the Max-Planck-Institut für Radioastronomie, Bonn, Germany. After correlation, we exported the data and constructed the interferometric visibilities using the calibration information provided by the staff at the observing antennas. We used the software package DIFMAP (Shepherd et al. 1995) to obtain images of both sources at each frequency (Fig. 1 only shows those at 8.4 GHz). The images clearly show the typical core-jet structure for these kind of radio sources, the brightest spot being usually associated with the flat-spectrum core, and the other components with steeper spectrum components of a jet. 1150+812 displays a southeastward jet, while 1803+784 displays a westward jet.

The goal of the data reduction is to obtain a data set of differenced phase-delays on which a weighted-least-squares algorithm can be applied to determine the relative position of the sources. To obtain this data set, we went through several steps, namely, phase connection (or correction of the 2π ambiguities of the phases; Shapiro et al. 1979), structure correction, and removal of opacity and propagation medium effects.

We used our hybrid maps to correct for the source structure contribution to the phase delays, using as reference the peak of the brightness distribution of each map. Most of the propagation medium effects are caused by the troposphere and the ionosphere. We used detailed monitoring of the weather parameters at each of the antenna sites to accurately track the tropospheric behaviour, which we modelled following Saastamoinen (1973). The dispersive character of the ionosphere makes its contribution to the phase delays scale as ν^{-2}. By carrying a dual-frequency experiment, we estimated and

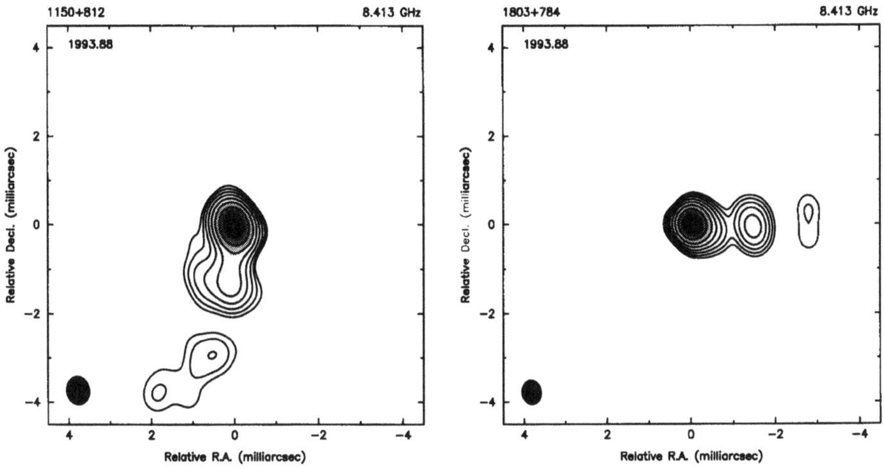

Figure 1. VLBI images of the radio sources 1150+812 and 1803+784, at 8.4 GHz. Contours are 0.5,1,2,4,8,16,32,64, and 90% of the peak of brightness (POB). The size of the natural restoring beam is a gaussian beam of $\sim$ 0.6 mas for 1150+812, and $\sim$ 0.5 mas for 1803+784. The angular distance determination reported here corresponds to the angular distance between the POB of each map. East is to the left, and North is up in both maps.

removed the latter. For it, the reference points at 2.3 GHz should be physically identified with those chosen at 8.4 GHz. We took into account opacity effects to obtain a reliable map registration (Guirado et al. 1998).

In this way, we obtained a set of (undifferenced) phase delays, free of structure and propagation medium effects. However, the existence of unmodelled systematic effects was evident from a visual inspection of the data. For that reason, we formed a set of differenced phase delays by subtracting the delay of each observation of 1150+812 from the delay of the previous observation of 1803+784. Although the sources are separated by 15°, this procedure largely cancelled out systematic effects still present in the undifferenced phase delays.

3. Relative position of the radio sources

We used the differenced phase delays in our weighted-least-squares fit to estimate the relative positions of the radio sources. The estimates (J2000.0 coordinate system) yielded by our observations, in right ascension and declination, are shown in Table 1. The error bars include formal uncertainties from a covariance analysis, and residual errors of the geometry of the interferometric array, source position, source opacity, and atmosphere. The largest contributor to the error budget in the relative separation of the sources is the a priori position of the reference radio source, 1803+784 (IERS, 1995). This contribution is particularly relevant for this pair of radio sources, given their special geometry (Pérez-Torres et al. 1999).

TABLE 1. Estimates of the angular separation of 1150+812 and 1803+784

$\Delta\alpha$	=	$6^h7^m33\overset{s}{.}18458$	$\pm$	$0\overset{s}{.}00013$
$\Delta\delta$	=	$-2^\circ30'25\overset{''}{.}13561$	$\pm$	$0\overset{''}{.}00030$

4. Conclusions

We have demonstrated the feasibility of using phase delays from dual-frequency VLBI measurements at 8.4 and 2.3 GHz, and determined with submilliarcsecond accuracy the angular separation of the pair of strong radio sources 1150+812 and 1803+784, separated by almost 15°. This accomplishment opens the applicability of the technique to a complete sample of radio sources (Ros et al. these proceedings), with the aim of building a quasi inertial celestial reference frame based on stationary radio source cores, with submilliarcsecond accuracy.

References

Guirado, J.C., Marcaide, J.M., Elósegui, P., Ratner, M.I., Shapiro, I.I., Eckart, A., Quirrenbach, A., Schalinski, C.J., Witzel, A., 1995, A&A, 293, 613

Guirado, J.C., Marcaide, Ros, E., Ratner, M.I., Shapiro, I.I., Quirrenbach, and Witzel, A., 1998, A&A, 336, 385

IERS, IERS Annual Report 1995, Observatoire de Paris (1996)

Lara, L., Marcaide, J.M., Alberdi, A., Guirado, J.C., 1996, A&A, 314, 672

Ma, C., Shaffer, D.B., de Vegt. C. Johnston, K.J., Russell, J.L., 1990, AJ, 99, 1284

Marcaide, J.M., Shapiro, I.I., 1983, AJ, 88, 1133

Pérez-Torres, M.A., Marcaide, J.M., Guirado, J.C., Ros, E., 1999 in preparation.

Ros, E., Marcaide, J.M., Guirado, J.C., Ratner, M.I., Shapiro, I.I., Krichbaum, T.P., Witzel, A., and Preston, R.A., 1999, submitted to A&A.

Ros, E., Marcaide, J.M., Guirado, J.C., Pérez-Torres, M.A., this volume

Shapiro, I.I., Wittels, J.J., Counselman III, C.C., Robertson, D.S., Whitney, A.R., Hinteregger, H.F., Knight, C.A., Rogers, A.E.E., Clark, T.A., Hutton, L.A., Niell, A.E., 1979, AJ, 84, 1459

Saastamoinen, J., 1973, Bull. Géodésique, 105, 279

Shepherd, M.C., Pearson, T.J., and Taylor, G.B. 1995, BAAS, 26, 987

GAIA: FUNDAMENTAL PARAMETERS FROM THE SPACE

J. TORRA, X. LURI, F. FIGUERAS AND C. JORDI

Departament d'Astronomia i Meteorologia. Universitat de Barcelona. Diagonal, 647, 08028 Barcelona, Spain

1. Introduction

GAIA is a project for an ESA's cornerstone astrometric mission at the 10 μas accuracy level satisfying the recommendation of the Survey Committee for the long-term plan "Horizon 2000+" (Battrick 1994). The main goal of the mission is to obtain positions, parallaxes and proper motions with an accuracy better than 10 μas for $V = 15$, as well as multicolour multiepoch photometry for the one billion stars brighter than $V = 20$. GAIA is also planned to perform a radial velocity survey, thus providing for the first time the determination of the three velocity components of the star's motion. The project takes advantage of the knowledge and experience of the Hipparcos mission (ESA 1997, Perryman et al. 1989) from which several concepts proven to be succesful have been taken for GAIA. As in Hipparcos, GAIA will measure large angles along a great circle scanned by the satellite. The motion of precession around the sun-satellite direction allows coverage of the whole sky. The scientific impact of the mission, the first statistically significant survey of the galaxy, will be enormous, especially for the study of the dynamics of our galaxy, its structure and evolution (Perryman and van Leeuwen 1995).

2. The objectives

The objective of the GAIA mission is to obtain positions, proper motions and parallaxes plus radial velocities and photometry for the stars inside a significant volume of the galaxy. We summarize the goals for the GAIA mission in the following table:

Astrophysics and Space Science is the original source of publication of his article. It is recommended that this article is cited as: *Astrophysics and Space Science* **263**: 315–318, 1999.

Limiting magnitude	$I \simeq 20 - 21^{\rm m}$
Completeness	$I \simeq 20^{\rm m}$
Number of objects	35×10^6 ($V < 15$)
	350×10^6 ($V < 18$)
	1.3×10^9 ($V < 20$)
Precision	4 μas ($V = 10$)
	10 μas ($V = 15$)
	0.2 mas ($V = 20$)
Radial velocities	3 – 10 km/s ($I < 16 - 17$)
Intermediate band photometry	6-colour ($I < 16 - 17$)
or Low Res. spectroscopy	$R = 20 - 40$Å ($I < 17 - 18$)
Broad band photometry	4-colour ($I < 20$)

10 μas precision is equivalent to a 10% precision in distance at 10 kpc and 10 μas/yr in proper motion corresponds to 0.5 km/s at this distance. The sets of broad and intermediate filters to be implemented for photometry have not yet been decided. The possibility of obtaining intermediate band photometry from low resolution spectroscopy is under discussion. In its present design and considering a lifetime of 5 years, each star will be observed by GAIA 100- 200 times, thus providing accurate multiepoch multicolour photometry.

3. Scientific Goals

GAIA is expected to have an unprecedented impact on astronomy. For the first time several fundamental parameters and functions will be defined in a homogeneous, precise and exhaustive form and *in situ* characterization of stellar population will be possible. Some of the topics in which GAIA will be relevant are:

– Luminosity Calibration: We can roughly compare the GAIA prospect with the Hipparcos results for different stellar populations:

	Hipparcos	GAIA
Limiting distances	within 100 pc	5 – 50 kpc
Halo	very few	large numbers
Spiral arms	very few	large numbers
Bulge	none	large numbers
White dwarfs	some	many
Mv limit	−4	supergiants
Open clusters	10	all galactic clusters
Globular clusters	none	many stars accesible
Cepheids and RRLyrae	few	all plus LMC and SMC

- Open clusters: Hipparcos observed only Hyades and Coma Berenices ($r < 100$ pc; $\epsilon_r < 10\%$), GAIA will observe 30 open clusters closer than 500 pc with precision of 1% in distance, allowing studies on cluster dynamics, testing of models of stellar evolution, evaporation, mass segregation, percentage of binaries, etc.
- Globular clusters: Internal motions, calibration of luminosities, ages, models of formation and evolution can be studied for about 20 clusters between 5 and 10 kpc with $\epsilon_r < 10\% - 20\%$
- Dynamics of binary systems: Hipparcos found ≈ 3000 new double systems. GAIA will resolve 1–2 mas, then, orbits and masses to 1% accuracy can be obtained for wide binaries out to 100 pc. Evolutionary history, mass transfer, etc. can be studied for interacting systems
- Structure and motion of the different components and populations of the galaxy
- Dynamical implications on disk and halo dark matter distribution
- Rotation curve of the galaxy and determination of $M_{\rm Gal}$, mass of the dark halo, etc.
- Proper motions in LMC/SMC
- Reference system: directly linked to quasars, very dense (30000 stars per Schmidt plate)
- Detection of planetary systems and brown dwarfs: precision of 20 μas allows detection of Jupiter-like planets up to 30 pc. This volume includes several thousands of candidates; all of them will be observed with GAIA. If the precision is of 2 μas, for bright stars, then the detection of Jupiter-like planets is possible up to 100 pc
- Test of General Relativity: light bending, gravitational lensing by MACHOs, detection of gravitational waves, etc.

A more detailed description of the scientific goals of GAIA can be found in Perryman and van Leeuwen (1995), Lindegren and Perryman (1996), and Gilmore et al (1998).

4. The Payload

In its present design the GAIA project considers a main instrument with astrometric capabilities and a second one devoted to intermediate band photometry or spectrophotometry and radial velocities. The astrometric measures are performed by two three reflection telescopes, 1.7m×0.75m entrance pupil and focal length 50 m, pointing to two regions of the sky separated by a large angle (106 degrees). In the focal plane of either telescope an array of 350 CCDs operating in TDI mode, covering about 0.26 deg^2, registers the image of a star as it crosses the focal plane as a consequence of the rotation (120 arcsec/s) of the satellite. As in Hipparcos, the measurement of the large angles between stars in either field of view gives the astrometric parameters after the whole sky has been scanned several times. The array of CCDs (pixel size 9 μm $\times$ 27 μm) is divided into three parts: the main part is devoted to the astrometric measurements, the second is used to detect the objects to be observed before they reach the astrometric area, and wide band photometry is performed in the third. The second instrument is aimed to provide the radial velocities and intermediate band photometry. Both measurements are performed by the same f=4.17 m three reflection telescope. The central region ($1^{\circ} \times 1^{\circ}$) of the field of view enters a slitless spectrograph, and two strips at either side of this region feed the narrow band photometer. Images through narrow band filters (or low resolution spectrograph) and spectra are collected in an array of CCDs (1 pixel = 10 μm), operated in TDI mode. The spectra will have a resolution of 0.75 Å and cover the 8470 – 8700 Å wavelength range. The magnitude limit for the spectrograph will be $I \simeq 16-17^{\rm m}$ and the estimated number of objects to be observed is 1.8×10^{8}. Photometry will be obtained for 3.5×10^{8} stars up to $I \simeq 17-18^{\rm m}$.

5. Present Status

In 1996 ESA decided to start system and technology studies for the candidate cornerstone missions within Horizon 2000+. In 1997, the GAIA concept and technology study was awarded to Matra Marconi Space. At the same time Alenia Spazio began a technological study for a Fizeau interferometer. A Science Advisory Group (SAG) was created in 1997 to determine the top level scientific goals, to follow the technological studies and to demonstrate the feasibility of the mission. The SAG is in charge of preparing the Red Report before the end of 1999. The results of the technological study have shown that the proposed payload and satellite concepts meet all the scientific requirements, that no fundamental technology development is needed. The technological activities needed and the development of the satellite can be achieved in the schedule for the cornerstone 5 mission, before 2009.

References

Battrick B., 1994, *Horizon 2000 Plus*, ESA, SP-1180
ESA, 1997, *The Hipparcos and Tycho Catalogues*, SP-1200
Gilmore et al., 1998, in preparation
Lindegren, L., Perryman, M.A.C., 1996, A&ASS, 116, 579
Perryman, M.A.C., Bernacca, P.L., 1997, *Hipparcos Venice'97*, ESA SP-402
Perryman, M.A.C., van Leeuwen, F., 1995, *Future Possibilities for Astrometry in Space*, ESA SP-379
Perryman, M.A.C., Hassan, H., et al., 1989, *The Hipparcos Mission*, ESA SP-1111

HIGH PRECISION ASTROMETRY WITH VLBI

FROM THE TRIANGLE 1803+784/1928+738/2007+777
TO THE COMPLETE S5 POLAR CAP SAMPLE

E. ROS
Max-Planck-Institut für Radioastronomie, Bonn, Germany
AND
J.M. MARCAIDE, J.C. GUIRADO AND M.A. PÉREZ-TORRES
Dept. d'Astronomia i Astrofísica, U. de València, Spain

1. Introduction

The Very Long Baseline Interferometry (VLBI) technique can image compact radio sources with a resolution of the order of the milliarcsecond and can determine astrometrically relative positions to precisions of the order of tens of microarcseconds. This technique is ideal to construct a precise celestial reference frame. Up to the present, the group delay observable has been regularly used in such a task. The use of the more precise phase delay observable should constitute an immediate improvement in accuracy. Furthermore, the phase-delay, if used differentially over radio source pairs, becomes the most accurate observable in astrometry. The technique of phase-delay differential astrometry has been applied to several source pairs, with separations ranging from 33″ (1038+528 A/B; Marcaide & Shapiro 1983) to 5°.9 (3C 395/3C 382; Lara et al. 1996). The pair 1928+738/2007+777 (4°.6 separation) has been also studied (Guirado et al. 1995, 1998) yielding precisions of $\sim$200 μas. Recently, we added a new source 1803+784 to the latter pair to take advantage of the constraints introduced by a triangular geometry in the determination of the angular separations (Ros et al., 1998). It represents a first step towards extending the differential phase-delay astrometry from pairs to a whole sky radio source frame.

2. The Triplet 1803+784/1928+738/2007+777.

We observed the radio sources of the triangle formed by the BL Lac objects 1803+784 and 2007+777, and the QSO 1928+738 on epoch 1991.89 with an intercontinental interferometric array simultaneously at the frequencies

Astrophysics and Space Science is the original source of publication of his article. It is recommended that this article is cited as: *Astrophysics and Space Science* **263**: 319–322, 1999.

of 2.3 and 8.4 GHz (Ros et al., 1998). We determined the angular separations among the three radio sources with submilliarcsecond accuracy from a weighted least squares analysis to the differential phases, after removing most of the contribution due to the geometry of the array and the atmosphere. The radio source structure contributions to the phase delays were also modeled using hybrid mapping images of the radio sources from the same observations. We checked the consistency of our astrometric determination through the use of the so-called Sky-Closure. The Sky-Closure was defined as the circular sum of the angular separations of the three radio sources, determined pairwise and independently. In our case the result was consistent with zero, and verified satisfactorily the data process followed. The final accuracy of the astrometric determinations was of 130 μas.

One important aspect in the astrometric work is the excess propagation delay due to the ionization of the propagation medium, mainly the ionosphere. The ionospheric contribution to the delays had been determined in the past from dual-band VLBI observations. Sardón et al. (1994) showed that the total electron content (TEC) of the ionosphere can be determined with high accuracy by using dual frequency Global Positioning System (GPS) data. We used their method to estimate the plasma contribution by using TEC estimates of the ionosphere obtained from data from different GPS sites neighbour to the VLBI stations (see Ros et al. 1998, 1999).

Proper Motions in 1928+738. The comparison of the measurements of the separations of the pair 1928+738/2007+777 with those presented by Guirado et al. (1995; 1998) for epochs 1985.77 and 1988.83 allows us to register adequately the absolute position of 1928+738 relative to 2007+777. We estimate the proper motion of components in 1928+738, and identify the position of the radio source core even though it is unseen at cm-wavelengths, as shown in Fig. 1. The average proper motion of the components emerging from the core region is of 0.30±0.15 mas/yr towards the South.

3. The S5 Polar Cap Sample.

The phase connection over a separation as large as 15° is critical for the success of the extension of the astrometry to larger samples. Pérez-Torres et al. (1999) have shown the phase-connection for such a pair, 1150+812/1803+784, to be possible. Therefore, the 13 radio sources of the complete S5 polar cap sample (Eckart et al., 1986), which have mutual separations less than 15° can be studied also astrometrically. These sources are the quasars 0016+731, 0153+744, 0212+735, 0615+820, 0836+710, 1039+811, 1150+812, and 1928+738, and the BL Lacertae objects 0454+844, 0716+714, 1749+701, 1803+784 and 2007+777. We observed this set of radio sources at 8.4 GHz

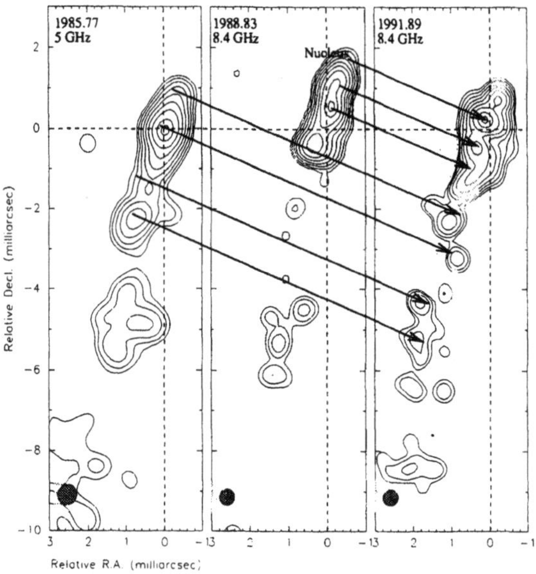

Figure 1. Astrometric alignment of the maps of 1928+738. We infer a proper motion of the components of about 0.3 mas/yr towards the south, and also that the nucleus of the radio source is near, but north of, the northernmost component of the 1988.83 map, at the position indicated in the central figure. The components have been convolved with circular gaussian beams (bottom, left) of diameters 0.7, 0.4, and 0.4 mas, showing slightly overresolved images.

over 24 hours on epoch 1997.93. We imaged the 13 radio sources using hybrid mapping techniques. On these images we defined reference points and then removed the structure contributions from the corresponding astrometric observables. After it, we used the differential phase-delays to obtain a global solution of all the source positions. Fig. 2 shows our preliminary results. A similar trend of systematic effects, which will cancel out when making the differences, is conspicuous.

With respect to the determination of the ionosphere contribution to the data, the density of the GPS network increased notably from 1991 to 1997, making the bias removal and the accuracy of the TEC determination much better. Now it is possible to have Global Ionospheric Maps from the Global Positioning System and thus estimate the ionosphere contribution to the astrometric observables of a single-wavelength VLBI observation and to remove the plasma effects from them with high accuracy.

4. Conclusions

The differential phase-delay astrometry has recently undergone important improvements. The phase-connection process has been extended to larger

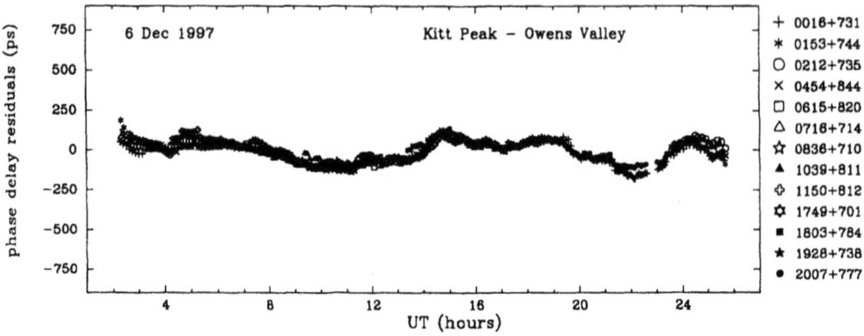

Figure 2. Preliminary results of the first epoch of 8.4 GHz observations of the complete S5 polar cap sample: postfit residuals of the (undifferenced) phase-delays after a weighted least-squares analysis which estimates the relative separations among the radio sources. One phase-cycle at the frequency of 8.4 GHz corresponds to 120 ps of phase-delay.

sets of radio sources with larger source separations, and the ionosphere contribution to the astrometric observables has been successfully removed using GPS data. We have determined with submilliarcsecond precision the relative separations in the triangle of radio sources 1803+784/ 1928+738/ 2007+777, and we have observed astrometrically the complete S5 polar cap sample, that among the 13 sources within 20° to the celestial North Pole includes the above 3. New observations are now underway in the framework of a long-term astrometric program to determine the absolute kinematics of radio source components in the S5 complete sample. This program, extended over 5 years, will reach a precision in the determination of the relative separations better than 0.1 mas and consequently in the proper motions of the radio source components.

References

Eckart, A., Witzel, A., Biermann, P., Johnston, K.J., Simon, R., Schalinski, C., Kühr, H., 1986, A&A, 168, 17

Guirado, J.C., Marcaide, J.M., Marcaide, J.M., Elósegui, P., Ratner, M.I., Shapiro, I.I., Eckart, A., Quirrenbach, A., Schalinski, C.J., Witzel, A., 1995, A&A, 293, 613

Guirado, J.C., Marcaide, J.M., Ros, E., Ratner, M.I., Shapiro, I.I., Quirrenbach, A., Witzel, A., 1998, A&A, 336, 385

Lara, L., Marcaide, J.M., Alberdi, A., Guirado, J.C., 1996, A&A, 314, 372

Marcaide, J.M., Shapiro, I.I., 1983, AJ, 88, 1133

Pérez-Torres, M.A., Marcaide, J.M., Guirado, J.C., Ros, E., 1999, this volume

Ros, E., Marcaide, J.M., Guirado, J.C., Ratner, M.I., Shapiro, I.I., Krichbaum, T.P., Witzel, A., Preston, R.A., 1998, A&A, submitted

Ros, E., Marcaide, J.M., Guirado, J.C., Sardón, E., Shapiro, I.I., 1999, in preparation

Sardón, E., Rius, A., Zarraoa, N., 1994, Radio Science, 29, 577

THE EVERSHED EFFECT IN SUNSPOTS: A THEORETICAL EXPLANATION

BENJAMIN MONTESINOS
LAEFF/INTA, P.O. Box 50727, E-28080 Madrid, Spain
IAA/CSIC, P.O. Box 3004, E-18080 Madrid, Spain

AND

JOHN H. THOMAS
Department of Mechanical Engineering, Department of Physics and Astronomy, and C.E.K. Mees Observatory, University of Rochester, Rochester, NY 14627, USA

Abstract. The siphon flow model, consisting in the simulation of a flow of gas moving along a thin magnetic flux tube and driven by the pressure drop between its footpoints, is proposed to explain the observational features of the Evershed effect, one of the longstanding problems in solar physics.

1. Introduction

The 'Evershed effect' is a wavelength shift and profile asymmetry observed in some absorption lines in the penumbra of sunspots which has been interpreted as a radial outflow of gas from the sunspot: the so-called 'Evershed flow'. It was discovered by John Evershed in 1909, and its theoretical interpretation has remained elusive to solar physicists for almost 90 years. Only recent observational results have clarified the geometry of the flow, leaving the challenge to theoreticians in order to find a picture fully consistent with the observations.

The penumbra is a deep structure, containing a significant fraction of the sunspot's emerging magnetic flux (see the review by Thomas & Weiss 1992). The penumbra consists of relatively dark radial filaments in which the magnetic field is nearly horizontal, alternating with relatively bright filaments, in which the field is inclined (by something like 40°) to the horizontal (Title et al. 1993, Lites et al. 1993).

Astrophysics and Space Science is the original source of publication of his article. It is recommended that this article is cited as: *Astrophysics and Space Science* **263**: 323–326, 1999.

The Evershed flow seems to be confined to the dark filaments (Title et al. 1993, Shine et al. 1994). There is in fact direct observational evidence that the flow occurs along elongated and elevated channels in the penumbra, oriented in the radial direction of the sunspot (Rimmele 1995, Stanchfield et al. 1997). The typical observed velocities reach up to ~ 6 km/s. The flow is also time dependent, with the rate of plasma flow along individual channels increasing and decreasing over times of the order of 10–15 minutes (Shine et al. 1994, Rimmele 1994).

2. The problem of the missing flow

Prior to 1997, very little magnetic return flux and gas flow, to the penumbra itself or to the surrounding photosphere, had been observed. Although the Evershed flow appeared to stop abruptly at the outer edge of the penumbra, the outflow itself must continue, though tracing its path proved to be difficult. This is the problem of the 'missing flow' that has been puzzling solar physicists for a number of years.

Rimmele (1995) and Stanchfield et al. (1997) found signatures of small, isolated patches of upflows in the penumbra and some patches of strong downflows outside it. Rimmele (1995) also showed that many of the Evershed flow channels extended outward across the penumbral boundary into the surrounding photosphere, sometimes ending in a 'porelike' structure.

Westendorp-Plaza et al (1997) showed that *a significant* part of the Evershed mass flux *flows back* into the Sun in the deepest atmospheric layers at the outer edge of the sunspot and its surroundings. Other, more elevated, magnetic tubes extend well outside the sunspot carrying a portion of the flow; these features are identified with the already known 'superpenumbral canopy' (Solanki et al. 1994).

3. The theoretical model

The most natural interpretation to this phenomenon arises from the identification of the upflows and downflows with the upstream and downstream footpoints of magnetic flux tubes. This, along with the evidence of the spatial correlation between the Evershed flow and the inclination of the magnetic field, suggests that the flux tube acts like a 'siphon' with the flow driven along the tube by the pressure drop between the endpoints of the tube.

This mechanism, first proposed by Meyer & Schmidt (1968), has been developed substantially in the recent years in a series of models (see Montesinos & Thomas,1997 and references therein). These models consist of

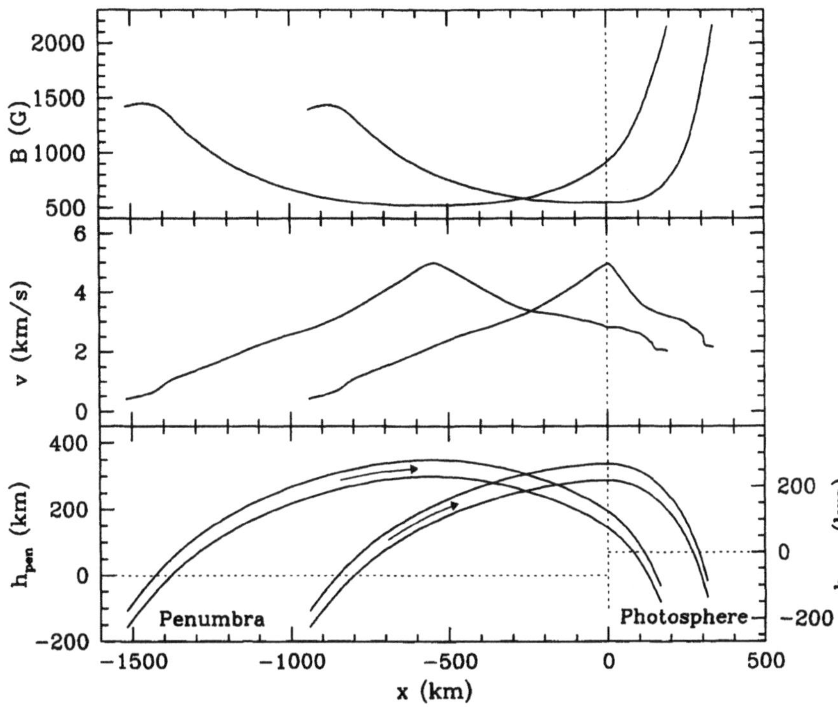

Figure 1. Two examples of computed siphon flows. The equilibrium path of the flux tubes, the velocity profiles and the magnetic fields are shown.

the numerical simulation of a steady siphon flow of a zero-viscosity gas along a thin, arched magnetic flux tube. The modelling allows the simultaneous calculation of the flow speed along the tube, the cross-sectional area of the tube, the magnetic field strength, the equilibrium path of the flux tube through the atmosphere, and all the thermodynamical variables. The calculations include the effect of radiative transfer between the flux tube and its surroundings, and variable ionization of the flowing gas. The only quantities that must be specified to calculate a full solution for a particular case are the initial conditions of the flow, namely, the speed, the radius of the tube, the plasma β ($8\pi p_g/B_0^2$), and the pitch angle, all specified at some reference point in the upstream branch of the flux tube.

The siphon flow model can be used in a variety of environments. In our case, since we have to model the behaviour of flux tubes that cross the interface between the penumbra and the surrounding photosphere, the ambient atmosphere consists of a magnetic penumbral atmosphere in mechanical equilibrium with an external field-free atmosphere representing the surrounding photosphere. For further details see Montesinos & Thomas

(1997).

4. Results

In Figure 1 we show two typical examples of computed siphon flows along magnetic flux tubes that emerge within the penumbra. One of them lies almost entirely within the penumbra, whereas the other crosses the penumbra-photosphere interface at the apex of the arch. The bottom panel shows the equilibrium path of the tubes through the atmosphere (height h versus distance x). Note that for the tube which extends well outside the penumbra into the photosphere, the curvature of the arch increases as the tube crosses the interface. Since the shape of the tube is due to the balance between magnetic buoyancy, magnetic tension and centrifugal forces, the tube is forced to bend more sharply in order to counterbalance the larger value of the buoyancy force in the photosphere. The middle and upper panels show the flow speed and the magnetic field strength along the tube. The behaviour of all these quantities is in good agreement with the results reported by Westendorp-Plaza et al. (1997). Their observations show that the magnetic flux returns to the photosphere at the edge of the spot with inclinations close to the vertical, as our model predicts. The velocities observed, of the order of a few km/s, are similar to the values obtained from the model. Finally, the magnetic field strengths at the downstream footpoints of the tubes in the surrounding photosphere, computed with the model, are typical of those observed in intense magnetic elements (Rüedi et al. 1992). In summary, the siphon flow model seems to account for all the important observational features of the Evershed effect.

References

Lites, B.W. et al., 1993, ApJ, 418, 928
Meyer, F., Schmidt, H. U., 1968, Zeits. Angew. Math. Mech, 48, 218
Montesinos, B., Thomas, J.H., 1997, Nature, 390, 485
Rimmele, T.R., 1994, A&A, 290, 972
Rimmele, T.R., 1995, A&A, 298, 260
Rüedi, I., Solanki, S.K., Rabin, D., 1992, A&A, 261, L21
Shine, R.A. et al., 1994, ApJ, 430, 413
Solanki, S.K., Montavon, C.A.P., Livingston, W., 1994, A&A, 283, 221
Stanchfield, D.C.H., Thomas, J.H., Lites, B.W., 1997, ApJ, 477, 485
Title, A.M. et al., 1993, ApJ, 403, 780
Thomas, J.H., Weiss, N.O., 1992, in *Sunspots: Theory and Observations*, eds. Thomas, J.H. & Weiss, N.O., 3, (Kluwer, Dordrecht)
Westendorp-Plaza, C. et al., 1997, Nature, 389, 47

3-D HELIOSEISMOLOGY: RINGS AND HORIZONTAL FLOWS

J. PATRÓN, I. GONZÁLEZ HERNÁNDEZ, A. JIMÉNEZ,
S.J. JIMÉNEZ REYES, I. MARTíN, P.L. PALLÉ,
F. PÉREZ HERNÁNDEZ, C. RÉGULO AND T. ROCA-CORTÉS
Instituto de Astrofísica de Canarias
C/Vía Láctea s/n, La Laguna, Tenerife SPAIN

1. Introduction

The study of 3-dimensional power spectra of the solar oscillations has provided in the last decade a method to determine the amplitude and direction of horizontal velocity flows in the upper layers of the convection zone. For high-degree modes ($l \geq 190$), a local plane wave approach for the decomposition of the modes is more appropriate than the usual spherical harmonics decomposition. By doing this, information about the two components of the horizontal wavenumber $\mathbf{k}$ can be obtained, and a 3-dimensional power spectrum in k_x, k_y and ω is constructed. Cuts at constant frequency of the distribution of power appear as individual rings for every radial order n; this is why the technique is known as a ring diagram analysis. A good description of this technique can be found in Hill (1988, 1995).

We can detect the presence of horizontal velocity flows beneath the solar surface using the 3-dimensional power spectrum $P(\omega, k_x, k_y)$. If no flows are present, the distribution of power around the origin $k_x = k_y = 0$ is axially symmetric, and the rings appear to be perfect circles centered at the origin. When a velocity flow field $\mathbf{U}$ is present below the solar surface, the wave front is advected by the flow and there is an apparent Doppler shift in the frequency of the mode $\Delta\omega$, given by $\Delta\omega = \mathbf{k} \cdot \mathbf{U} = k_x U_x + k_y U_y$, where U_x and U_y are the components of $\mathbf{U}$ in an arbitrary heliographic coordinate system such as latitude and longitude. In this case the rings of power will no longer be centered at $k_x = k_y = 0$ and by measuring the position and shape of these rings, we can estimate U_x and U_y . In addition, once we have $U_x(\omega)$ and $U_y(\omega)$, we can use an inversion procedure to infer $U_x(r)$ and $U_y(r)$, where r is depth in the solar interior. Finally, by performing this analysis at several heliographic positions, we can estimate the variation of

Astrophysics and Space Science is the original source of publication of his article. It is recommended that this article is cited as: *Astrophysics and Space Science* **263**: 327–330, 1999.

the horizontal velocity field not only with depth but also with position on the solar surface.

This information is of great interest for solar physics, not only because a general view of large scale flow fields can be obtained by the combination of the local results, but also because the spatial structure and temporal evolution of flows in the solar convection zone beneath the photosphere play an important role in the solar cycle problem, solar convection zone physics and the dynamo problem.

2. Data processing and recent results

The typical set of images are either relatively high resolution ($\sim$ 2 arcsec /pixel) intensity or Doppler velocity images. The reduction process starts by tracking and remapping a particular region of the solar surface onto a great circles reference system. The region is tracked at the rotation rate of the center of the region during a time span from one to three days, with an image taken every 60 s. Then, a 3-dimensional image is created and an FFT is applied in three dimensions; the result is a 3-dimensional power spectrum as a function of k_x, k_y and ω. The peaks on the power spectra are fitted to a model where the two components of the horizontal velocity flows are estimated. Finally, an inversion procedure is applied to recover the depth information out of the velocity distribution obtained for the set of modes available in each power spectra (Patrón et al. 1995).

Here we show the latest results obtained with this technique: an estimation of the differential rotation at latitudes 0, 15 and 30 degrees, in both hemispheres (figure 1), and evidences of the presence of meridional flows at least down to the first 20Mm of depth (figure 2). These results have been obtained from a whole set of local data covering 360 degrees in longitude and about 75 degrees in latitude with a total of 120 regions analyzed. The data consist of a set of full-disc dopplergrams taken with the Michelson Doppler Imager (MDI) on board SOHO during May and June 1996 (González Hernández 1998).

The results shown in figure 1 have been obtained by averaging the longitudinal component of the velocity flows for all the regions studied at the same latitude (24 regions for every latitude) and then averaging the results for the southern and northern hemispheres. The solid lines in figure 2 correspond to the average of the latitudinal component of the flows for all the regions at the same latitude. The dash-dot-dot-dot line in this figure shows also the same average results but for an average of only two adjacent regions and also using a different subset of latitudes. The solid line give more stable results since we are averaging both in time and position.

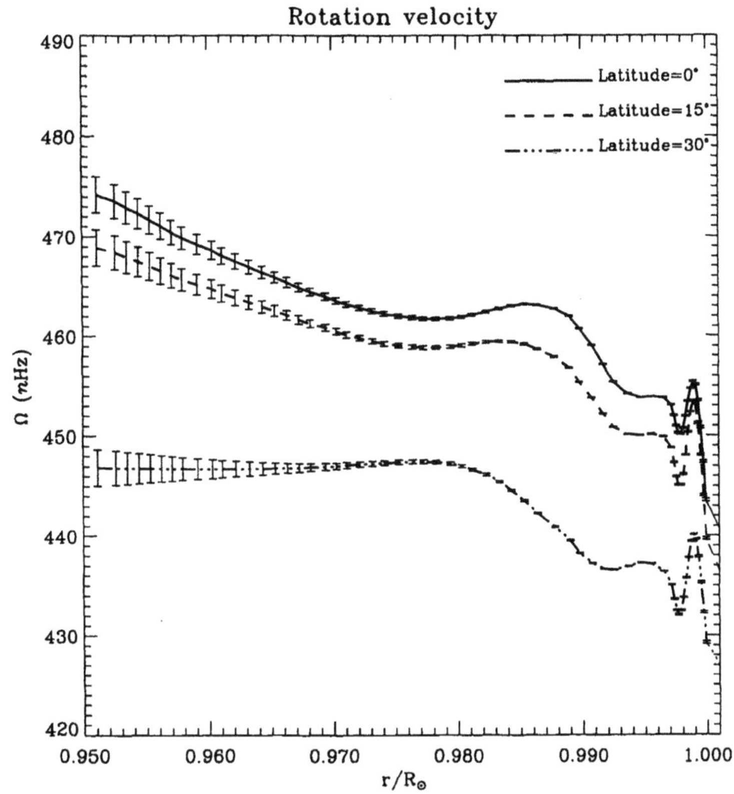

Figure 1. Rotational velocity versus depth at latitudes $0°, 15°$ and $30°$.

3. Conclusions

The results obtained with this technique show similar general behaviours: organized structures in depth and spatial coherence. The presence of convergent and divergent zones may be an indication of the existence of large scale convective structures such as giant cells and meridional flows. These kind of results are very interesting when studying the interactions between movements of material in the Sun and the presence of magnetic activity.

References

González Hernández, I., 1998, PhD. Thesis, University of La Laguna, Spain

Hill, F., 1988, ApJ, 333, 996

Hill, F., 1995, in *GONG'94: Helio- and Astero-Seismology from the Earth and Space*, ASP Conf. Ser., 76, eds. R.K. Ulrich, E.J. Rhodes, Jr. and W. Dappen, 484

Patrón, J., Hill, F., Rhodes, E. J., Jr., Korzennik, S. G., Cacciani, A., 1995, ApJ, 455, 476

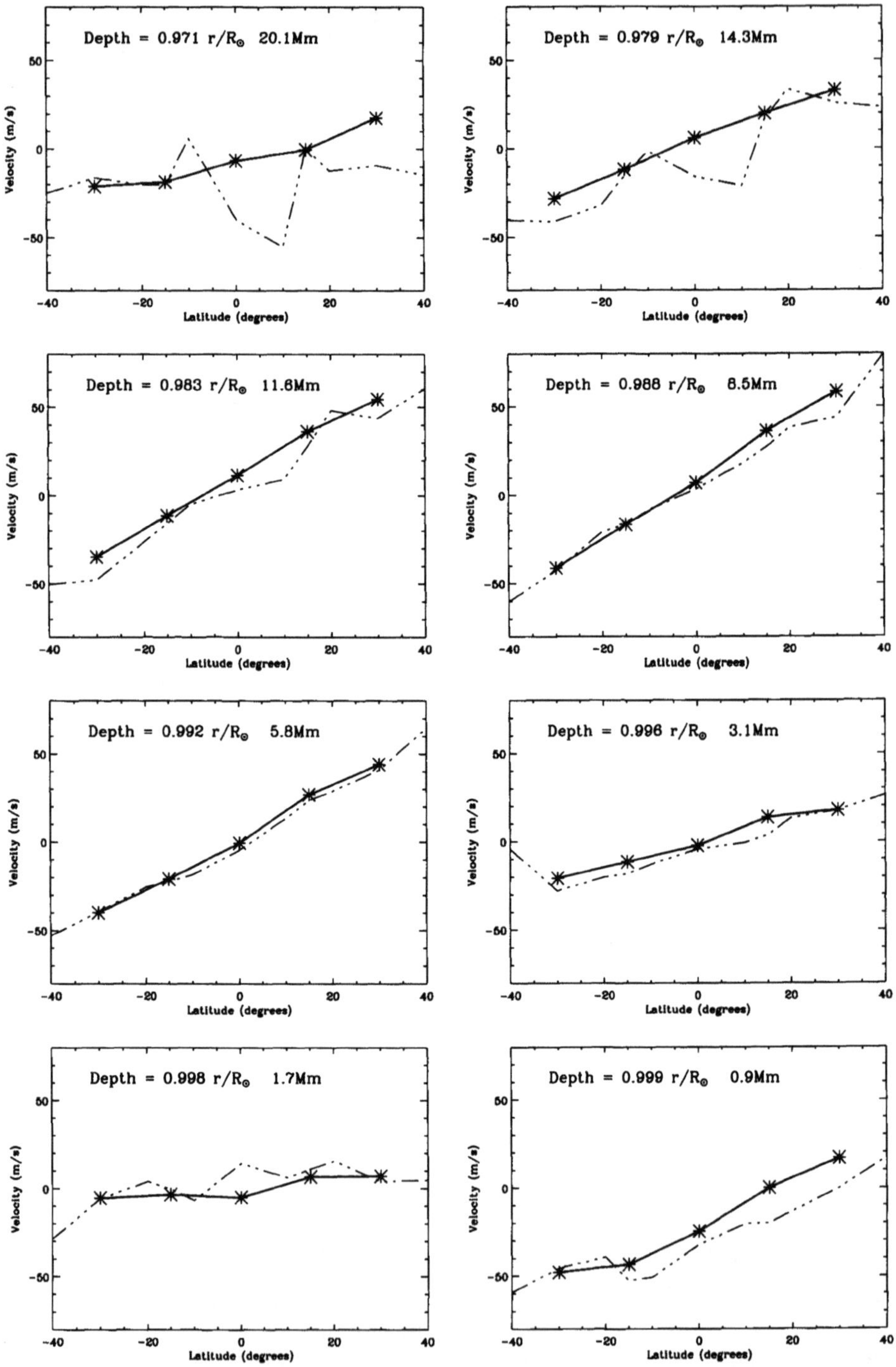

Figure 2. Meridional flows at several depths.

SIR: AN INVERSION TECHNIQUE OF SPECTRAL LINES

B. RUIZ COBO
Instituto de Astrofísica de Canarias, E-38200 La Laguna, Spain

Abstract. During the last years, inversion techniques have become one of the most powerful tools to obtain, from spectropolarimetric data, the stratification of physical quantities (temperature, pressure, velocity fields, magnetic field, etc.) describing different structures in the solar atmosphere.

The SIR code (Stokes Inversion based on Response functions), developed at the Instituto de Astrofísica de Canarias (IAC), is internationally acknowledged nowadays as the most sophisticated and accurate inversion method. The main results derived from its application to the study of the solar photosphere, are presented. Two recent extensions are reported as well: inversion under NLTE conditions, and the MISS code (Multiline Inversion of Stellar Spectra).

1. Introduction

The so-called spectroscopic *inversion* methods are aimed at obtaining the model atmosphere which best reproduces a set of observed spectral line profiles. The availability of high quality data (observed full Stokes spectra and accurate line parameters) and modern computers have made these inversions possible by taking into account the information contained in the whole line profile, which can potentially provide not only the temperature (T) stratification in the atmosphere, but also insights into other physical quantities such as line of sight velocity (V) and magnetic fields ($\vec{B}$).

The basic idea of inverting Stokes spectra is to use the misfit between synthetic and observed profiles to modify a guess model atmosphere until the synthetic spectrum matches the observed one. This process is complex because of the nonlinearity of the radiative transfer equation. To keep the problem tractable, approximations in one or another direction were often made in the past. In 1987, Skumanich & Lites introduced their inversion

Astrophysics and Space Science is the original source of publication of his article. It is recommended that this article is cited as: *Astrophysics and Space Science* **263**: 331–334, 1999.

code based on the assumption of Milne-Eddington atmospheres. Later, Ruiz Cobo & del Toro Iniesta (1992) presented the first inversion code (SIR: Stokes Inversion based of Response functions) capable of dealing with arbitrary stratifications of the physical parameters in the atmosphere. Inversion methods of spectropolarimetric data have recently been reviewed by Wood & Fox (1995) and del Toro Iniesta & Ruiz Cobo (1995, 1996, 1997).

2. The SIR code

The sole restrictions of this inversion method are plane-parallel geometry and Local Thermodynamical Equilibrium (LTE).

The inversion itself proceeds by minimizing a χ^2 merit function by means of a nonlinear least-squares Mardquart's algorithm and a modified singular value decomposition (SVD) method (see Ruiz Cobo & del Toro Iniesta 1992). The partial derivatives of the synthetized Stokes profiles with respect to the various model parameters, the so-called Response Functions (RFs), are fundamental quantities for Marquardt. It is important to keep the number of free parameters as low as possible in order not to crowd the χ^2–hypersurface with local minima. For this reason, the stratifications of the depth-dependent physical quantities are sampled in a coarse spatial grid (containing typically of the order of 2–7 points called *nodes*). Once the perturbations of the physical quantities at the nodes which lead to a smaller χ^2 value are provided by Marquardt, the whole stratifications in the original spatial grid can be retrieved by means of cubic-spline interpolation.

3. Some applications of SIR

SIR has been succesfully applied to a variety of solar scenarios, briefly reported in the following:

- *Empirical models of sunspots.* The first empirical models of sunspots including the stratification of the magnetic field strength were published by Collados et al. (1994). Based on the analysis of Stokes I and V, the authors found significant differences between large and small sunspots. Large spots turn out to be cooler and with larger magnetic fields than small spots. These results confirm to a large extent evidences previously reported by several authors from other kind of analyses.
  In addition, the optical tomography of a sunspot (i.e., maps of $\vec{B}$ and other physical quantities like T or V at several optical depths) has been obtained after applying SIR to Advances Stokes Polarimeter (ASP) observations (Westendorp Plaza et al. 1997a). New light is cast on topics like penumbral "corrugated" fields and superpenumbral canopies. Westendorp Plaza et al. (1997b) have reported on a signature of the

Evershed flow tied to the magnetic field lines returning back into the solar surface at the continuum forming layers.

- *Structure of unresolved solar magnetic elements.* SIR has been modified to implement the simplest two-component model of solar magnetic elements, the thin flux tube model (Bellot Rubio et al. 1997, 1998, 1999). By considering velocity gradients within the tubes, the authors have been able to match the full ASP Stokes profiles observed in a plage region near disk center. This is the first time the asymmetries of observed profiles have been properly reproduced. In addition, the thin flux tube scenario predicts the shift of the Stokes V zero-crossing wavelengths of more than one hundred of FeI and FeII lines observed with the Fourier Transform Spectrometer.
- *Granular-intergranular model atmospheres.* From inversion of 2D high spatial resolution solar spectroscopic data obtained with the aid of the IAC's Correlation Tracker, the runs with height through the quiet photosphere of T, V, pressure and density have been obtained (Rodríguez Hidalgo et al. 1995). The model atmospheres corresponding to individual spatial points have been averaged according to their continuum intensity. From this sequence of empirical models, the thermal and dynamical structure of a typical solar granule has been built (Ruiz Cobo et al. 1996).
- *Stratification of the 5-minute oscillation through the solar photosphere.* From inversion of a time series of KI 7699 Å line spectra at the solar disk center, the runs with optical depth and, for the first time, with geometrical height (Lagrangian and Eulerian coordinates) of the T and V oscillation have been obtained (Ruiz Cobo et al. 1997; Rodríguez Hidalgo et al. 1999). The increase in the V fluctuations amplitude with height has been confirmed and quantified, and the first quantitative estimate of the T oscillations (which is delayed about 90° through the photosphere, up to the temperature minimum height) has been presented.
- *NLTE inversion.* Combining SIR with robust and efficient solution methods for the NLTE multilevel radiative transfer problem, a NLTE inversion code of spectral lines has been developed (Socas-Navarro et al. 1998a,b). For the fast approximate evaluation of NLTE RFs, the fixed departure coefficients (FDC) approximation brings the inversion procedure into affordable CPU times. As the first step to exploit the code as a diagnostic tool for the solar chromosphere, it has been extensively tested for various groups of chromospheric lines (e.g. CaII H and K, MgI b_1 and b_2).
- *Multiline Inversion of Stellar Spectra (MISS).* MISS has been adapted from the original SIR code, to recover LTE, 1D model photospheres

for late-type stars from very high resolution, high signal-to-noise ratio stellar line profiles. It has proved able to find a LTE temperature stratification for the solar photopshere, reproducing the solar continuum and limb darkening (Allende Prieto et al. 1998).

References

Allende Prieto, C., Ruiz Cobo, B., García López, R. J., 1998, ApJ, 502, 951

Bellot Rubio, L.R., Ruiz Cobo, B., Collados, M., 1997, ApJ, 478, L45

Bellot Rubio, L.R., Ruiz Cobo, B., Collados, M., 1998, A&A Lett., in press

Bellot Rubio, L.R., Ruiz Cobo, B., Collados, M., 1999, ApJ, in press

Collados, M., Martínez Pillet, V., Ruiz Cobo, B., del Toro Iniesta, J.C., Vázquez, M., 1994, A&A, 291, 622

Rodríguez Hidalgo, I., Ruiz Cobo, B., del Toro Iniesta, J.C., Collados, M. Sánchez Almeida, J., 1995, JOSO Ann. Rep., 162

Rodríguez Hidalgo, I., Ruiz Cobo, B., Collados, M., 1999, in preparation

Ruiz Cobo, B., del Toro Iniesta, J.C., 1992, ApJ, 398, 375

Ruiz Cobo, B., del Toro Iniesta, J.C., Rodríguez Hidalgo, I., Collados, M., Sánchez Almeida , J., 1996, in *Cool Stars Stellar Systems, and the Sun, 9th Cambridge Workshop*, eds. R. Pallavicini & A.K. Dupree, ASP Conf. Ser., 109, 155

Ruiz Cobo, B., Rodríguez Hidalgo, I., Collados, M., 1997, ApJ, 488, 462

Skumanich, A., Lites, B.W., 1987, ApJ, 322, 473

Socas-Navarro, H., Trujillo Bueno, J., Ruiz Cobo, B., 1998a, ApJ, 507, 470

Socas-Navarro, H., Trujillo Bueno, J., Ruiz Cobo, B., 1998b, in *2nd International Workshop on Solar Polarization*, in press

del Toro Iniesta, J.C., Ruiz Cobo, B., 1995, in *La Polarimétrie, outil pour l'étude de l'activité magnetique solaire et stellaire*, eds. N. Mein & S. Sahal-Bréchot, 127

del Toro Iniesta, J.C., Ruiz Cobo, B., 1996, Solar Physics, 164, 169

del Toro Iniesta, J.C., Ruiz Cobo, B., 1997, in *Forum THEMIS*, eds. N. Mein & S. Sahal-Bréchot, 93

Westendorp Plaza C., del Toro Iniesta, J.C., Ruiz Cobo, B., Martínez Pillet, V., Lites, B.W., Skumanich, A., 1997a, in *1st Advances in Solar Physics Euroconf. Advances in the physics of sunspots*, eds. B. Schmidier, J.C. del Toro Iniesta & M. Vázquez, ASP Conf. Ser., 118, 202

Westendorp Plaza C., del Toro Iniesta, J.C., Ruiz Cobo, B., Martínez Pillet, V., Lites, B.W., Skumanich, A., 1997b, Nature, 389, 47

Wood, K., Fox, G.K., 1995, Inverse Probl., 11, 795

PROBING THE SOLAR SUBPHOTOSPHERIC LAYERS USINC RING ANALYSIS

I. GONZÁLEZ HERNÁNDEZ
Instituto de Astrofísica de Canarias
c/Vía Láctea s/n, La Laguna, Tenerife SPAIN;
Astronomy Unit, School of Mathematical Sciences, Queen Mary and Westfield College, Mile End Road, London E1 4NS, UK

AND

J. PATRÓN, T. ROCA-CORTÉS , F. PÉREZ HERNÁNDEZ, A. JIMÉNEZ, S.J. JIMÉNEZ REYES , I. MARTíN, P.L. PALLÉ AND C. RÉGULO
Instituto de Astrofísica de Canarias
c/Vía Láctea s/n, La Laguna, Tenerife SPAIN

1. Introduction

The knowledge of the material behavior in the solar convection zone will cast some light in more complex phenomena such as the solar dynamo (DeLuca & Gilman 1991). Local helioseismology gives an opportunity to probe the subphotospheric layers using medium-high ℓ solar acoustic modes. The modes penetrate inside the Sun from the surface to a returning point depending on ω/ℓ, and for the case of medium-high ℓ modes this point remains near the surface.

Rings diagram analysis is one of these local helioseismology techniques. The horizontal velocities in the region where the waves propagate, produce a Doppler shift in their frequencies (Hill et al. 1991), displacing the surfaces in a 3-dimensional (3-D) power spectrum of the solar p modes (k_x-k_y-ω). Figure 1 shows the rings of power produced by cutting a real 3-D spectrum at two different temporal frequencies, ν = 3500 and 4000 μHz. The frequency displacement ($\Delta\omega = \vec{k_h} \cdot \vec{U_h}$) can be deduced by fitting the spectrum to a model that takes it into account (Patrón et al. 1995), and so the components of the horizontal velocity fields can be found.

The application of local helioseismology has been retarded up to now by the quantity and the quality of the available data. The recent develop-

Astrophysics and Space Science is the original source of publication of his article. It is recommended that this article is cited as: *Astrophysics and Space Science* **263**: 335–338, 1999.

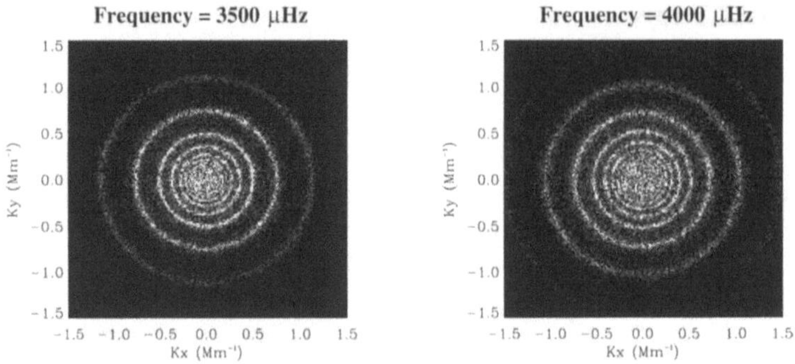

Figure 1. Rings of power produced by cutting a 3-D spectrum at temporal frequencies, $\nu = 3500$ and 4000 μHz. The spectrum have been built using SOI/MDI data for a region of about 15×15 degrees over the solar surface.

ment and success of projects like SOI/MDI (Scherrer et al. 1995), on board SOHO, have provided uninterrupted series of high spatial resolution data.

In this paper we will show a sample of the horizontal velocity vectors obtained for five latitudes and 24 longitudes at several depths above $0.97R_{\odot}$. About 30000 Doppler solar images, with a pixel size of ~2 arcsec, have been used to complete this work. The images were taken between May and June of 1996 during the first *Dynamic Program* of the SOI/MDI project.

2. Data reduction and analysis

A total of 120 sections of about 15×15 degrees have been tracked during a time span of 1536 minutes (~ 25 hours) using a superficial rotation rate with the coefficients proposed by Snodgrass (1984). The sections were chosen at latitudes $0°, \pm 15°, \pm 30°$ and every 15 degrees in longitude covering an entire solar rotation. The first longitude was 75° of Carrington rotation 1909. The tracked regions were then mapped onto a great-circle coordinate system and the spectra were obtained by performing a 3-D Fourier transform in both spatial directions and time. The spectra have been fitted to get the horizontal velocity flows affecting the different modes to the model given by Patrón et al. (1995). Finally, these velocities were inverted using a least-square with second-derivative smoothing inversion method to infer the depth dependence.

Figure 2 shows the velocity vectors for the 120 sections at five positions in depth of the whole range studied. Theoretically, with the modes considered in the analysis, $183 < \ell < 999$ and $0 < n < 7$, it is possible to cover a depth range from the surface to r = $0.95R_{\odot}$. However, due to the noise in the data, the errors become very large for the zone beneath r = $0.97R_{\odot}$. Since the studied regions have been tracked, the surface rotation should not appear in the results, but the radial differential rotation remains. We assumed this radial differential rotation to be the mean longitudinal velocity of all the sections located at the same latitude (Patrón et al. 1999), then we subtracted this mean from each individual region at that latitude.

The latitudinal component of the velocity vectors show a meridional behavior of the flows for most depths, with material flowing from the equator toward the poles (Patrón et al. 1999, González Hernández et al. 1999). The longitudinal component seems to play a less important role. It changes with longitude at a fixed latitude, but we do not find any systematic behavior with these data. There is a minimum in the velocities at about 2Mm depth, then the meridional behavior appears again until near the surface where the flows are mostly disorganized. For detailed information see González Hernández (1998).

3. Conclusions

The presented map shows the possibility of probing the upper layers of the convection zone by using ring analysis. This first map gives an idea of the information that can be obtained using the method. Nevertheless, more work providing higher spatial resolution as well as temporal studies of the obtained horizontal flows are necessary before further conclusions.

We thank Dr. Rick Bogart for introducing us to the use of the SOI/MDI data. Part of the reduction process was carried out on the Jet Propulsion Laboratory Caltech Cray supercomputer. The supercomputer used in this investigation was provided by funding from the NASA offices of Mission to Planet Earth, Aeronautics, and Space Science. IGH thanks ESA for its post-doctoral fellowship.

References

DeLuca, E.E. and Gilman, P.A., 1991, The Solar Interior and Atmosphere, ed A.N. Cox, W.C. Livingston & M.S. Matthews, (The University of Arizona Press), 275

González Hernández, I., 1998, PhD. Thesis, University of La Laguna

González Hernández, I., Patrón, J., Bogart, R.S. and the SOI Ring Diagram Team 1999, ApJ Letters, 510, L153-L156.

Hill, F., Deubner R.-L., Issak, G., 1991., The Solar Interior and Atmosphere, ed. A.N. Cox & W.C. Livingston, (The University of Arizona Press), 479

Patrón, J., Hill, F., Rhodes, E.J. Jr., Korzennik, S.G., Cacciani, A., 1995, ApJ, 455, 476

Patrón, J., González Hernández, I., Jiménez, A., Jiménez Reyes, S.J., Martín, I., Pallé, P.L., Pérez Hernández, F., Régulo, C., Roca Cortés, T., 1999, this volume

Scherrer, P.H. et al., 1995, Solar Physics, 162, 129

Snodgrass, H.B., 1984, Solar Physics, 94, 13

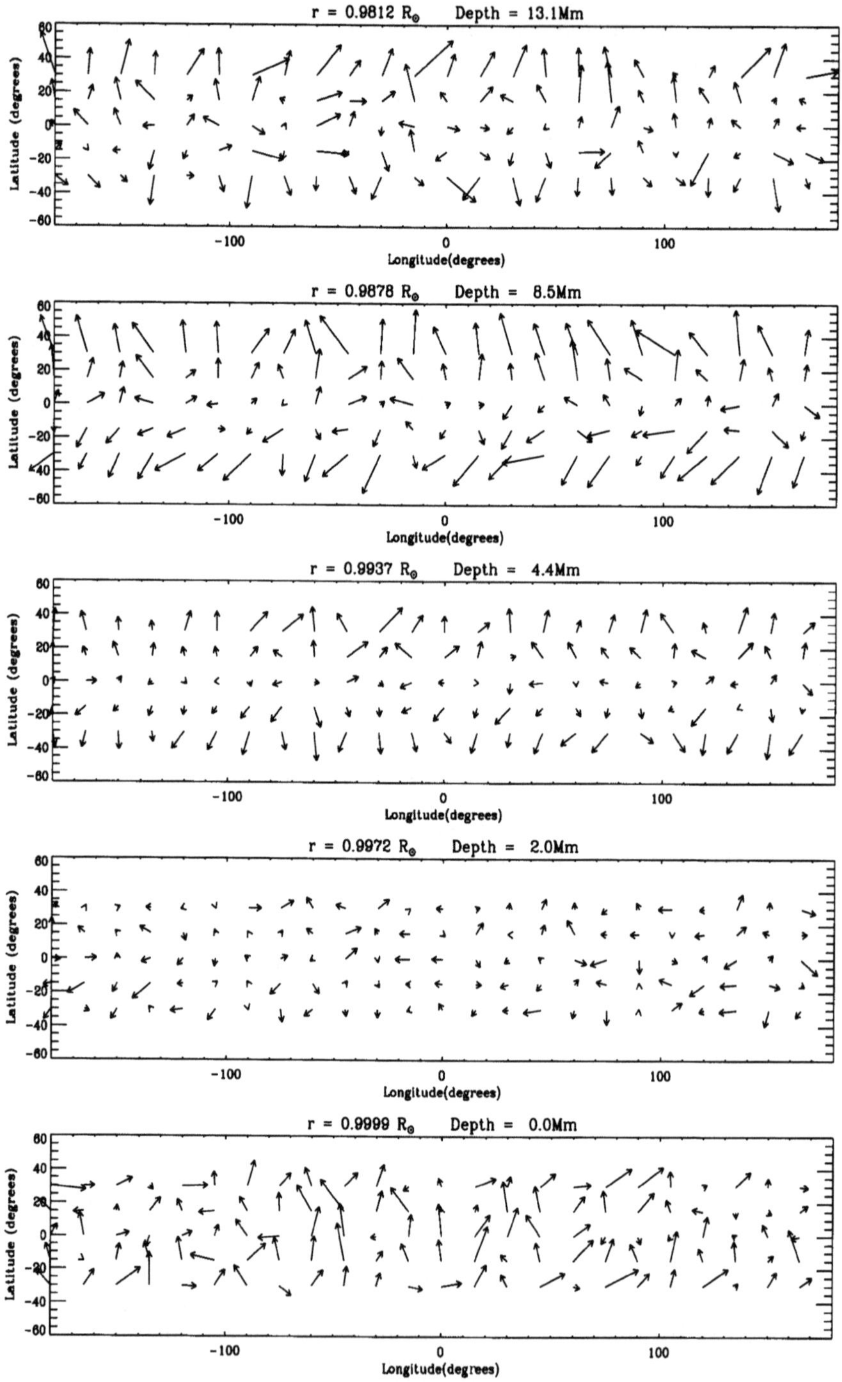

Figure 2. Horizontal velocity flows of the 120 regions at five different depths.

A NEW OBSERVATIONAL STRATEGY TO SEARCH FOR SOLAR G-MODES

I. MARTÍN MATEOS, I. GONZÁLEZ HERNÁNDEZ, A. JIMÉNEZ,
S.J. JIMÉNEZ REYES, J. PATRÓN, P.L. PALLÉ,
F. PÉREZ HERNÁNDEZ, C. RÉGULO AND T. ROCA-CORTÉS
Instituto de Astrofísica de Canarias
C/Vía Láctea s/n, La Laguna, Tenerife SPAIN

1. Introduction

The study of the g-modes could turn out to be one of the best ways of knowing about the deep solar interior. The g-modes are transversal waves whose principal component is tangential to the propagation of the oscillation. These waves are trapped in the interior of the Sun. The biggest problem on the observational study of g-modes is their small amplitudes outside the convection zone, they remain below the noise level.

In order to develop an adequate study of the g-modes, it is necessary to have data continuity with a high duty cycle, high sensitivity and stable instrumentation. The ideal approach would be the use of a combination of the highest quality data, plus theoretical models, plus the new techniques. In this paper we describe one of these new observational options and the results obtained from its application.

2. Observations and Data

The data used for the analysis were obtained at the Observatorio del Teide (Tenerife, Spain) using CANON, a resonant scattering spectrophotometer. The line-of-sight velocity was determined by measuring the Doppler shift of the absorption line KI λ7699Å. The new observational strategy is characterized for taking data with alternative continuous measurements at six different positions uniformly distributed over the solar disc (Figure 1).

Astrophysics and Space Science is the original source of publication of his article. It is recommended that this article is cited as: *Astrophysics and Space Science* **263**: 339–342, 1999.

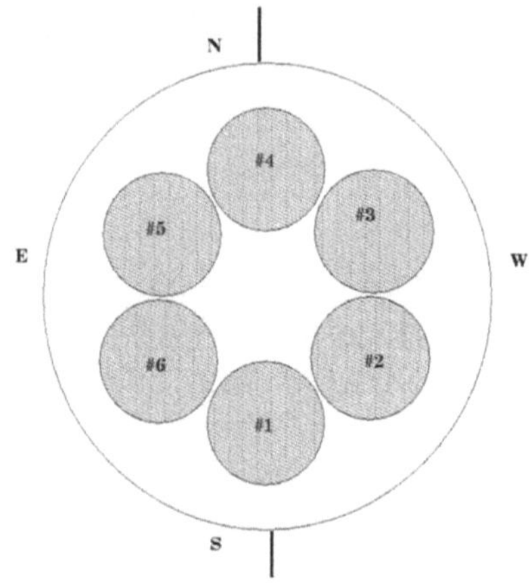

Figure 1. Schematic representation of the six regions scanned over the solar surface during the observations.

The advantage of observing at the solar periphery is that this increases the sensitivity to horizontal velocity fields, the most important component for gravity modes. According to the phases of the modes, adequate combinations of the different apertures allow us to filter out the signal from some particular modes (and also some of the noise) and to enhance others.

The data used in the analysis are the best series obtained between 1993 July 5 and September 29, in which almost 704 hours of good data were recorded, the duty cycle being 33 %.

3. The New Strategy

The analysis of the spectra is focused on the detection of significant frequency peaks that could be assigned to solar g-modes.

For this analysis we have taken advantage of the spatial position of the apertures and considered the possibility of symmetries. We have chosen the combination giving the best resolution and identifications.

The combinations considered by taking into account the East-West and North-South symmetries are:

$S_{\mathrm{I}} = \frac{1}{2}(S_{S_{\mathrm{EW}}} + S_{N_{\mathrm{EW}}})$ where $S_{S_{\mathrm{EW}}} = \frac{1}{2}(S_2 + S_6)$

$S_{\mathrm{II}} = \frac{1}{2}(S_1 + S_4)$ $S_{N_{\mathrm{EW}}} = \frac{1}{2}(S_3 + S_5)$

$S_{\mathrm{III}} = \frac{1}{2}(S_{S_{\mathrm{EW}}} - S_{N_{\mathrm{EW}}})$

$S_{\mathrm{IV}} = \frac{1}{2}(S_1 - S_4)$

$S_{\mathrm{IV}} = \frac{1}{2}(S_1 - S_4)$

The interval from 50 to 250 μHz was selected because the signal-to-noise ratio was higher as well as the expected g-modes density. The power spectrum was obtained by means of the ISWF procedure.

4. Peak Selection

The reference level, 3σ was determined by the noise level in the frequency range considered. This threshold is strongly frequency dependent due to the slope of the solar background. Because of this, the frequency range considered here, from 50 to 250 μHz, has been divided into ten different intervals of width 20 μHz. A noise level has been estimated for each of these intervals as the average power value.

The phase is also considered, so that only peaks with a phase difference between the two series less than $\pm 20°$ or $\pi \pm 20°$ will be interpreted as common to both series.

The time evolution of the selected peaks has also been checked by looking for their continuous presence over time. For this purpose, the initial series, with 3 months of data, has been divided into 13 series of 2 months of data. The first of the new series starts on the first day of the total series (July 5) and finishes two months later; the second series starts on the third day of the total series and finishes two months later, and so on, until the last series, which starts on the 26th day of the total series and finishes at the end of it (Sept. 29), so that each of the series is displaced by two days.

For each temporal sub-series the combined series, $S_{\rm I}$, $S_{\rm II}$, $S_{\rm III}$ and $S_{\rm IV}$, have been obtained with similar symmetry characteristics. The same criterion has been applied for the peak selection, but in the final selection we consider only those peaks present in all the sub-series.

In order to work with totally independent series two new ones were formed: one with data from odd days and the other with data from even days. Nine peaks, selected previously, are present in both series with the same properties.

Studying the probability of appearance of the selected peaks, one hundred different random redistributions of the present data in the original combined series were created obtaining the new position by a Gaussian scramble.

To make a comparison with data from other instruments, a data series with the same temporal window, obtained with the Mark-I instrument (working with whole disk integrated light in the potassium line), has been analyzed. Another series was constructed with random data, keeping the same window and also the same standard deviation. No peaks were found.

5. Discussion

The selected peaks (Table 1) were compared with the ones selected from different works done by Stanford (Delache and Scherrer, 1983; Henning and Scherrer, 1988) and Tenerife (van der Raay, 1988; Pallé, 1986) groups. It was not possible to find clear similarities between the results from the works compared, but it is clear that the lowest noise level is obtained from the CANON data.

Looking for more information about the selected peaks, those were compared with different sets of frequencies provided by 3 different solar models (Christensen-Dalsgaard *et al.*, 1993). There are more identifications with the first two models (72%), but it is not possible to reach a conclusive identification.

In summary, 7 peaks have been selected which exhibit the characteristics of g-modes with strict limits, and 18 with less stringent limits.

TABLE 1. Selected peaks (resolution 0.01 μHz)

Frequency (μHz)	Period (min)	Amplitude (cm s^{-1})	Noise (cm s^{-1})
53.35	312.4	107.69	7.5
58.39	285.43	60.01	4.52
59.28	281.4	67.65	4.52
70.35	236.9	103.94	6.16
70.79	235.43	75.94	4.54
71.08	234.47	96.08	6.16
79.79	208.88	121.76	6.16
105.13	158.53	67.94	5.87
114.03	146.16	62.18	4.77
116.75	142.75	45.89	4.77
129.41	128.78	45.68	4.77
130.54	125.58	28.67	3.57
135.91	122.60	26.65	3.57
164.14	101.53	38.57	2.98
170.65	97.66	31.42	2.96
182.57	91.28	24.81	2.96
194.16	85.83	35.17	2.91
232.41	70.53	18.35	2.46

References

Christensen-Dalsgaard, J., Proffitt, C.R., Thompson, M. J., 1993, ApJ, 403, L75
Delache P., Scherrer P. H., 1983, Nature, 306, 651
Henning, H. M., Scherrer, P. H., 1988, in *Seismology of the Sun and Sun-like Stars*, ed. E.J. Rolfe, Paris, ESA SP-286, 419
Martín Mateos, I., Pallé P. L., 1998, Solar Physics, submitted
Pallé P. L., 1986, *Thesis*.. University of La Laguna
van der Raay, H. B., 1988, in *Seismology of the Sun and Sun-like stars*, ed. E.J. Rolfe, Paris, ESA SP-286, 339

CENTRE-TO-LIMB VARIATION OF THE SOLAR GRANULATION

M. SÁNCHEZ CUBERES, J. A. BONET AND M. VÁZQUEZ
Instituto de Astrofísica de Canarias
38200 La Laguna, Tenerife, Spain

AND

A. D. WITTMANN
Universitäts- Sternwarte, D-37083 Göttingen, Germany

Abstract. An excellent series of images of solar granulation was taken during the partial solar eclipse of 1994 May 10 at the Swedish Vacuum Solar Telescope at Roque de los Muchachos Observatory. Making use of the lunar limb profile, images at different heliocentric positions were corrected for instrumental and atmospheric effects. The centre-to-limb variation of the granulation contrast was calculated and compared with previous determinations.

1. Introduction

The lower layers of the solar photosphere are dominated by the fluctuations of temperature, ΔT, and velocity, ΔV, associated with granulation. The study of the centre-to-limb variation (hereafter CLV) of the granulation contrast in the continuum is one of the methods used to get information about the variation of ΔT with altitude in the atmosphere.

A common problem is how to get information about the degradation introduced by the telescope and the atmosphere. This was possible by means of the lunar limb profile exposed simultaneously with the granulation images (see Figure 1).

2. Observations and data analysis

Images of solar granulation at different $\mu = \cos\theta$ positions on the solar disc were obtained at the Swedish Vacuum Solar Telescope (SVST, Observatorio del Roque de los Muchachos, La Palma, see Scharmer et al., 1985)

Astrophysics and Space Science is the original source of publication of his article. It is recommended that this article is cited as: *Astrophysics and Space Science* **263**: 343–346, 1999.

during the 1994 May 10 solar eclipse. This annular solar eclipse (see Espenak & Anderson 1993 for general information), which was observed as a partial eclipse from the Canary Islands, began at 17:53 UT and reached the maximum at 19:05 UT.

Image acquisition was performed with a KODAK Megaplus 8-bit CCD camera (1360×1035 pixels) with a exposure time of 13 ms. An interferential filter at $\lambda6708 \pm 20$ Å selected a spectral range with a reduced number of spectral lines. The frames were taken with an automatic selection system (the best two images within selection periods of 10 s were recorded). The image scale was $0.''124$/pixel.

The images were pre-treated with flat-fielding and dark-current subtraction. For this purpose sets of dark currents and flat-fields were recorded before the solar eclipse.

The method of correction of the images for instrumental and atmospheric effects is described in detail in Sánchez Cuberes et al. (1999)

Figure 1. Image of the solar granulation together with the lunar limb.

A simulation of the eclipse (Figure 2) was performed using the astronomical ephemerides, which allow the determination of the relative positions of the Sun and the Moon throughout the eclipse. They included calculations of the differential astronomical refraction taking into account the local atmospheric conditions. This information was used to calculate the heliocentric

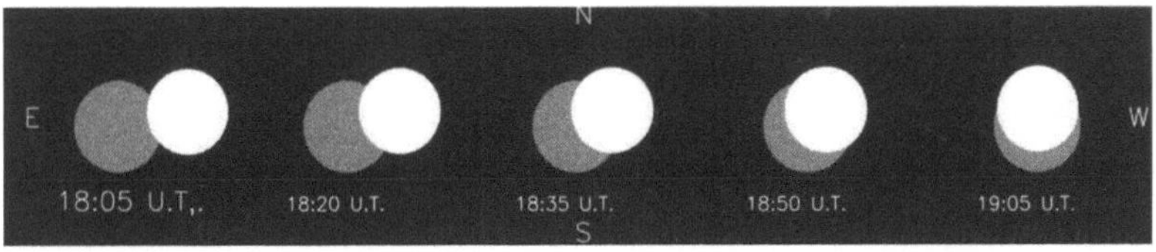

Figure 2. Eclipse simulation. The Moon is the white circle and the Sun is the grey one.

position of every frame by fitting the observed lunar limb to the simulated one.

3. Results

Figure 3 shows the centre-to-limb variation of the granulation contrast, ΔI_{rms}. In the same plot, normalized to the value at the disc centre, we show the results obtained by other authors in different observing conditions and at different times.

Calculations of the response function to temperature fluctuations (Kneer 1984) show that a steep ΔT gradient in deep photospheric layers leads to a slow and monotonic decrease in the CLV of ΔI_{rms}. In contrast, a smaller ΔT gradient would give rise to a CLV with a maximum somewhere between the disc centre and the limb. Our observations clearly support the first model.

Apart from differences due to different observing conditions and/or correction procedures, two main causes are invoked for explaining the scatter in the results: a) different wavelengths imply different heights of formation of the continuum, and b) a possible variation of the granulation contrast and its CLV over the solar cycle.

References

Druesne,P., Borgnino, J., Martin, F., Ricort, G., Aime, C., 1989, A&A, 217, 229
Durrant, C.J., Mattig, W., Nesis, A., Schmidt, W., 1983, A&A, 123, 319
Espenak, F., and Anderson, J., 1993, NASA RP-1301
Kneer, 1984, in *Small-scale dynamical processes in quiet stellar atmospheres*, Sac Peak Workshop, ed. S.L. Keil, 110
Rodríguez Hidalgo, I., Collados, M., Vázquez, M., 1992, A&A, 264, 661
Sánchez Cuberes, M., Bonet, J.A., Vázquez, M., Wittmann, A.m, 1999, in *High Spatial Resolution in Solar Physics*, eds. T. Rimmele, R. Radick & K. Balasubramanian, ASP Conf. Ser., in press
Scharmer, G.B., Petterson, L., Brown, D.S., Rehn, J., 1985, Applied Optics, 24, 2558
Wiesmeier, A., Durrant, C.J., 1981, A&A, 104, 207
Wilken, de Boer, C.R., Denker, C., Kneer, F., 1997, A&A, 325, 819

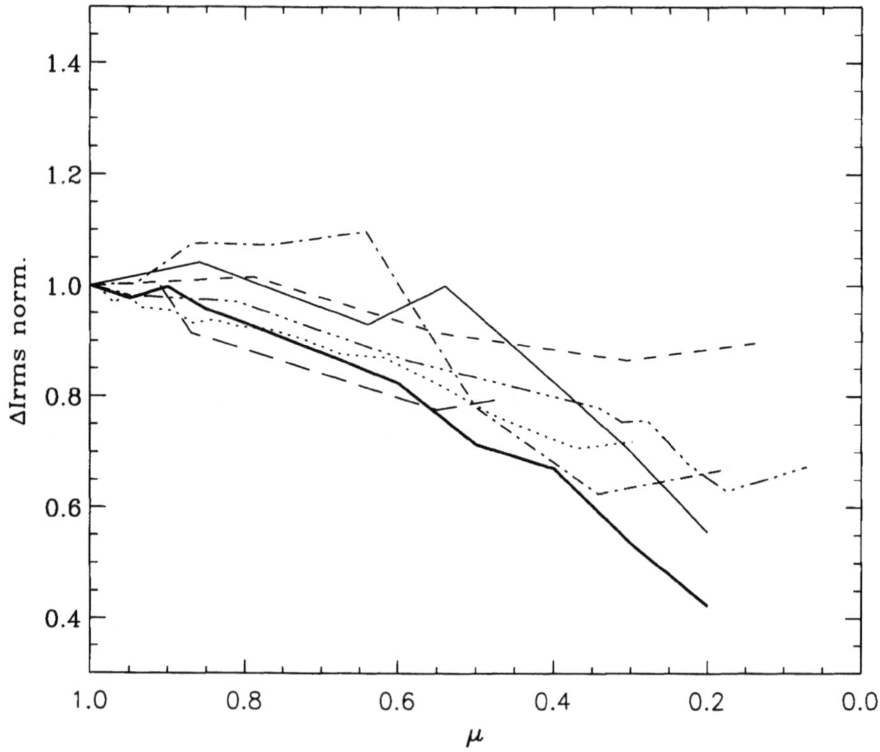

Figure 3. Measurements of the CLV of the solar granulation. (Thick line): this work (λ6708); (- · · -): Rodríguez Hidalgo et al. (1992)(λ4686); (- · · · · -): Wilken et al. (1997) (λ5500); (short dashed): Durrant et al.(1983) (λ5560); (· · ·): Druesne et al. (1989) (λ5840); (thin solid line): Spektrostoskop data Wiesmeier & Durrant (1981) (λ5560); (long dashed): Eclipse data Wiesmeier & Durrant (1981) (λ4220)

THE SOLAR ROTATION RATE $\Omega(R, \theta)$

A. EFF-DARWICH AND S.G. KORZENNIK
Harvard-Smithsonian Center for Astrophysics
60 Garden St., Cambridge, MA, 02139, USA

AND

T. ROCA CORTÉS, F. PÉREZ HERNÁNDEZ AND P. PALLÉ
Instituto de Astrofísica de Canarias
C/ Via Láctea, s/n, La Laguna, Tenerife, 38205, Spain

Abstract.

In recent years, the capability to detect and analyse solar oscillation acoustic modes has greatly improved. The development of ground based networks like GONG or BiSON and the use of space platforms like SOHO has allowed us to study the structure and dynamics of the Sun with unprecedented precision.

In this work we explore the distribution of the rotation in the solar interior, specially in the core. This kind of information is essential in order to study the physical processes involved in the evolution of the Sun and many other stars.

1. Statement of the problem

The oscillations of the Sun may be described by its eigenfrequencies $\nu_{n\ell}$. The solar rotation splits the degeneracy of oscillation modes in a multiplet with $2\ell + 1$ components $\nu_{n\ell m}$.

Hansen *et al.* (1977) derived an analytical relation between the splittings of the frequencies $\delta\nu_{n\ell m}$ (the separation between the components of a multiplet) and the rotation rate $\Omega(r, \theta)$ in the Sun, that can be written as:

$$\delta\nu_{n\ell m} = \int_0^R \int_0^\pi K_{n\ell m}(r, \theta)\Omega(r, \theta) dr d\theta \qquad (1)$$

Astrophysics and Space Science is the original source of publication of his article. It is recommended that this article is cited as: *Astrophysics and Space Science* **263**: 347–350, 1999.

where $K_{n\ell m}$ are known as the kernels of the inverse problem defined by equation (1), and are given explicitly in Christensen-Dalsgaard (1997). The independent variables r and θ correspond, respectively to the radial and angular (colatitude) locations in the Sun.

Our goal is to infer the rotation rate $\Omega(r,\theta)$ in the solar interior, specially in the core ($r < 0.5R_\odot$). To do this, we used variations of traditional numerical techniques that allows us to invert relation (1), where the resulting estimate of the rotation rate $\Omega(r,\theta)$ is computed from a linear combination of the splittings $\delta\nu_{n\ell m}$. The reliability of the rotation rate depends on how well the specific inversion technique works, but of course on the quality of the splittings.

In this case we implemented a two-dimensional inversion code that will allow us to infer the radial and angular distributions of the rotation rate. The method is based on the Optimal Mesh Distribution technique (Eff-Darwich & Pérez Hernández, 1997), a varitiaon of the Regularized Least Squares method (Craig & Brown, 1992).

We have used data (splittings) coming from different instruments and different analysis techniques, and the data sets cover a wide range of degrees and frequencies. The instruments selected were SOI/MDI (Scherrer *et al.*, 1998), LOI and GONG (Rabello-Soares & Appourchaux, 1998), GOLF (Lazrek *et al.*, 1997), BiSON (Chaplin *et al.*, 1996) and IRIS (Gizon *et al.*, 1997) . In Fig. 1 we present a sample of splittings taken from SOI/MDI ($\ell > 2$) and GOLF ($\ell \leq 2$). It is also indicated the depths for which the different data sets provide specific information. Although the GONG instrument provides observation up to $\ell = 250$, in this work we only used the estimates for $\ell = 1, 2$ obtained by Rabello-Soares & Appourchaux (1998).

It is evident from Fig. 1 that the number of modes sensitive to the core is very small when compared to the number of modes sensitive to the upper layers (*i.e.* the upper part of the radiative region and the convective zone). Note also that the uncertainties associated to the low degree modes are significantly larger than the uncertainties associated to the high degree modes. Moreover the time spent by the modes that penetrate the core is only a small fraction of their total travel time, hence these modes are only partially sensitive to the dynamics of the core. It is very important and necessary to have data of exceptional quality in order to extract from them the small amount of information they can provide on the dynamics of the core.

2. The solar rotation rate

We have carried out several inversions of the SOI/MDI data set (for $3 \leq \ell \leq 250$) complemented with the low degree splittings ($1 \leq \ell \leq 2$) provided by

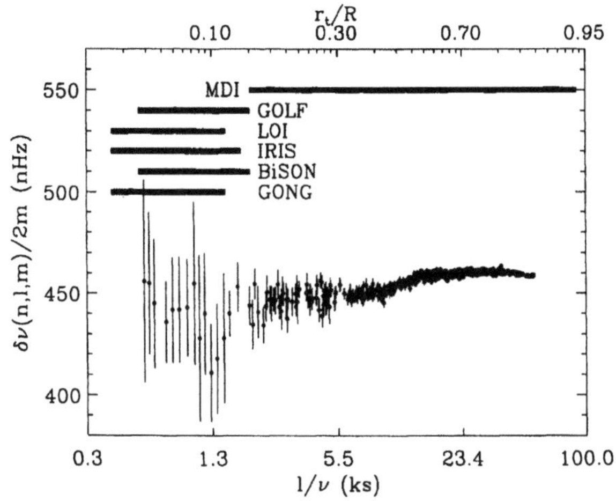

Figure 1. Sample of splittings taken from SOI/MDI ($\ell > 2$) and GOLF ($\ell \leq 2$). The splittings are plotted versus $\ell/\nu_{n\ell}$ that is proportional to the inner turning point of the modes (shown in the upper axis). The thick horizontal lines represent the regions where the different instruments provide information about the dynamics of the solar interior.

GOLF, GONG, LOI and BiSON. The resulting profiles are shown in Fig. 2. The error bars correspond to the inversion of the combined SOI/MDI and GONG data set. The rotation rate in the convective and outer radiative zones ($0.5 \leq r/R_\odot \leq 1.0$) does not depend on the low degree splitting sets. This is the reason we only show the inversion of the SOI/MDI and GONG data. Filled circles, open circles, crosses and asterisks represent the rotation rates at 0, 25, 40 and 50 degrees, respectively. We only give the rotation rate in the solar core ($0.2 \leq r/R_\odot \leq 0.5$) for the equatorial belt because the equator is the only region sensitive enough to the data down there. The rotation rate in the core (its upper limit is marked with dashed lines) is presented in filled circles for the inversion of GOLF data, in open circles for LOI, in crosses for GONG and in asterisks for BiSON.

It is clear in Fig. 2 that there is a transition from a differential rotation distribution in the convective zone ($r > 0.71R_\odot$) to a solid rigid rotation in the outer radiative zone ($0.5 \leq r/R_\odot \leq 0.71$). In the convective zone the equator rotates faster than the polar regions. The gradient in the rotation rate at the base of the convective zone plays a very important role in the solar dinamo theory. The study of the width and angular distribution of this gradient is one of our tasks in the near future.

All the inversions in the inner radiative zone (the solar core) are compatible within the error bars with a core rotating at the same rate that the outer radiative envelope. The inversions are reliable (in terms of the inversion technique) down to $0.2R_\odot$. From these results we conclude that

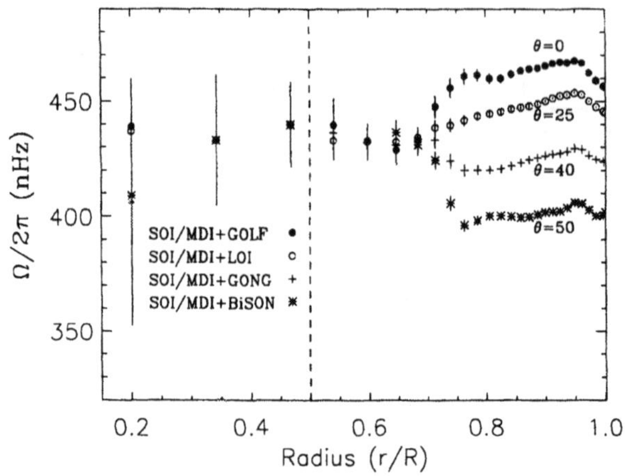

Figure 2. Rotation profiles obtained from the 2D inversion of SOI/MDI and GONG data for the outer radiative and convective zones. The filled circles, open circles, crosses and asterisks represent the rotation rates at 0, 25, 40 and 50 degrees, respectively. The rotation rate in the solar core is only shown for the equator . In this case we present the rates obtained using the four different low degree splitting sets: GONG (crosses), LOI (open circles), BiSON (asterisks) and GOLF (filled circles).

the existence of a super-rotator in the solar center (keeping memory of the high rotation speed of the young Sun) is very unlikely.

Nevertheless, both BiSON and GONG data are also compatible with a core rotating slower than the radiative envelope. These differences between the four low degree data inversions are due to the different locations and capabilities of the instruments. Indeed, some of the instruments are on board SOHO (namely LOI, GOLF and SOI/MDI) while the others are ground based (BiSON and GONG). Some of them have spatial resolution (SOI/MDI, LOI and GONG) while the rest observe the Sun as a star. Finally, even the same data are analyzed with different techniques, leading to different results. We believe the small differences in the inversions in the core will dissapear with a deeper understanding of the effects of the analysis and observational techniques in the data.

References

Chaplin, W.J., Elsworth, Y., Howe, R., 1996, MNRAS, 280, 849
Christensen-Dalsgaard, J., 1997, *Lecture Notes on Stellar Oscillations*
Craig, I.J.D., Brown, J.C., (1992) *Inverse Problems in Astronomy: A Guide to Inversion Strategies for Remotely Sensed Data* A. Hilger, Bristol
Eff-Darwich, A., Pérez-Hernández, F., 1997, A&ASS, 125, 1
Gizon, L., Fossat, E., Lazrek, M., 1997, A&A, 317, L71
Hansen, C., Cox, J., van Horn, H., 1977, ApJ, 217, 151
Lazrek, M., Baudin, F., Bertello, L., 1997, Solar Phys., 175, 227
Rabello-Soares, C., Appourchaux, T., 1998, Private communication
Scherrer, P., Schou, J., Bogart, R. S., 1998, ApJ, in press

AN OVERVIEW OF SATURN'S EQUATORIAL STORMS: 1990 - 1997

A. SÁNCHEZ-LAVEGA
Dpto. Física Aplicada I, Universidad del Pais Vasco, Bilbao, Spain

J. R. ACARRETA
L.A.E.F.F. - I.N.T.A.,Madrid, Spain

R. HUESO AND J. F. ROJAS
Dpto. Física Aplicada I, Universidad del Pais Vasco, Bilbao, Spain

J. LECACHEUX
DESPA, Observatoire de Meudon, Paris, France

F. COLAS
Bureau des Longitudes, Paris, France

AND

J. M. GÓMEZ
Grup d'Estudis Astronomics, Barcelona, Spain

Abstract.

We present an overview of the studies we have performed of the unusual storm activity that is occurring in the Equatorial Zone (EZ)of Saturn since 1990 based on a long-term CCD imaging (261-953 nm).

1. Introduction and Observations

Late September 1990 a giant storm, historically known as Great White Spot or abbreviated GWS, erupted in Saturns Northern Equatorial Zone (Sanchez-Lavega et al. 1991;Beebe et al. 1992). This is a very rare phenomena that has been recorded in just five occasions in the history of the telescopic observations of Saturn in the following years and latitudes: 1876 (8°N), 1903 (36°N), 1933 (2°N), 1960 (60°N) and 1990 (4°N)(Sanchez-

Astrophysics and Space Science is the original source of publication of his article. It is recommended that this article is cited as: *Astrophysics and Space Science* **263**: 351–354, 1999.

Lavega, 1982; 1994). The GWS are the most impressive and magnificent meteorological phenomenon that can be observed in this giant planet. It was not active during the epoch of the Pioneer 11 (1979) and Voyager 1 and 2 (1980 1981) close encounters with the planet (Burke et al., 1980; Smith et al., 1981, 1982). The 1990 event was probably the most spectacular among the five GWS registered, and the best imaged since its first steps, because of the new CCD cameras available in ground-based telescopes and to the HST capabilities.

Most CCD images were obtained with the 1-m planetary dedicated telescope at Pic-du-Midi Observatory (French Pyrenees) covering the spectral range from 0.4-1.0 μm. Broad and narrow -band interference filters including a set centered in the methane absorption bands at 0.619, 0.725, 0.890 μm were usually used (see e.g. Sanchez-Lavega et al., 1994). These observations were supported occasionally with CCD images taken with 0.4 -m telescopes by I. Miyazaki and D. Parker. We also used in this research archived images obtained during this period with the HST facility (Sanchez-Lavega et al., 1996). Typically we observed during about 40-50 nights each apparition spanning in time $\sim$ 8 months each year (the hole of $\sim$ 4 months corresponds to the Saturn's solar conjunction period).

2. Storms cloud morphology and dynamics

The 1990 GWS initiated at 5° N latitude with the formation of a very bright (from blue to red) conspicuous spot that grew in few days to a maximum horizontal size $\sim$ 20 000 km, (onset stage). Outflow from the central spot took place with radial velocities $\sim$ 30-40 ms^{-1} (Sanchez-Lavega et al.,1991;Beebe et al.,1992). This spot represents in fact the core of the storm and survived for $\sim$ 2-3 months. One week following the outbreak, the northern and southern extremities of the spot expanded zonally, with the bright clouds moving away from the storm center along parallel circles, with extreme values of $\sim$ 450 ms^{-1} at 9°N and $\sim$ 220 ms^{-1} at 16°N. The disturbance reached a planetary scale (mature stage) in about 1-2 months following the outburst when the clouds encircled the whole equatorial band from 2°S (limit imposed by ring shadowing) to 22° N.

This stage was followed by a gradual decay in the bright cloud activity in 1991 and later in 1992, although bright cloud elements survived several months in 1991 (Sanchez-Lavega et al., 1993). The homogenization of the equatorial cloud cover at the resolution reached by ground-based telescopes took place in about 2.5 years (decay stage), ending in 1993 when no signs of discrete cloud activity were observed in the Equator. However, in July 1994, a new huge cloud system formed at 12°N (Sanchez-Lavega et al., 1996). This feature, although large in size ($\sim$ 27 000 km) and lifetime ($\sim$ 1 year) showed a different morphology and behavior to the previously discussed 1990 GWS. Its most significant aspect was its steadiness in morphology and its slow motion relative to the normal zonal wind profile (as measured with the Voyagers). During 1995 the cloud activity decreased again (except by the presence of the 1994 feature). However in 1996 and 1997 new bright medium sized storms ($\sim$ 10 000 km) formed this time in the southern Equatorial Zone (at 12°S). This change in hemispheric activity within the equatorial area was coincident with a pronounced increase in insolation at these latitudes in accordance with the shadowing effect produced by the rings (Sanchez-Lavega et al., 1998).

Absolute photometric measurements from the ultraviolet to near infrared (including different intensity methane absorption bands at 619, 725 and 890 nm) of the diffuse light reflected by the storm clouds and surroundings (Sanchez-Lavega et al., 1994), were used to constraint radiative transfer models of the vertical structure of the cloud in the 1990 GWS during the different phases between 1990 and 1991 (Acarreta and Sanchez-Lavega, 1998). The radiative transfer equation was solved using the 'doubling an adding method' for different cases of haze and cloud particles and vertical altitude, including absorption by methane gas and Rayleigh scattering by the H_2. The simplest models that fit the whole spectral range contain three layers: pure gas + haze (mixed in gas) + semi-infinite cloud (mixed in gas). Synthetic double Henyey-Greenstein functions and Mie phase functions were tested to simulate the scattering and absorption properties of the particles.

The altitude distribution of each layer was studied for three different periods: undisturbed EZ in October 1990, mature phase in November 1990 and evolved phase in July 1991. The most significant result is that the main cloud forming the storm, presumably composed of ammonia crystals, elevated $\sim$ 60 km above its normal level at a pressure $\sim$ 1.4 bar, suggesting vigorous upward ascending motions. This cloud descended 28 km in 1991 during the evolved phase. Another interesting result is that the EZ particles were brighter during the mature storm phase in the blue than during the undisturbed and evolved phases. If the haze particles are assumed spheres, then their mean size should be $\sim 0.8\ \mu$m.

The GWS onset can be explained by a wet convective disturbance occurring in the deep water clouds (pressure levels $\sim$ 10 bar) as first proposed by Sanchez-Lavega and Battaner (1987). We have recently developed one-dimensional time-dependent dynamical models using as a reference the vertical temperature profile determined from the radio-occultation experiment in Voyager 1 (Lindal et al., 1985) extrapolated adiabatically in depth, together with different hydrogen distributions and abundances of ammonia and water (Hueso and Sanchez-Lavega, 1998). These models predict maximum vertical velocities w $\leq$ 80 ms^{-1} so the ascending buoyant parcels can finally reach the level P $\sim$ 300 mbar as calculated from the radiative transfer models. The subsequent evolution of the GWS disturbance involves probably a wave dynamical phenomenon (Sanchez-Lavega et al., 1994). But more efforts are needed to interpret the interaction of the convective disturbance with the strongly sheared zonal winds.

Time tracking of the longitude changes of the cloud elements forming the storms are transformed to wind velocities relative to the assumed internal rotation period of Saturn (System III, 10 hr 39 min 22 s). A direct comparation between the wind profiles in the Equator using the Voyager 1 and 2 images with the data available for some historical storms and from the large data set coming from the 1990-1997 storm features shows quite similar profiles, considering the large time span of the observations (more than 100 years). This suggest that the wind flows are deep seated although there are some departures in the wind velocity of the biggest storm cores.

Understanding Saturn meteorology and atmospheric dynamics will gratly improve with the arrival of the Cassini mission to Saturn in 2005.

This work has been supported by Gobierno Vasco research grant PI 1997-34. J. R. A. acknowledges the support from Laboratorio de Astrofísica y Física Fundamental (LAEFF-INTA, Madrid). R. Hueso acknowledges a fellowship from Gobierno Vasco.

References

Acarreta, A., Sanchez-Lavega, A., 1998, Icarus, (in the press)
Beebe, R.F., C. Barnet, Sada, P.V., Murrell, S. M., 1992, Icarus, 95, 163
Burke, J.J., Gehrels, T., Strickland, R.N., 1980, J. Geophys. Res., 85, 5883
Hueso, R., Sanchez-Lavega, A., 1998, Proceed. IX Rencontres Blois (in the press)
Lindal, G.F., Seetnam, D.N., V.R. Eshleman, 1985, Astron. J.,90, 1136
Sanchez-Lavega, A., 1982, Icarus, 49, 1
Sanchez-Lavega, A., Battaner, E., 1987, A&A, 185, 315
Sanchez-Lavega, A., Colas, F., Lecacheux, J., Laques, P., Miyazaki, I., Parker, D., 1991 Nature, 353, 397
Sanchez-Lavega, A., Lecacheux, J., Colas, F., Laques, P., 1993, J. Geophys. Res., 98, 18857
Sanchez-Lavega, A., Lecacheux, J., Colas, F., Laques, P., 1994 Icarus, 108, 158
Sanchez-Lavega, A., 1994, Chaos, 4, 341
Sanchez-Lavega, A., Lecacheux, J., Gomez, J.M., Colas, F., Laques, P., Noll, K., Gilmore, D., Miyazaki, I., Parker, D., 1996, Science, 271, 631
Sanchez-Lavega, A., Lecacheux, J., Colas, F., Rojas, J. F., Gomez, J. M., 1998, Planet. Space Science (in the press)
Smith, B. A. et al., 1981, Science, 212, 163
Smith, B. A. et al., 1982, Science, 215, 504

THE GTC: AN ADVANCED 10M TELESCOPE FOR THE ORM

J.M. RODRíGUEZ ESPINOSA, P. ALVAREZ AND F. SÁNCHEZ
Instituto de Astrofísica de Canarias
C/ Vía Láctea s/n, La Laguna, Tenerife, Spain

Abstract. The Gran Telescopio CANARIAS (GTC) is a high performance 10-m class telescope whose construction has been promoted by the IAC (Instituto de Astrofísica de Canarias). It will be installed at the Roque de los Muchachos Observatory (ORM), in the island of La Palma. First light is planned for end-2002. The key science drivers for the project are image quality, operational efficiency and reliability, as emphasized in the Conceptual Design Document which was finished in mid-97. The Preliminary Design is now proceeding on all aspects of the project. The GTC Project is presently funded at the 70% level by Spain. The scientific drives behind the GTC project are described here, as well as the current technical, managerial, and operational baseline.

1. Introduction

The GTC Project is a Spanish initiative lead by the IAC (Instituto de Astrofísica de Canarias) to build a large telescope at the ORM (Observatorio del Roque de los Muchachos), in the island of La Palma. The GTC Project was approved in 1996 and is funded by the Office of Science and Technology, dependent on the Spanish Presidential Office and the Regional Government of the Canary Islands (through the local Consejería de Educación, Cultura y Deportes). They are advocating that this project becomes an international partnership. Thus, international partners are invited to share up to 30% of the project with Spain. The Conceptual Design phase of the project was finished during the second half of 1997. A document (Alvarez et al. 1997) summarizing the Conceptual Design was finished in August 1997. The major results of this phase and the current status of the project are presented here. The authors refer the interested reader to the Conceptual Design Document (Alvarez et al. 1997) for details.

Astrophysics and Space Science is the original source of publication of his article. It is recommended that this article is cited as: *Astrophysics and Space Science* **263**: 355–360, 1999.

2. The GTC Requirements

It is realised that the science programmes to be made with the GTC are photon starved with the current instruments as well as spatial resolution driven. They therefore require a large collecting area and excellent image quality. Besides, the available observing time on the GTC is limited and hence its scientific use has to be optimized. These goals are to be achieved by making design decisions preventing the loss of observing time during operation and through the use of adequate observing modes. Thus the GTC Project is driven by the following simple, clear and ambitious requirements:

- A major improvement in limiting magnitude between 0.3 and 25 microns beyond what is currently possible on a 4-m class telescope.
- Excellent image quality over a reasonably wide field of view (FOV).
- The simultaneous availability of imaging and spectroscopic capabilities over a wide wavelength range (0.3 to 2.5 microns).
- High operational efficiency.
- High reliability.

The following sections explain how the GTC will accomplish these top level requirements.

3. Improving the Limiting Magnitude

A minimum of 10 meters in diameter is needed to achieve, at least, 1 magnitude improvement in limiting magnitude over the existing 4-meter class telescopes. This is the diameter selected for the GTC in a classical Ritchey-Chrétien configuration. A larger diameter results in an unacceptably high cost if the specified image quality is to be maintained. The segmentation scheme adopted is similar to that of the Keck telescope: 36 hexagonal segments. The side length of each hexagon being 936 cm and its thickness 80 mm. In the infrared, achieving low telescope emissivity is important to keep a low background thereby improving the sensitivity. The GTC emissivity will be kept below 5% with fresh coating and maintained below 8% with frequent "in-situ" cleaning. To achieve this emissivity level the entrance aperture stop is situated at the secondary mirror and will have a central hole to avoid radiation from telescope elements inside the central obscuration.

4. Excellent Image Quality

The GTC is to be situated in a top class observatory enjoying excellent seeing, low water vapor and a high number of photometric nights per year (e.g. Murdin 1985; Muñoz-Tuñón et al. 1998 and references therein). The GTC will be erected at the Roque de Los Muchachos Observatory (ORM), in the island of La Palma, located at 28 N and 17 W, and at an altitude of 2400 meters. This is well above a stable inversion layer generally situated between 800 and 1500 meters. It is a well proven site for astronomical observations (Murdin 1985; Brandt & Righini 1985; Vernin & Muñoz-Tuñón 1994; Muñoz-Tuñón et al. 1998), where many first class telescopes have been installed and successfully exploited for the last 15 years. A Spanish Law protects the Observatory from any kind of pollution which can affect its quality for astronomical observations. The median seeing value obtained after several site testing campaigns is 0.64 arcsec, but values as low as 0.2 arcsec have been routinely measured. Specific measurements have been carried out to improve the knowledge of the site for IR observations. Integrated atmospheric water vapour column measurements have been obtained since 1996 using different methods. A median value of 4 mm of water-vapour column has been obtained, and more than 25% of the nights measured have yielded values below 2 mm with a significant number of nights having been below 1 mm (Picirillo et al. 1997; Kidger et al. 1998). The quality of the atmosphere above the ORM allows establishing a tight image quality requirement for the GTC. The quantitative requirement on image quality that has been adopted for the GTC is that a free-air seeing in the visible of 0.4 arcsec (FWHM) should not be degraded to more than 0.45 arcsec. This implies that the combined image degradation from dome seeing, mirror seeing and optics should be under about 0.18 arcsec on axis (FWHM, assuming that it adds in quadrature), or that θ_{80} is 0.84 arcsec. The following paragraphs present the design guidelines followed to achieve the above requirement.

The GTC will have a well ventilated enclosure to facilitate the unimpeded flow of air through it. This will grant that any warm air accumulated inside the dome is carried away, specially in the proximity of the primary mirror where the damage on the image quality is maximum. The ventilation level will be under control to avoid wind buffeting on the telescope structure. Additionally, the sources of heat inside the telescope chamber will be minimized and, when possible, the heat generated will be exhausted. To avoid the warming up of the telescope structure and the mirror blanks during daytime, the telescope area will be kept closed and actively cooled during the day.

The segmented primary mirror ($f/D = 1.65$) is the major source of error on the image quality. A complex system of actuators and sensors, working in real time, will keep the overall primary figure under specifications. It will probably be the most complex system ever made for a telescope mirror. Tip-tilt and piston correction on the segments will be made with actuators and sensors similar to those used by Keck. These work well on Keck and no additional efforts are necessary except to increase their resolution and sensitivity due to the GTC tighter requirements. To correct the low frequency errors on the figure of every segment and to keep the polishing requirements well matched with the capabilities of the optical manufacturers, an active force actuators system is placed behind each segment. This system will permit to correct lateral position errors, curvature errors, and errors due to thermal and gravitational deformations of either the segments or their support systems. The control system of the primary mirror will use a set of nested control loops with different bandwidths. The inner loop, at the level of every segment, will work between 50 to 100 Hz to achieve high stiffness. A wavefront sensor in the A&G box will measure the quality of the primary surface and close the loop with the control system. When sources for guiding are not available, the system will work in open loop using a telescope model previously calibrated which provides the required accuracy. The secondary mirror will be controllable in five degrees of freedom to keep the alignment between the primary mirror and the science instrument and to perform focus corrections. This secondary mirror is a lightweighted piece of glass (ULE^{TM} or Zerodur^{TM}) of about 60 to 65 kg. Its size will be 1.17 meters with a serrated outer shape and a central hole, similar to the primary. The secondary mirror constitutes the entrance stop of the system, resulting in an F17 effective f-ratio for the telescope, and a plate scale of 1.21 arcsecond mm^{-1}.

A deployable tertiary mirror will be used to feed the Nasmyth and folded Cassegrain foci. The available field of view (FOV) will be 25 arcmin in diameter at the Cassegrain and Nasmyth foci. The unvignetted FOV is 8 arcmin in diameter. For the folded Cassegrain foci the FOV will be 9 arcmin. The GTC is designed to have prime focus available. The FOV at Prime will be 40 arcmin in diameter. Image motion has to be limited to achieve the requirement on image quality. This sets requirements on the dynamic behavior of the telescope structure and driving mechanisms. The image motion residuals will be corrected by tip-tilt capabilities on the secondary mirror. With these tip-tilt capabilities, up to frequencies of 40 Hz, the secondary will be used to correct structure vibrations coming from wind buffeting on the structure or residuals of the telescope driving system. The fast tip-tilt capability has the additional advantage of being capable of correcting atmospheric image-motion at 4.8 micron or in good seeing even at shorter wavelengths. For IR observations the secondary will have chopping capabilities up to frequencies of 5 Hz with a chopping throw of 15 arcsec and 80% duty cycle. An adaptive optics (AO) system for the GTC is planed for 3 years after Day 1, hence the GTC is being designed to elegantly accomodate the AO system. To increase the sky coverage of the AO system a future laser guide star is planned for the GTC. The GTC design requirement for the AO system is that it should be able to achieve essentially diffraction-limited images at 2.2 microns (0.8 Strehl Ratio in 0.5 arcsec visible seeing).

5. High Operational Efficiency

To maximize the scientific return from the GTC, high operational efficiency is needed. Operation modes, which minimise telescope time losses like changes in telescope configuration or telescope setup, and which make optimum use of the prevailing conditions (e.g. seeing, humidity, instruments available), will be implemented. These will be achieved through the use of queued service observing. In this mode the actual observations to be undertaken are not determined until immediately before the observations take place. Therefore, the conditions are known and those observations that require the best conditions have access to them. The GTC will be designed to facilitate this observing mode. As such, a single secondary configuration has been adopted for visible and IR wavelengths which avoids changes between secondaries, which are wasteful of telescope time and expensive if done during the night. This scheme also requires multiple focal stations (Cassegrain, two Nasmyth and four folded Cassegrain foci) available and several instruments ready to be used. A deployable tertiary mirror that is capable of switching between these foci when deployed, or, if parked, of feeding the Cassegrain focus will be used. Day one instruments will always be available. They will permit to cover a wide wavelength range with multiple configurations so that the current conditions can be scientifically exploited in queue-observing mode.

A strategy to do the alignment, phasing and shaping of the primary mirror segments during daytime is being studied. Only the calibration of the segment shape is left for the night, therefore requiring less telescope time. Queued service observing will facilitate the standardization of observing protocols to increase the usefulness of the data archive. This archive, fully integrated in the Control System, will be a powerful tool to achieve the maximum scientific return from the GTC.

6. High Reliability

The GTC is being designed to be a highly reliable telescope. The maximum percentage of downtime due to failures must not exceed 2% of the useful time. The total non-availability of the telescope is established to not exceed 5% including maintenance and downtime due to failures. This reliability budget is a design requirement for the GTC.

7. Budget and Schedule

The estimated cost of the project, based mainly on the information obtained directly from potential suppliers of subsystems, on the experience of other telescope projects and the costs of constructions carried out at the Canarian observatories, is 93.5 Millions US $ (MUSD) (1997). The operation budget is estimated to be 6.1 MUSD where development of new instruments and system updates are included. The project is scheduled to have first light at the end of 2002 and starts full scientific operations at the end of 2003. Major contracts will be awarded mainly in 1999.

8. Project Status

Intensive work is proceeding on the preliminary design of the different subsystems of the GTC. The Civil work and Enclosure are the more advanced subsystems. The Preliminary Design Review (PDR) of the Civil Work already happened in mid 98. Ground breaking is planed for mid 99. Several development programs are running to optimise the polishing and verification procedures for the primary segments. A prototype of the support of the primary mirror segments is being fabricated. It will permit to test the control algorithms and the daytime alignment and phasing subsystem. An Announcement of Opportunity requesting conceptual designs of two Day One Science Instruments was issued in summer of 1998. The two instruments will be selected early in 1999 to be supplied in 2003. These instruments will be developed by consortia of research institutions. Finally, contacts and negotiations with different organizations around the world are being held to arrive to the signature of formal agreements for joining the GTC project. Meanwhile, the Spanish administration (the Central and the Regional Governments) guaranty the 30% of the cost of the project that is expected to be covered by non-Spanish partners.

References

Alvarez, P., et al., 1997, *Gran Telescopio CANARIAS. Conceptual design*, GTC Project Document GEN/STMA/0012-L
Brandt, P.N., Righini, A. 1985, Vistas in Astronomy, 28, 437
Kidger, M., Rosario, J.C., 1998, *Water measurement Campaign for the GTC*, GTC Project Document
Muñoz-Tuñón, C., Varela, A.M., Mahoney, T., 1997, *Estudio del sitio para el Gran Telescopio. Informe final*, GTC Project Document EXT/IAC/0008-R
Muõz-Tuñón, C., Vernin, J., Varela, A.M., 1997, A&AS, 125, 183
Murdin, P., 1985, Vistas in Astronomy, 28, 449
Picirillo, L., et al., 1997, ApJ, 475, L77
Vernin, J., Muñoz-Tuñón, C., 1994, A&A, 284, 311

THE EMIR NEAR-INFRARED MULTI-OBJECT SPECTROGRAPH FOR GTC

M. BALCELLS
Instituto de Astrofísica de Canarias
38200 La Laguna, Tenerife, Spain

Abstract. EMIR is a near-infrared (NIR) multi-object spectrograph (MOS) proposed for first-generation common user instrument of the Gran Telescopio Canarias GTC. It has three observing modes: imaging, multiobject spectroscopy and integral-field spectroscopy, in the NIR Z, J, H and K bands. The design uses cryogenic multislit masks to allow simultaneous spectroscopy of up to approximately 50 objects. The spectral resolution R=5000 allows for effective "see through" the OH line forest that dominates the J and H sky backgrounds. EMIR addresses one of the fundamental goals of large telescopes, the spectroscopy of large quantities of faint sources. Science applications include young stellar objects embedded in molecular clouds, brown dwarfs, stellar clusters, resolved population in external galaxies and distant galaxies. EMIR is being designed by a consortium of Spanish, UK and French Institutes. Project updates and information will be posted at `http://www.ucm.es/info/emir/`.

1. Introduction

The Spanish astronomical community has expressed a strong interest for the procurement of a NIR spectrograph for GTC. Desirable qualities include a multi-object capability reaching the K band, over a field as wide as possible; such functionality is not available in any other MOS existing or planned for a large telescope, hence it will open a number of research niches to GTC observers. In addition, high efficiency, high multiplex factor and the ability to provide imaging are obvious design goals. An international team of scientists and instrumentalists has been set up to design and build such an instrument. Grantecan funded the initial proposal written by R. Guzmán, for its Feasibility Study Phase.

Astrophysics and Space Science is the original source of publication of his article. It is recommended that this article is cited as: *Astrophysics and Space Science* **263**: 361–364, 1999.

2. Science programs

In order to set the top-level science requirements of this common-user instrument, the definition team enquired the Spanish astronomical community and identified a broad range of science applications that would benefit from the NIR multi-object spectroscopic, IFU and imaging capabilities of EMIR. These range from the chemical composition of planets in the Solar System, to the study of distant galaxies in the early Universe. Applications abound in stellar astronomy, eg. the determination of the low-end of the stellar luminosity function, and young stellar objects embedded in molecular clouds. NIR spectroscopy provides as well key diagnostics on the nature and evolution of Supernovae. Studies of nearby, unresolved galaxies will benefit from the mapping of NIR CO and Si lines, giant/supergiant discriminators, and from the ability to determine kinematics and composition in dust obscured galaxy regions. Both the MOS and the IFU will allow for dynamical studies of resolved stellar systems such as stellar clusters and resolved external galaxies.

A key driver behind EMIR, and indeed of most 10m telescopes, is to open new windows onto the study of the formation and evolution of galaxies. Current spectrographs in the visible wavelentth allow the study of galaxies out to $z \sim 1$. NIR MOSs operating in the J, H bands (AUSTRALIS, NIR-MOS) will soon extend this domain to $z \sim 2$. Beyond that, distant galaxies are identified by the Lyman limit entering the visible range at $z \sim 5$. By providing MOS at the 2.2μm K band, GTC will open the field of study of galaxies in the range $z = 2 - 5$. A team of Spanish, British and American astronomers (the COSMOS consortium) are working already now for the exploitation of this unique capability that EMIR will provide to GTC.

3. The instrument

To accomplish the science goals described in the previous section, we propose a multi-slit spectrograph. In order to perform at K band, masks will be cooled at -100°C. At this temperature their thermal background count rate is well below the dark current of the HgCdTe Rockwell detectors. We ruled out fiber robot systems given the technical difficulties in their implementation at cryogenic temperatures. The spectrograph optics are entirely reflective except for a field lens and the corrector plate in the Schmidt camera. This makes for a particularly simple optical design which is highly efficient and is free from chromatic aberrations. We are studying the use of Volume-Phase Holographic gratings (VPHGs) to replace the current ruled reflection gratings, once the IR performance of VPHGs is established. VPHGs will improve the efficiency by a factor above 2, allowing for spectroscopy with efficiency similar to that of imaging. EMIR characteristics are sumarized in

Figure 1. Mechanical layout of **EMIR**.

Table 1. The optical design of EMIR is shown in Figure 1.

4. The science group and instrumental consortium

EMIR poses a number of technological challenges. These derive from the instrument's large size and the need to work at cryogenic temperatures. An Instrument Definition Team (IDT) has been set up to define the main science drivers, to study the performance of the instrument, to establish its working requirements, to build an instrument simulator and a data-reduction pipeline. This team will guarantee that EMIR is ready to produce useful science from day one. Members of the IDT include Balcells (IAC), Carter (JMU), Gallego (UCM), Guzmán (Yale), James (JMU), Lowenthal (UMass), Mas-Hesse (LAEFF), Parry (Cambridge), Peletier (Durham), Pello (OMP), Prieto (IAC), Rauscher (Durham), Sharples (Durham). The EMIR science team is led by M. Balcells (PI, IAC) and R. Guzmán (co-PI, Yale).

EMIR will be built by an instrumental consortium which includes Durham, Cambridge and Liverpool Universities in the UK, Observatoire Midi-Pirennées in France, and IAC, UCM, LAEFF/INTA and CD6 in Spain. A novel approach to the work breakdown is being plan ned, in which Spanish, British and/or French groups contribute together to the design development of major components such as mechanics, optics and control electronics. IAC is in charge of the overall management of the project, with Jesús Patrón (IAC) as Project Manager. Parry (Cambridge) and Manescau (IAC) contribute to the optical design. Fuentes (IAC) and Gellatly (TTL) are in

Table 1: Technical characteristics of **EMIR**

Focal Station	Nasmyth
Wavelength range	0.9-2.5 microns
Optimization	Entire wavelength range
Observing modes	Multi-object spectroscopy Integral-field spectroscopy Imaging
Spectral resolution	5000 for 0"75 (3-pixel) wide apertures 7500 for 0"50 (2-pixel) wide apertures
Spectral coverage	One observing window (Z, J, H and K) per single exposure
Array format	2 Rockwell Hawaii-2 2048x2048 HgCdTe
Scales at detector	0.25 arcsec / pixel
OH supression	In software
Cryostat temperature	77 K
Multi-object mode:	
FOV	6'x6', up to 45 slitlets (0.6"×8")
Sensitivity	K=19.5, t=3600s, S/N=5 per resolution element (continuum) F=2x10^{-18} erg/s/cm^2/Å, t=3600s, S/N=5 per resolution element (Halpha line at z=3)
Imaging mode:	
FOV	6'×6'
Image quality	>80% of energy within 0.2"
Sensitivity	K=22.3, t=3600s, S/N=10 in one arcsec box
Integral Field mode:	
FOV	7.5"×7.5"
Fibre bundle/mirror slicer	TBD

charge of the overall mechanical design. Sharples and Rauscher (Durham) provide the cold-mask unit, while the IFU will be developed by Allington-Smith (Durham).

5. Current status

The EMIR Feasibility Study and Conceptual Design will be submitted to Grantecan in December 1998. EMIR will compete with other proposals for first science IR instrument for GTC. GTC instrument selection will take place in early March 1999. Commissioning for the GTC first generation NIR instrument is planned for the middle of 2003.

AN OPTICAL INTERMEDIATE DISPERSION SPECTROGRAPH FOR THE GTC

ODIN: Espectrógrafo Óptico de Dispersión INtermedia

R. J. GARCÍA LÓPEZ
Instituto de Astrofísica de Canarias
E-38200 La Laguna, Tenerife, Spain

Abstract. ODIN is an international consortium constituted to design and build a first-light multi-purpose optical intermediate dispersion spectrograph for the *Gran Telescopio Canarias* (GTC, 10-m) to be installed at the Observatorio del Roque de los Muchachos (La Palma). This instrument should fulfill an important fraction of the needs of the Spanish astronomical community in optical spectroscopy performed at large telescopes, and must be highly competitive in a variety of research fields. We are proposing a spectrograph that will be capable of providing multi-object observations by means of optical fibers, integral field spectroscopy, and also multi-integral fields. It will be mounted at the Nasmyth focus of the GTC. One of the main drivers of our proposal is to cover the whole visible range allowed by the atmospheric cut-off with the highest efficiency, particularly at the UV end.

1. The ODIN Consortium

This international consortium is promoted by the Instituto de Astrofísica de Canarias and is constituted by the following researchers and institutions:

- *Principal Investigator*: Ramón J. García López[1]

- *Operations Manager*: Luis Cuesta[1]

- *Instrument Definition Team* (in alphabetical order):

Astrophysics and Space Science is the original source of publication of his article. It is recommended that this article is cited as: *Astrophysics and Space Science* **263**: 365–368, 1999.

Santiago Arribas[1] Jacques Baudrand[2]
Phil Charles[3] Paul Felenbok[2]
Jean Guérin[2] Jean Pierre Picat[4]
Mauro Pucillo[5] José Luis Rasilla[1]
Justo Sánchez del Río[6] Antonio Talavera[7]

– *Science Advisory Team* (in alphabetical order):

Carlos Abia[8] John Beckman[1]
Jorge Casares[1] Héctor Castañeda[9]
Miriam Centurión[5] Gavin Dalton[3]
Ascensión del Olmo[6] César Esteban[1]
Ignacio González Martínez-Pais[1] Artemio Herrero[1]
Eduardo Martín[10] Minia Manteiga[11]
Evencio Mediavilla[1] Enrique Solano[12]
Geneviève Soucail[4] Ana Ulla[11]
José M. Vílchez[1] Giovanni Vladilo[5]

1: Instituto de Astrofísica de Canarias (Spain); 2: Observatoire Paris-Meudon (France); 3: University of Oxford (UK); 4: Observatoire Midi-Pyrenées (France); 5: Osservatorio Astronomico di Trieste (Italy); 6: Instituto de Astrofísica de Andalucía (Spain); 7: Instituto Nacional de Técnica Aeroespacial (Spain); 8: Universidad de Granada (Spain); 9: Observatorio Astronómico Nacional, UNAM (Mexico); 10: University of Berkeley (USA); 11: Universidad de Vigo (Spain); 12: Universidad Europea de Madrid (Spain).

2. Instrument Description

The instrument proposed is an intermediate dispersion spectrograph, supplying resolving power in the range 5000 to 40000, which will work from 310 nm (with an acceptable throughput) to 1 μm. Low order and echelle gratings will be used to cover the proposed range of spectral resolution. The instrument will provide a fiber multi-object option with a multi-integral field capability. A single large integral field unit (IFU) and a simulated long slit will also be supplied. The spectrograph will be located on one of the Nasmyth platforms to increase the spectral stability and thermal and mechanical insulation.

The spectrograph unit is a classical one, with mirror optics and silica correctors. Taking advantage of the large-size detectors available (a mosaic of two 2k$\times$ 4k pixels thinned CCD chips), we propose also to have several orders available simultaneously on the detector with a limited number of objects. The design will be optimized to yield the largest spectral coverage in the higher resolution modes while providing a very large spectral coverage for all the fibers in a single exposure when the lower resolution modes are used. An optical image slicer will be provided for the highest resolution.

The fiber positioner will be attached at the Nasmyth rotator. To join the probes to the spectrograph slit the fibers should not exceed a length of 10 m, with an internal transparency of 75% at 310 nm. We are considering a classical positioner for handling around 60 arms. The advantage of this type of positioner over the robot one for a small number of fibers, is the high configuration speed and the ability to move the fibers in real time during the observations. It is less expensive to build all arms identical in their mechanical and electronic design. At the tip of each arm a bundle of nine fibers will be located, fitted with micro-lenses to act as image slicer and allowing the highest spectral resolution with a grating of reasonable size. The micro-lenses will handle the focal ratio degradation by changing the input F/17 beam to a smaller value (about F/3), increasing the spectral stability by injecting the beam in the pupil mode.

The spectrograph will also allow integral field spectroscopy. Two bundles, providing different fiber configurations in the telescope focal plane will be used for observations of single extended objects. These bundles will be located at the center of the focal plane using a specially designed mechanism. It is also planned to implement a multi-integral field capability. About 20 arms among the 60 available at the fiber positioner will carry the 'small' integral field units, which will be made of an array of several fibers. At the fiber output, the exit beam will pass through a linear micro-lens array to recover an aperture which matches the entrance focal ratio of the spectrograph. At the spectrograph slit, the different fiber modes will be exchanged by simple commutation, and the spectrograph will work in the same way for all of them. The main characteristics of ODIN are summarized in Table 1.

3. Scientific Drivers

The scientific programmes for which ODIN is intended to be used cover fields related to stellar formation, structure and evolution, chemical evolution of the Galaxy, chemical evolution at high redshifts, constraints on primordial nucleosynthesis models, interstellar medium, galactic dynamics, extragalactic objects, etc. All of them would have specific and clear impacts on a wide variety of astrophysical fields of research. We must emphasize that ODIN is equipped with features which make it unique among similar instruments which (will) operate on other large telescopes. Reaching the at-

TABLE 1. ODIN Specifications

Focal station	Nasmyth
Wavelength range	0.31 nm to 1 μm
Detector array	4096 × 4096 pixels
Field of view	12 arcmin in diameter (maximum vignetting of 4% for the field outside 8 arcmin)
Spectral resolution	$R = 5000$ to 40000
Simultaneous spectral coverage	5000 Å at $R = 5000$ 2000 Å at $R = 10000$ 1000 Å at $R = 20000$ 500 Å at $R = 40000$
Field of a single fiber (+microlens) in the sky	0.38 arcsec
Multi-object mode	60 arms
Multi-IFU mode	20 arms
Central IFU mode: High spatial resolution Low spatial resolution	 12 arcsec × 12 arcsec (minimum) 1 arcmin × 1 arcmin (minimum)
Simulated long slit	2 to 3 arcmin

mospheric cut-off at intermediate-high spectral resolution with multi-object or multi-IFU configurations, in an observatory which has demonstrated its high transparency in this wavelength range, will not be possible with other instruments.

4. Present Situation

At the moment of writing these proceedings (November 1998), ODIN is in its conceptual design phase and will compete with other proposals to be one of the two first-light instruments to be installed at the GTC. The final decision about these instruments will be made by the GTC Project Office at the beginning of March 1999. All efforts are being directed within the consortium to make our proposal a successful one and to provide the community with a new workhorse in optical spectroscopy at the end of 2003. For more information, please contact R. J. García López at *rgl@ll.iac.es*.

OSIRIS

Imaging and Spectroscopy for the GTC

JORDI CEPA
Departamento de Astrofísica
Facultad de Física, Universidad de La Laguna,
E-38071 La Laguna, Tenerife, Spain

AND

Instituto de Astrofísica de Canarias
E-38200 La Laguna, Tenerife, Spain

Abstract.
The characteristics of OSIRIS (Optical System for Imaging and low-Resolution Integrated Spectroscopy), an instrument for imaging and low resolution spectroscopy proposed for the GTC 10m telescope, are described. OSIRIS is a multi-purpose instrument that incorporates in a "classical design" several innovations, already tested in 4m-class telescopes, that will provide a powerful first light instrument for the GTC.

1. Introduction

OSIRIS represents a new generation of instrument observing techniques, that includes the concepts of tunable filters (TFs), charge shuffling on the CCD detectors, volume phase holographic gratings, etc, already tested and used in 4m-class telescopes. OSIRIS is an imaging system and a low-resolution long-slit and multi-object spectrograph covering the wavelength range 0.33-1.1 μm, in a field of view of 8 arcmin in diameter (Table 2).

OSIRIS can be coordinated with any instrument in the Nasmyth focus, providing parallel observing modes and therefore allowing simultaneous imaging and spectroscopy, in the same or different spectral ranges. With this facility of parallel observing modes, OSIRIS will allow to obtain simultaneous optical imaging and high resolution spectroscopy of the same FOV, or imaging/spectroscopy of the same or different FOVs in the optical and the NIR (Table 3).

Astrophysics and Space Science is the original source of publication of his article. It is recommended that this article is cited as: *Astrophysics and Space Science* **263**: 369–372, 1999.

TABLE 1. OSIRIS expected throughput. The expected efficiencies are of OSIRIS without GTC, including losses due to reflection, absorbtion and obstruction but ignoring slit losses and losses due to filters.

Imaging mode	
4000 Å	60%
6000 Å	55%
8000 Å	40%
Spectroscopic mode (at R=600)	
4000 Å	55%
6000 Å	50%
8000 Å	35%

2. Instrument Description

2.1. IMAGING

Broad band imaging in the Johnson, Strømgren and Gunn systems will be provided. Two tunable filters located in the pupil of the optical system will cover the full optical range, providing an unlimited number of narrow and intermediate band interference filters with FWHM between 1 and 70 Angstrom. Depending on the wavelength and the FWHM selected, orders can be blocked by broad band filters or by an additional set of interference filters of 250-400 Å FWHM. Both order-sorting and TFs will be tiltable up to 20°. Several advantages for this instrument are: to avoid the expense to manufacture a large number of narrow band filters, uniformity of the instrumental response function for all wavelengths and all band-passes, and the use of the same optical path for all the configuration. Table 1 shows the expected performance of OSIRIS in imaging mode.

2.2. CHARGE SHUFFLING

High quality CCD detectors, as they are produced now, allow to synchronize TFs frequency switching and movement of charge in a CCD. TFs can be tuned with rates in excess of 100 Hz, and its change can be synchronized with the movement of charge over large areas on a CCD. In this form it is very straightforward to obtain simultaneously images of object and continuum in a single exposure or to carry out time-series photometry in the same or in different wavelengths with timescales of less than one second. Also, surveys for high redshift objects with emission lines can be efficiently performed allowing, for example, a fast cluster membership identification.

2.3. SPECTROSCOPY

Low-resolution spectroscopy (R=600, 2000 and 5000) will be available, for the entire wavelength range, by using grisms. An adjustable long-slit will be available, together with multi-object spectroscopy in a standard mode. Given the queue-observing schedule for the GTC, a multi-slit unit will be available although it will be also possible, upon request, to use a mask of slits. Volume Phase Holographic gratings will provide, in the near future, grisms with efficiencies larger than 90% over a wider spectra range, with equal response in S and P polarizations. Table 3 shows the expected performance of OSIRIS in spectroscopic mode.

2.4. DETECTOR

On first light a CCD 4k×4k (or two buttable 2k×4k), of $\sim 15\mu$m pixel size type, equivalent to some $0.08''$ pix^{-1} on the sky, sensitive from 0.33 through 1.1 μm will be used. Charge shuffling and raster scan capacity will be available.

3. Summary

Given the innovations provided by OSIRIS with respect to similar instruments existing or planned in 10m-class telescopes: TFs combined with charge shuffling and simultaneous observing modes with other instruments, OSIRIS will be a competitive instrument for the first decade of the XXI century. It is expected to be one of the most used GTC instruments, not only for its innovations, but due to the wide range of astronomical projects that can be covered with such a multi-purpose instrument. The success of similar instruments like LRIS, to which KECK II is almost fully dedicated, is a good example of the scientific achievements that can be obtained with OSIRIS.

Note added while in proof

The concept of the instrument has changed drastically and is now intended as a first light, standalone instrument for imaging and low and intermediate resolution spectroscopy, that incorporates volume phased holographic gratings and belongs to the new class of "tunable spectrographs".

TABLE 2. OSIRIS specifications

General specificacions	
Focal Station	Nasmyth
Wavelength range	0.33 – 1.10 nm
Detector Array	4096 × 4096 pixels (or two buttable 2K×4K)
Scale at detector	0.08 arcsec/pixel
FOV	8 arcmin in diameter
Spectroscopic mode	
Spectral resolution	R= 600, 2000, 5000 (over all wavelength range)
Long Slit	Extending over all the FOV and adjustable
Multi-slit	Up to 15 slitlets of 10 arcsec length
Mask of slits	Up to 120 targets at R=600
Integral Field Unit	Yes
Imaging mode	
Blocking Filters	Yes
Broad band	Johnson, Strømgren and Gunn
Narrow Band	Tunable filters (FWHM: 1 to 70 Å over all the wavelength range)

TABLE 3. Simultaneous use of OSIRIS and other instruments in the same Nasmyth focus. In its simultaneous use with a NIR instrument with the same FOV, since the dicroic will not be cooled, the NIR wavelength range will be limited to ~1.6 μm. This does not happen when used simultaneously in different FOVs.

Parallel modes	
OSIRIS	Other Instruments
Imaging	High-res optical spectroscopy different wavelengths, same FOV NIR imaging same or different FOV NIR spectroscopy same or different FOV
Spectroscopy	NIR imaging same or different FOV NIR spectroscopy same or different FOV

SITE-TESTING RESULTS AT LA PALMA OBSERVATORY

A.M. VARELA, C. MUÑOZ-TUÑÓN AND T. MAHONEY
Instituto de Astrofísica de Canarias, E38200-La Laguna Tenerife, Spain

1. Introduction

In view of the high cost of large telescopes and the advances in imaging techniques, it is essential to know the seeing potential of both future and already existing sites. Systematic site testing is justified not only by the need to find the best sites but also to take the most of the already classified as excellent.

In view of this concern a workshop on *Site Properties of the Canarian Observatories* was organized recently by the Subcommittee for Site Properties of the International Scientific Committe of the Canarian Observatories in which the latest results concerning water vapour content, atmospheric transparency and site testing at different telescopic installations, adaptive optics, etc., were presented and analysed. The proceedings of this conference are just about to appear in a special edition of *New Astronomy Reviews*. We present here a summary of the results obtained from several site-testing campaigns at the Roque de los Muchachos Observatory (ORM), following the classical scheme of simultaneous optical and meteorological monitoring, which now confirm the ORM to be one of the best observing sites.

2. Site testing campaigns: methods and instrumentation

In Figure 1 we give a schematic view of the site-testing equipment operated by the IAC at the Observatory. Seeing is referred to the zenith taking into account the appropiate air mass correction, the zenithal angle being small ($\leq 30°$). The FWHM is accurate to better than $0.1''$, and under reasonably good weather conditions it is possible to monitor the seeing with a temporal sampling rate of less than a half a minute for several hours thanks to the automatic guiding. Meteorological parameters are sampled continuously once each minute.

Astrophysics and Space Science is the original source of publication of his article. It is recommended that this article is cited as: *Astrophysics and Space Science* **263**: 373–376, 1999.

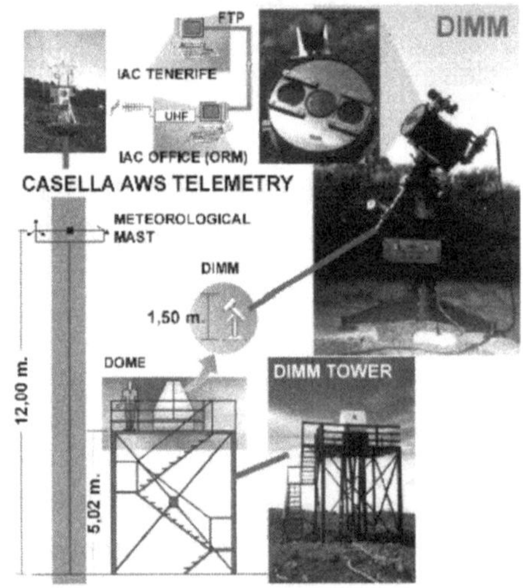

Figure 1. The site-testing equipment includes the Differential Image Motion Monitor (Vernin & Muñoz-Tuñón 1994), an instrument for measuring the astronomical quality of the image, or seeing , a 12-m mast equipped with standard meteorological sensors (wind speed, wind direction, air temperature, temperature of soil and sub-soil, humidity, pressure, etc.), and the necessary infrastructure to support the instruments. To avoid the surface layer influence the DIMM is located on a 5-m tower (Vernin & Muñoz-Tuñón 1992).

3. Site assessment campaigns at the TNG, WHT and GTC sites

Since June 1994 the DA/IAC DIMM has been in operation at the Roque de los Muchachos Observatory to begin a site testing campaign for the *Telescopio Nazionale Galileo* (TNG). The mean seeing at the TNG site during the entire campaign was 0.76″, the median was 0.64″ and 7% of the time the seeing was better than 0.3″. The best seeing measured was 0.17″. A seasonal variation was noticeable, the best seeing conditions appearing during the summer, which corresponds to the occurence of a well-defined inversion layer due to the high prevalence of trade winds (see Muñoz-Tuñón, Vernin and Varela 1997).

From these statistics for the TNG slope, there is no doubt concerning the outstanding quality of the area. The 4.2-m William Herschel Telescope (WHT) area has also been extensively studied, due to the long-standing presence of large telescopic facilities. The NOT site, has also been demonstrated to be excellent. For an appropiate comparison of the different sites with standard techniques, a seeing monitor was installed close to the WHT.

From September 1995 to December 1996, a site-testing programme led by the IAC was carried out to select the site for the future *Gran Telescopio*

Canarias (GTC; Alvarez et al. 1998). The choice between two sites preselected, GTC1 and GTC2 (Muñoz-Tuñón, Varela and Mahoney, 1998) was made not only on the basis of the seeing, but also taking into account the meteorological variables. The proximity of GTC1 to La Caldera required continous statistical meteorological tests over a period of, at least, one year, in order to ensure a seasonal sample. At each site we installed a DIMM (5m above the ground) and a 12-m mast equipped with standard meteorological sensors. Any local fluctuation in the thickness of the surface layer due to La Caldera is detectable by the seeing statistics since DIMMs yield a temporal sampling rate of 1/2 minute. This large and updated database will allow us to study the relationship between both the optical quality and meteorology at the ORM (see Muñoz-Tuñón, Varela and Mahoney, 1997).

Table 1 shows the total statistical results of the seeing data gathered during different campaigns at several sites at the ORM.

	GTC1	GTC2	TNG	WHT
N data	160119	87978	86385	135332
Min (″).	0.15	0.18	0.17	0.11
Mean (″)	0.72	0.75	0.76	0.81
std (″)	0.55	0.40	0.47	0.45
Median (″)	0.72	0.65	0.64	0.69
$< 0.5''$	19%	22%	28%	21%
$< 1''$	76%	84%	82%	74%
$> 2''$	5%	2%	3%	3%

In Figure 2 we present the cumulative frequency of the seeing data presented in Table 1. Comparing GTC1 and GTC2, slightly better conditions at the lower site (GTC2) are measured (see Muñoz-Tuñón, Varela & Mahoney, 1998). That has played an important role in the final decision on the future location of the GTC.

To summarize, the excellent seeing results obtained confirm the ORM among the best astronomical sites (Paranal and La Silla, Chile) for which seeing statistics are available, the ORM presenting better behaviour, with the best median values, and slightly larger percentage of seeing values below $0.3''$. The parameters corresponding to the best month are also slightly better.

The meteorological statistics show similar temperature, humidity and wind speed conditions at GTC1 and GTC2. The overspill of clouds from La Caldera and the instability of inversion layer occurred with the same frequency. The percentage of useful observing hours was the same.

The prevailing winds are from the north east as is shown in Figure 3 (Mahoney, Muñoz-Tuñón & Varela, 1998). There seems to be a correlation between wind direction and seeing behaviour (for moderate wind speed).

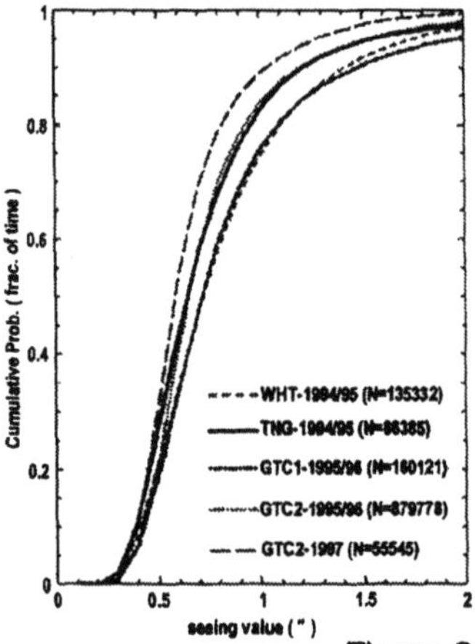

Figure 2

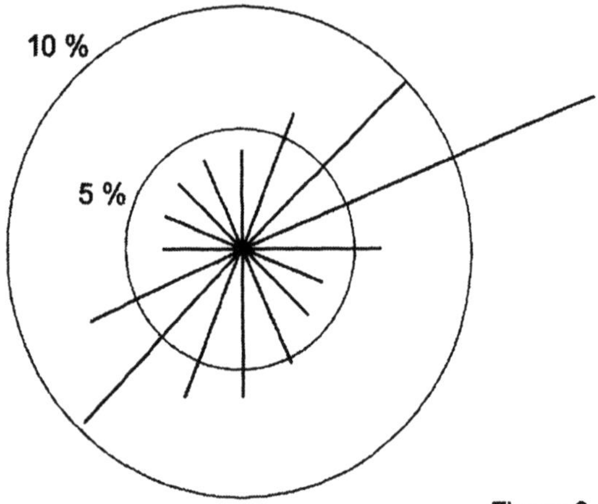

Figure 3

Figure 2. Cumulative probability plot of seeing values obtained for all data gathered at the GTC1, GTC2, WHT and GTC sites. The whole data set at every location spans almost one year and therefore includes summer and winter.

Figure 3. Nocturnal windrose for February 1995 to December 1996. The prevailing winds are from the north east and there seems to be a correlation between wind direction and seeing behaviour (for moderate wind speed). This is the most comprehensive windrose published for the ORM.

The study of these possible relationships between image quality and wind is the next aim of our group.

References

Alvarez-Martín, P., Rodríguez-Espinosa, J.M., Sánchez. F., 1998, New Rev. in Astron., ed. C. Muñoz-Tuñón

Mahoney, T., Muñoz-Tuñón, C., Varela, A.M., 1998, New Rev. in Astron., ed. C. Muñoz-Tuñón

Muñoz-Tuñón, C., Vernin, J., Varela, A.M., 1997, A&AS, 125, 183

Muñoz-Tuñón, C., Varela, A.M., Mahoney, T., 1997, Campaña de Prospección Astronómica del Lugar Idóneo para la Instalación del Telescopio de 10 m, Informe Final

Muñoz-Tuñón, C., Varela, A.M., Mahoney, T., 1998, New Rev. in Astron., ed. C. Muñoz-Tuñón

Vernin, J., Muñoz-Tuñón, C., 1992, A&A, 257, 811

Vernin, J., Muñoz-Tuñón, C., 1994, A&A, 284, 311

STATUS AND RESULTS FROM THE HEGRA AIR SHOWER EXPERIMENT

V. FONSECA (HEGRA COLLABORATION)
Dept. Física Atómica, Fac. Ciencias Físicas
Universidad Complutense. 28040 Madrid, Spain

Abstract. The HEGRA experiment investigates the relativistic (non thermal) universe by measuring air showers initiated in the atmosphere by photons and charged particles from the cosmos, using the imaging atmospheric Cherenkov light technique (> 500 GeV) and the showerfront sampling technique (> 20 TeV). New results concerning the search and study of gamma ray sources above 500 GeV, the measurement of the energy spectrum and chemical composition of cosmic rays above 1 PeV and the search of TeV counterparts of Gamma Ray Bursts are briefly described.

1. Introduction and Experimental Method

The HEGRA (High Energy Gamma Ray Astronomy) air shower experiment explores processes of the violent, non thermal universe. The measurement of the cosmic primary particle is indirect, via the showers generated in the atmosphere, through the cascading process. A direct measurement with detectors on board of satellites of the cosmic primary photons and nuclei with energy higher than several GeV is not possible, due to their small detection area in view of the low cosmic ray rates involved. HEGRA exploits two detection methods: the imaging atmospheric Cherenkov technique with 6 atmospheric Cherenkov telescopes ($E > 500$ GeV) and the shower front sampling technique ($E > 20$ TeV) with three types of arrays (scintillator, wide angle Cherenkov counter AIROBICC and Geiger-towers). HEGRA has pioneered new methods in both detection techniques. The experiment is located at the Observatorio del Roque de los Muchachos on the Canary island of la Palma (2200 m a.s.l., 28.8°N, 17.9°W) covering and area of about 40000 m^2 (Figure 1). The experiment is run by a collaboration of German, Spanish and Armenian research institutes and universities.

Astrophysics and Space Science is the original source of publication of his article. It is recommended that this article is cited as: *Astrophysics and Space Science* **263**: 377–380, 1999.

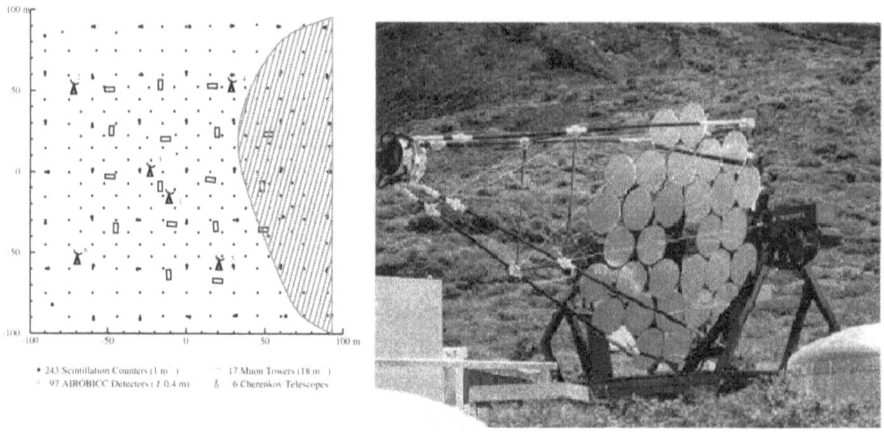

Figure 1. Left: The layout of HEGRA. The hatched region was destroyed by a bush fire in October 1997. Right: One of the HEGRA atmospheric Cherenkov telescopes used for the stereoscopic observations.

1.1. THE IMAGING AIR CHERENKOV TELESCOPES TECHNIQUE

The showers in the atmosphere, initiated by the primary photons or nuclei, generate atmospheric Cherenkov light which is recorded during the night using a tesselated mirror which focuses the Cherenkov light onto a multipixel (photomultipier) camera in the focal plane of the mirror. The image given by each air shower can be analyzed and the direction and energy determined as well as gamma induced showers can be separated from hadron initiated ones (which constitute in general a severe background). The Cherenkov light pool at detection level has a radius of about 100 m and thus the detection area is of the order of 3×10^4 m^2. HEGRA has six telescopes. Five are identical and measure simultaneously the images of the same air shower from different viewing angles (stereoscopic observation). With this unique technique HEGRA achieves an angular resolution of $< 0.1^\circ$, an energy resolution better than 20% and an excellent separation of gamma from hadron initiated showers, resulting in a flux sensitivity of $\mathrm{SIGNAL}/\sqrt{\mathrm{BACKGROUND}} = 5\sigma$ in one hour of observation for a flux corresponding to 1/4 the Crab flux. These telescopes have a segmented mirror of 8.5 m^2 and a 271 pixel (photomultiplier) camera of 0.25°, resulting in a 4.3° field of view.

The remaining telescope (built in 1991) runs in stand-alone mode and has been upgraded from 5 to 10 m^2 in 1997 using a new type of all Aluminum mirrors.

1.2. SHOWERFRONT SAMPLING TECHNIQUE

At energies of about 20 TeV the secondary particles of an air shower reach ground distributed on a thin disk (the shower front). The arrival time and the particle density can be sampled with an array of scintillation counters and the direction and energy of the primary particle can be obtained. This is the classical method of airshower observation. At HEGRA, 243 scintillation counters have been placed on a 200×200 m^2 area in a 15 m grid. Also the atmospheric Cherenkov photons are concentrated in the shower front and the Cherenkov light is sampled by an array of open photomultipliers (20 cm diameter) placed in a 30 m grid (AIROBICC). The energy thresholds of the arrays for vertical incident primary photons is 20 (11) TeV requiring 14 (6) counters in coincidence for scintillators (AIROBICC). The angular resolution is 0.8° (0.3°) for events triggered with the corresponding arrays. For part of the time, data were collected also with 17 Geiger towers (measuring the electromagnetic energy at the detection level and the direction of muons). After a bush fire in October 1997, which destroyed about 30% of the arrays, there are now, after a restauration, 97 AIROBICC counters and 182 scintillation counters in operation.

2. Results and Discussion

2.1. IMAGING ATMOSPHERIC CHERENKOV TECHNIQUE

The Crab nebula was observed by the imaging atmospheric Cherenkov technique for the first time in 1989 (Whipple, $E > 300$ GeV)). Since then Crab has been observed by several experiments and it is now the standard candle of TeV gamma astronomy. At present, there are eight sources identified as TeV gamma emitters. In the northern sky, the Crab nebula (galactic source) and the extragalactic BL Lac objects Mrk 421 and Mrk 501 ($z \approx 0.03$) have been observed by HEGRA with high significance ($> 5\sigma$). In 110 hours of observation between March and October 1997 an excess of 38000 events from the direction of Mrk 501 have been measured by the HEGRA telescopes system. Also, during partial moonshine, measurements were done with the prototype telescope that allowed to record the most complete sample of the flaring states of Mrk 501. During this period Mrk 501 showed strong outbursts, being the strongest gamma source in the TeV sky (up to 10 times more intense than the standard candle, the Crab nebula, during the largest flare in 26/27 June, 1997). A detailed study of the time scales of the flux variability from night to night and also on sub-day time scales have been performed with the HEGRA data of 1997 and variation time scales between

5 and 15 hours have been obtained. The shape of the energy spectrum of Mrk 501 in the energy range from 500 GeV up to 30 TeV has also been measured. Correlated measurements of Mrk 501 with the ground Cherenkov telescopes and X–ray satellites are done. The Rossi X-ray Timing Explorer satellite (RXTE), which has monitored Mrk 501 since January 1996 with the all sky monitor, showed a correlation between the 2–12 KeV count rate of RXTE and the TeV flux amplitudes measured by HEGRA.

2.2. SCINTILLATOR AND AIROBICC ARRAYS

The showerfront sampling method with the arrays permits the monitoring of a large fraction of the sky (1 sr) simultaneously ($E > 20$ TeV). Two types of research is performed: gamma astronomy and cosmic ray physics.

Up to now no clear signal of a gamma ray source with a significance above 5σ has been observed in the region of the sky accesible from la Palma. The Crab nebula has been detected with a significance of about 3.5σ. A search for TeV counterparts of gamma ray bursts has also been performed. Except for one case, no clear excess is found and the flux upper limits obtained are in agreement with recent measurements in the optical waveband.

The measurement of the energy spectrum and composition of charged cosmic rays (nuclei) in the energy range from 0.4 PeV to 10 PeV (1 PeV $= 10^{15}$ eV) is a central theme of the cosmic ray physics with arrays. This energy range is only accessible to these type of detectors. Preliminary results show that the energy spectrum steepens in the PeV region and that the mass composition does not change abrupty in this energy range.

Acknowledgements

We thank the Instituto de Astrofísica de Canarias for supplying excellent working conditions at la Palma. HEGRA is supported by the BMBF (Germany) and CICYT (Spain).

References

For a list of publications of the HEGRA collaboration see the web address www.gae.ucm.es. See also the HEGRA contributions to the Proceedings of the 25th International Cosmic Ray Conference 1997, Durban (South Africa) and proceedings of the 16th European Cosmic Ray Symposium 1998, Alcalá (Spain).

SPANISH PARTICIPATION IN THE MILLIMETER ARRAY

J.M. TORRELLES AND A. ALBERDI
Instituto de Astrofísica de Andalucía, CSIC

A. RIUS
Institut d'Estudis Espacials de Catalunya, FCR-CSIC-UB-UAB-UPC

J.M. PAREDES
Departament d'Astronomia i Meteorologia, UB

X. BARCONS
Instituto de Física de Cantabria, CSIC-UC

F. BRIONES
Instituto de Microelectrónica de Madrid, CSIC

J. CEPA AND J. VILCHEZ
Instituto de Astrofísica de Canarias

C. EIROA
Departamento de Física Teórica, UAM

J.M. MARCAIDE
Departament d'Astronomia i Astrofísica, UV

AND

J.M. MAS
Laboratorio de Astrofísica Espacial y Física Fundamental, INTA

Abstract. The Spanish participation as an Associate Member in the Millimeter Array (MMA) project – led by the U.S. National Radio Astronomy Observatory (NRAO) – has been recently proposed to the Spanish Science and Technology Bureau (OCYT). This request to the Spanish Government is based on two key factors: (1) The high scientific interest and support of the Spanish scientific community, for whom major new fields of research—inaccessible with other instruments—will become available. (2) An industrial return that will clearly benefit Spanish companies developing cutting-edge technology and providing them with a high international profile.

Astrophysics and Space Science is the original source of publication of his article. It is recommended that this article is cited as: *Astrophysics and Space Science* **263**: 381–388, 1999.

1. Introduction

The MMA project (see http://www.mma.nrao.edu) consists of 36 interconnected transportable 10-meter telescopes, which can be extended over a length of up to 10 km. Sensitive to polarized emissions, the interferometer will have an observable frequency range of 30–850 GHz ($\lambda \simeq$ 10–0.3 mm). It will be possible to arrange the 36 telescopes in four different configurations, such that both wide and narrow fields can be observed in great detail. In its most compact configuration the MMA will have an extensive field of view, which will be used to obtain images of objects with a large angular size, such as galaxies, molecular clouds and the Sun's disk, whereas in its widest configuration the interferometer will reach its highest resolution, around 10 milliarcseconds—one order of magnitude greater than that achieved by the Hubble Space Telescope at optical and infrared wavelengths. There are well layed hopes that the MMA will help solve a whole host of scientific conundrums.

The MMA will be installed in the Atacama Desert, Chile, at an altitude of 5,000 meters (16,400 feet), where the atmosphere is extremely dry and highly transparent at millimeter/submillimeter wavelengths. The interferometer will be in use 24 hours a day, and the telescopes will be completely exposed to the elements. Observations will be made on a remote basis.

Due to its high complexity, the MMA project will be one of the biggest technological challenges facing observational astronomy in the years to come. The NRAO has offered other countries the opportunity to participate in the MMA as "Associate Partners," provided they meet the criteria of knowledge and experience required to ensure that the project is a success. The minimum capital contribution to be made by each Associate Partner is 10% of the total budget. The conditions for joining the project as a partner will be established by an agreement to be made with the NRAO. The NRAO has already obtained funding from the U.S. Government to launch the three-year development and prototype program (1998–2001). The construction phase will begin in 2001 and last until 2006–2007, when the MMA will finally be operational.

Spanish participation in the MMA as an Associate Partner was first discussed on October 9, 1995, within the Network of Astrophysics Centers of the *Consejo Superior de Investigaciones Científicas* (Spanish Scientific Research Council; CSIC). The members of this network recommended that a national committee be set up to study the viability of Spanish participation in the project. This committee—made up of researchers from a wide range of institutions—was duly formed, and funding was obtained from the *Dirección General de Investigación y Desarrollo* (Directorate General for Research and Development; DGID). The committee's report, submitted to the DGID in July, 1997, concluded that Spanish participation as an MMA Associate Partner is perfectly viable, and—given the great interest and widespread support for the project shown by the Spanish scientific community, and the considerable industrial return that can be expected—was to be strongly recommended. Spain's status as an MMA Associate Partner would entitle the Spanish scientific community to a proportion of 10% of the instrument's observation time.

The *Centro para el Desarrollo Tecnológico Industrial* (Industrial Technology Development Center; CDTI)—part of the Spanish Industry and Energy Ministry (MINER)—looked more deeply into the question of Spain's possible participation in the MMA project. Moreover, the CDTI organized a workshop at its Madrid headquarters in June 1998 where the NRAO presented the project to potential Spanish industrial participants. The CDTI's report concludes that the involvement of Spanish industry in the MMA project is a viable proposition, not only with an assured industrial and technological return on the country's investment, but also providing an excellent opportunity for Spanish companies to gain valuable experience in an area of the electromagnetic spectrum that could be the basis for a number of commercial applications in the near future. The NRAO, meanwhile, has verified that Spanish industry has the know-how, experience and facilities required for Spain, as an Associate Partner, to make a major contribution towards the overall success of the MMA project.

2. Use of the MMA by the Spanish scientific community

Today, the observational branch of astronomical research is inconceivable without exploring every possible window of the electromagnetic spectrum—from gamma- and X-rays, through ultraviolet, visible and infrared light, to radio waves—because our view of any given astronomical object may vary greatly according to the wavelength at which we observe it, depending on the physical conditions of the object and its surrounding environment.

At an official national level, Spain is currently involved in the ESA's space astronomy projects (covering mainly the FIR, UV, X-ray, and gamma-ray ranges), collaborates with NASA's exploration work using the radio antennas at the Madrid Deep Space Communication Complex (MDSCC) at Robledo de Chavela, has access to optical (at the Roque de los Muchachos, Teide and Calar Alto observatories) and radio-astronomy facilities (NRAO, open to astronomers from all countries; Very Long Baseline Interferometry [VLBI] within the EVN), and is preparing to build the 10-meter optical telescope *Gran Telescopio de Canarias* (Canary Islands Large Telescope; GTC). There is a pressing need to extend the band accessible to Spanish astronomers to include the millimeter/submillimeter range, since this is one of the least-explored observational windows. This is precisely the observational window that will be covered by the MMA, enabling a wide range of scientific goals—ones that would be unthinkable with the instruments currently available—to be pursued in great detail. Since the scientific goals made viable by the MMA project are so wide-ranging (see below), it is hard to find a single subfield of astrophysical activity in Spain that will not be catered for. In other words, the project's scientific interest and possibilities for future exploitation for all the Spanish astronomical community are almost literally limitless. Even more, it should also be stressed that access to the MMA will not be limited to astronomers, but will also be open to other physicists (e.g., theoretical physicists interested in using cosmology to study elementary particles), chemists (e.g. astrochemical researchers) and geologists (e.g. those wishing to make detailed studies of the Solar System). The following list shows some of the scientific goals that could be pursued with the facilities offered by the MMA:

a) Solar System

- Providing us with new information on the nature and origin of comets, with the detection of new molecular isotopes and high angular resolution imaging of the comet's structure. This is important if we want to determine the conditions that were dominant when the Solar System first formed.
- Enabling us to learn the structure of planetary atmospheres through their temperature and molecular composition.
- Solar eruptions can be studied with the aid of coronographs and other auxiliary instruments, especially in the phase when electrons and ions are accelerated to energies of the order of MeV. Also, we will be able to study the structure and dynamics of the solar chromosphere by studying solar oscillations.

b) Star Formation: Protoplanetary Disks

- Detecting thousands of objects that have recently formed or are still forming and are still enveloped in large amounts of molecular gas and dust.
- Studying the physical evolution, dynamics and chemistry of star-formation regions at linear scales ranging from the nuclei of molecular clouds (0.1 pc) to inner circumstellar disks (of the order of 5 Astronomical Units). The MMA will be a unique instrument to map the densities and velocity fields of protoplanetary disks (objects believed to be some 100 AU in size, associated with star formation, from which planets are subsequently formed). This will help us to make great strides forward in our knowledge of the origin of our own Solar System.
- Determining the orientation and evolution of the magnetic fields of these disks by the polarized thermal dust emission, analyzing their role in star-formation processes.

• Direct detection of protoplanets around other stars with masses of the order of the mass of Jupiter in star-formation regions at distances of hundreds of parsecs (e.g., at similar distances to the clouds of Taurus, Ophiucus, etc.). It will be possible to detect the presence of hollows in protoplanetary disks, which may be associated with planet formation. It will also enable us to study the chemistry of these inner regions.

• Studying the physical conditions that might have led to the formation of binary systems, establishing their role in the evolution of protoplanetary disks.

c) Galactic Structure, Stellar Evolution and Physics

• Studying the strong winds associated with evolved stars, in order to determine how much material is ejected from the stars to the galaxy as a function of the galactocentric radius. It will be also able to study the formation and evolution of planetary nebulae.

• Determining C/O and isotope quotients in circumstellar structures, thus obtaining information on nuclear processes and the star-formation rate as a function of the galactocentric distance.

• Measuring photospheric or lower-chromosphere emissions from several thousand stars with an integration time per star of under 10 minutes. It will be able to measure the photospheric radii—and so directly obtain effective temperatures—of hundreds of stars.

• For the first time, the MMA will enable us to make direct studies of the formation of dust at distances of a few stellar radii from evolved stars. We will even be able to observe the "levitation" of material over the surface of the stars concerned. In addition, it will detect stars in any part of the H-R diagram.

• Studying the distribution of diffuse molecular clouds in our galaxy via molecular absorption lines (e.g, CO, HCO^+, HCN).

• Detecting maser emissions of SiO in Mira and OH/IR stars, enabling us to study stellar dynamics near the galactic center ($\leq$ 20 pc).

• Studying SgrA*, the compact object at the center of our galaxy. The MMA will enable us to learn the kinematics of neutral gas at linear scales of the order of 10^{16}cm, together with the continuum emission of the accretion disk around the possible black hole SgrA*.

d) Molecular clouds

• Making detailed studies of the chemistry of the interstellar medium, both in our own galaxy and in others. Imaging the spatial variation in molecular abundances throughout shock-affected regions.

• Obtaining high-quality images of galactic molecular clouds, in order to determine the nature and origin of their filament and turbulent structure and the way in which these structures evolve, giving rise to denser nuclei which eventually form stars by a process of gravitational collapse. It will also enable us to map the magnetic fields of molecular clouds.

e) Galaxies and Cosmology

• Detecting the Sunyaev-Zeldovich effect. This effect will be readily detectable by the MMA where it is most significant (in the millimeter/submillimeter band), especially in the 9-mm band. Since the gas that causes the Sunyaev-Zeldovich effect also emits X rays in clusters, combined measurements by the MMA and, for example, AXAF or XMM will be able to provide us with precise measurements of the Hubble constant or even the deceleration parameter.

• Imaging anisotropies in the background microwave radiation up to angular scales of 2 arc minutes with a sensitivity of 2 μK. The MMA will be a unique instrument, capable of measuring bursts or "seeds" of galaxy formation in the primitive Universe (redshifts of the order $z \sim 1000$).

• Analyzing galaxy formation times by imaging dust emissions from galaxies with luminosities of over 10^{11} solar luminosities, reaching redshifts of up to $z = 20$.

• Studying the earliest evolution of galaxies by analyzing their CO emissions, both in ultraluminous galaxies (up to $z \simeq 3$) and in normal galaxies (up to $z \simeq 1$). These studies will enable us to extend the Tully-Fisher relation to much greater redshifts than those reached with HI studies.

• Obtaining high-resolution images of the Magellanic Clouds (with linear resolutions of 0.25 pc). This will enable us to discover the roles played by metallicity, radiation pressure and external pressure in the structure of the clouds, together with their fragmentation, star-formation processes, etc.

• Obtaining relevant information on active galactic nuclei (AGNs): properties of gas and dust at linear scales of the order of 1 pc, the formation and "fueling" of AGN jets, the study of circumnuclear disks, the interaction between jets and the circumstellar medium, etc.

• Studying non-thermal processes in the lobes and jets of extensive radiogalaxies, measuring their magnetic fields, the energy distribution of electrons, the reacceleration processes (if any) of electrons, and the depolarization mechanisms of radio lobes.

• Studying molecular absorption lines in intermediate galaxies located in front of distant radio sources. The optical counterparts of some distant radio sources are obscured by the presence of dust and molecular gas in intermediate galaxies. This material can be detected by means of molecular absorption lines. Most radio sources are very weak at millimeter wavelengths. Thanks to the high sensitivity of the MMA, we will be able to increase the sample of radiogalaxies that can be studied, and enhance the characterization of intermediate material.

3. Industrial return to Spain

Besides its scientific interest, the MMA also offers major opportunities for Spanish industry. The main interest in the MMA on the part of Spanish industry would lie in the field of receiving millimeter waves, in a range of the spectrum that has until now been inaccessible to those outside scientific laboratories. The design and manufacture of telescopes and structural parts using carbon fiber is also a priority area, given the experience of certain companies already identified in this sector. The positioning and monitoring systems of the telescopes and the presentation, transmission and analysis of data are further potential areas in which Spanish firms could play a key role. Other activities, such as the telescopes' support structures and transport systems, and the preparatory civil works at the site in Chile could potentially also make a significant contribution to the industrial return on Spain's investment.

It is considered to be of the utmost importance for Spain to be involved in the MMA from the initial stages—i.e., the design and development of prototypes, gaining a favorable position to play a major role in the production of the set of telescopes.

Spanish participation in the MMA as an Associate Partner would call for a capital investment of $20M (10% of the total budget), to be spread over the period 1998–2007. The exact year-by-year breakdown of this sum would depend on the specific terms of any agreement reached with the NRAO. It is possible that Spain could always decide to participate in the MMA project as an Associate Partner by making a contribution that exceeds the minimum sum required, since there is no doubt that Spanish industry is quite capable of achieving a return of considerably more than $20M to guarantee the success of the project.

4. Possible merger LSA/MMA

Since 1992 several European institutions, including the European Southern Observatory (ESO), the *Institut de Radioastronomie Millimétrique* (IRAM), the Onsala Space Observatory (OSO), and the Netherlands Foundation for Research in Astronomy (NFRA) have discussed the design of a millimeter interferometer, known as the "Large Southern Array" (LSA).

In June, 1997, the ESO and the NRAO reached an agreement, signed by the ESO Director, R. Giacconi, and the NRAO Director, P. Vanden Bout, to consider the possibility of ultimately merging their respective large-array projects—the LSA and the MMA—into a single project, while forging links to collaboration on ensuring compatibility between the two designs. If an agreement is finally reached with the ESO, and an expanded LSA–MMA array is constructed instead of just the MMA alone, Spain, as an MMA

Partner, would have acces in any case to that LSA/MMA array and would be in a very good position to enter the LSA collaboration by providing the experience acquired with the MMA.

5. Conclusions

Spanish astronomy has undergone a spectacular development in recent years, owing to the installation of innovative facilities in Spain (e.g., at the observatories of the *Instituto de Astrofísica de Canarias* [Canary Islands Astrophysics Institute], the *Centro Astronómico Hispano-Alemán* [Hispano–German Astronomy Center] at Calar Alto, the NASA facilities at Robledo de Chavela, and the 30-m IRAM telescope), free access to the facilities of the VLA, VLBA (NRAO) and VLBI (EVN), participation in space astrophysics projects in association with the ESA, and the scientific training and know-how of Spanish astronomers. If we are not to be left behind by the technological and scientific challenges that we face at the threshold of a new millennium, we cannot afford to pass up on an opportunity to be involved in one of the most important projects for astronomy in the twenty-first century: a major interferometer working at millimeter and submillimeter wavelengths, and the MMA (LSA/MMA) is a cutting edge observational instrument working within exactly that range of the electromagnetic spectrum. Spanish participation as a MMA Associate Partner will require only a modest financial investment, if we consider how much Spanish science and industry stand to gain. To sum up:

- There is a large scientific community in Spain with the potential to make optimal use of the MMA (LSA/MMA) and ensure that highly useful research results are obtained. Spanish scientists will enjoy free access to the corresponding proportion of the observation time of the large array, based only on the scientific merit of the proposals submitted. There is widespread support for the Spain's participation in the MMA among the Spanish scientific community.
- The industrial return will be highly beneficial to Spanish industry, since it will not only provide the firms involved with access to the very latest technology, but is also bound to reflect positively—and globally—on the corporate image of any company involved.

SIXE: A PAYLOAD FOR MINISAT–02

JORDI GÓMEZ-GOMAR
Institute for Space Studies of Catalonia (IEEC)
Gran Capità 2-4, 08034, Barcelona (Spain)

JORDI ISERN
IEEC & Consejo Superior de Investigaciones Cientificas (CSIC),
Barcelona (Spain)

EDUARDO BRAVO
IEEC & Departament of Physics and Nuclear Engineering (UPC)
Barcelona (Spain)

AND

FRANCO GIOVANNELLI AND CESARE DE LA PADULA
Istituto di Astrofisica Spaziale (IAS)
Via Fosso del Cavaliere, 00133 Rome (Italy)

Abstract. A description of the X-ray instrument SIXE proposed for the MINISAT–02 mission is presented. The description includes the scientific goals and instrument performances.

1. Introduction

The Spanish Italian X-ray Experiment (SIXE), is a project fruit of the collaboration between the Institute for Space Studies of Catalonia (IEEC) and the Instituto di Astrofisica Spaziale (IAS) which was presented in response to the call for proposals for the Spanish satellite MINISAT–02. According to the philosophy of MINISAT–0X program, SIXE was conceived with three ideas in mind: 1) the main goal of the experiment had to be its scientific outcome, 2) the requirements of the instrument should be compatible with the performance of MINISAT, and 3) an important part of the design and manufacturing of the instrument should take place in Spain. SIXE was one of the 4 proposals selected for phase-A study. The results of the study have shown that the experiment is scientifically and technologically feasible within the framework of the MINISAT programme.

Astrophysics and Space Science is the original source of publication of his article. It is recommended that this article is cited as: *Astrophysics and Space Science* **263**: 389–392, 1999.

2. Scientific goals

Previous space-borne instruments (Turner et al. 1989; Swank et al. 1994) have shown that many interesting sources, most of them of variable nature, emit appreciably in hard X-rays (HXR). These sources are mainly related to accreting compact objects, and include galactic and extra-galactic phenomena. The most valuable information about the nature of these phenomena has been gained through the study of their rich variability signatures (bursts, pulsations, quasi-periodic oscillations, etc) and their spectral properties. Accordingly, SIXE is conceived as an excellent tool for providing variability monitoring and moderate energy resolution spectroscopy of HXR targets, while at the same time proposing an innovative philosophy to observe these sources.

All major HXR instruments share a limitation when observing targets. These missions have to satisfy a great number of observers and, in many cases, the requirements of several instruments. Thus, long-lasting exposures of individual objects are excluded of their observing programmes. However, a plenty of questions about HXR emitters require either a long-term monitoring or very long integration times, and longer exposures are demanded. Such observations will be possible for a small mission devoted to a reduced number of targets as proposed for SIXE. During 2 years, the experiment will observe $\sim$ 20 targets for accumulated periods of $\sim$ 2 months. Targets will be selected among the brightest candidates of different classes of phenomena, and taken as their prototypes. Observations carried out by SIXE will provide a unique database which shall give answer to some of these open questions. At the same time, SIXE is designed to reach the best time resolution and accuracy attained up to now in X-rays. With this combination of capabilities, SIXE will allow to study the variability in an unprecedented range of time-scales going from 10^{-6} to 10^{7} s, which covers from the characteristic time-scales of instabilities in the accretion flow onto neutron stars up to that of AGN variability. Moreover, an agreement has been reached with the team developing the Optical Monitoring Camera (OMC) on INTEGRAL, to include a modified version of the OMC on MINISAT–02 (Más-Hesse et al. 1998). The combined operation of both instruments will provide an unprecedented continuous multi-band campaign, which will enhance the scientific outcome of the mission.

It is planned to set up a scientific committee formed by experts in the different classes of sources to be observed, which will be responsible for the definition of the observing program. A preliminary list of phenomena to be studied is shown below:

- *Galactic:* pre-main sequence stars, flaring stars, X-ray pulsars, low-mass X-ray binaries, high-mass X-ray binaries, black hole candidates

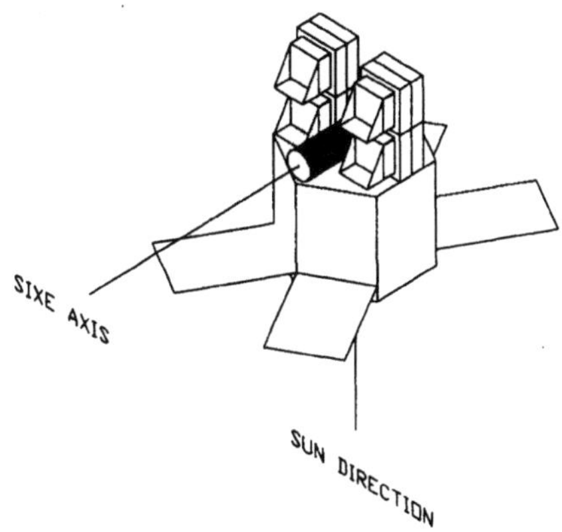

Figure 1. SIXE modules and OMC camera mounted on MINISAT's payload tray. The collimator and sun shields of SIXE can be distinguished in front of the detectors.

and microquasars.

- *Extragalactic:* cuasars, Seyfert-1 galaxies, Seyfert-2 galaxies, BL-Lac objects and clusters of galaxies.

3. The instrument and the mission

SIXE will be formed by four identical detectors (Fig. 1) based on the MCPC technology (Multi Cell Proportional Counter) (Ubertini et al. 1989). Each module will be filled with a mixture of Xe, Ar and Isobutane, and will comprise 70 detection cells defined by a distribution of cathodes-anodes at a relative potential of $\sim$ 2000 V. X-ray photons penetrating through the detector window will interact with the gas inside, generating a charge avalanche which will be collected on one of the anodes allowing to determine the energy and time of arrival of the photon. The characteristics of the detector window and the gas mixture have been selected to optimize the efficiency in the nominal range 3–50 keV. The combination of the four modules will give a total sensitive area $\sim$ 2300 cm^2 and the energy resolution will be:

$$\frac{\Delta E}{E} = 0.19\sqrt{\frac{5.9\mathrm{keV}}{E}}\ (E \leq 35\mathrm{keV}), \qquad \frac{\Delta \mathrm{E}}{\mathrm{E}} \leq 0.05\ (\mathrm{E} > 35\mathrm{keV}) \quad (1)$$

Special features have been implemented in the design of SIXE to reduce the internal background by more than a 95% and to improve the efficiency at high energies (internal anti-coincidence and fluorescence gating). In order to avoid the confusion of sources, while being compatible with MINISAT's

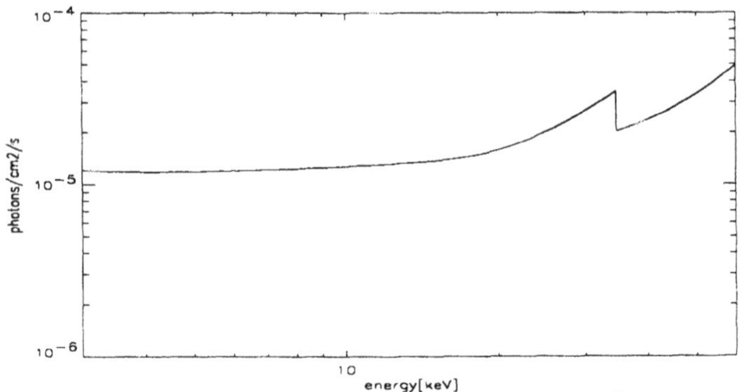

Figure 2. Narrow line sensitivity for SIXE (3σ) for a 10^5 observation.

pointing capability, each detector will have a collimated field of view of $\sim 1^{\circ} \times 1^{\circ}$. The modules will be mounted on two different banks disaligned by 3.5°. This configuration will allow to simultaneously observe X-ray sources and a blank sky field to monitor the background in real-time. Thus, the systematic errors in the determination of the internal background that have affected previous experiments and have limited their sensitivity to long-term variabilities, will be eliminated (Hayashida et al. 1989; Jahoda et al. 1999). In Figure 2, the narrow line sensitivity of the instrument for a 10^5 s observation is shown. Particular attention has been paid to the design of the detector front-end electronics and experiment data system, in order to reduce the expected dead time of the detector ($\sim$ 25 μs) and attain a timing accuracy of 2 μs.

The preferred orbit for SIXE is a low earth orbit with i $\sim 20^{\circ}$, h550–600 km and $\epsilon = 0$ (similar to MINISAT–01 orbit). The instrument will be pointed at angles $> 60^{\circ}$ respect to the Sun requiring 6 months for a full sky coverage.

References

García Primo, G., 1997, Revista del Instituto de Navegación de España, 1, 31

Hayashida, K., et al., 1989, PASJ, 41, 373

Isern, J., Bravo, E., Gómez-Gomar, J., Giovannelli, F., Bausells, J., Cabestany, J., De la Padula, C., García-Berro, E., García-Senz, D., Gutiérrez, J., José, J., Hernanz, M., Madrenas, J., 1998, SIXE phase-A report, Barcelona.

Jahoda, K., Swank, J.H., Rossi, B., 1999, in preparation

Más-Hesse, J.M., et al., 1998, Physica Scripta, T77, 44

Swank, J.H. et al., 1994, *The X-ray timing explorer: progress and science prospects*, GSFC, Greenbelt.

Turner, M.J.L., Thomas, H.D., Patchett, B.E., et al., 1989, PASJ, 41, 345

Ubertini, P., et al., 1989, IEEE Transactions on Nuclear Science, 36, 921

EURD: FIRST YEAR RESULTS

CARMEN MORALES AND JOSÉ F. GÓMEZ
Laboratorio de Astrofísica Espacial y Física Fundamental, INTA
Apartado de Correos 50727, 28080 Madrid, Spain

JOAQUíN TRAPERO
Laboratorio de Astrofísica Espacial y Física Fundamental, INTA
Apartado de Correos 50727, 28080 Madrid, Spain
Present address: Universidad SEK, Cardenal Zúñiga, s/n
40003 Segovia, Spain

AND

STUART BOWYER, JERRY EDELSTEIN AND ERIC KORPELA
Space Sciences Laboratory, University of California, Berkeley
CA 94720-5030, USA

Abstract. EURD (Espectrógrafo Ultravioleta extremo de Radiación Difusa), one of the instruments onboard the Spanish satellite MINISAT-01, is a spectrograph specially designed to detect diffuse radiation, covering the wavelength range of 350-1100 Å, with a spectral resolution of 6-8 Å. Its main scientific objectives are the detection of the emission line spectrum from the hot phase of the interstellar medium and the spectrum of the upper atmospheric airglow. In order to reduce geocoronal noise, EURD always observes in the anti-sun direction and only when the satellite is in orbital eclipse. After more than one year of observation we have obtained the best spectrum of the upper atmospheric nightglow in this wavelength range, the spectrum of 15 OB stars and the spectrum of the full Moon throughout the year.

1. The observations

EURD is a collaboration between the University of California, Berkeley, and the Instituto Nacional de Técnica Aeroespacial (INTA). It is on board the Spanish satellite MINISAT-01, launched from the Canary Islands on April

Astrophysics and Space Science is the original source of publication of his article. It is recommended that this article is cited as: *Astrophysics and Space Science* **263**: 393–396, 1999.

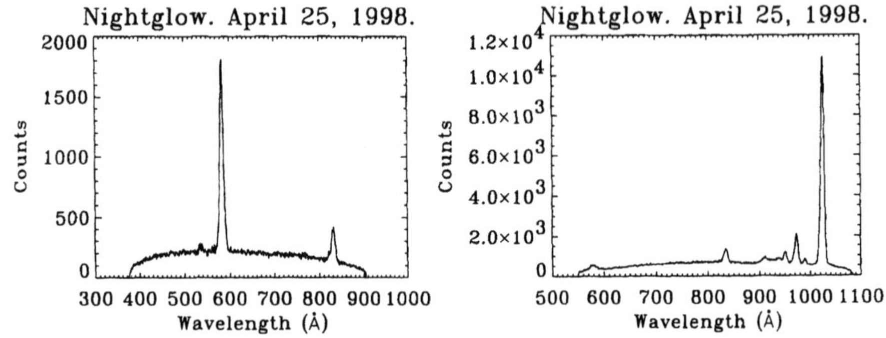

Figure 1. Typical spectra of one day of observation.

21, 1997, which has a circular orbit at 595 Km altitude and 151 degree inclination. INTA's Satellite Tracking Station at Maspalomas is the facility used for MINISAT-01. Once per day MINISAT memory is downloaded and tecnichal data are transmited to the INTA's Headquartes in Torrejón de Ardóz (Madrid) were the Stellite Operations Center is located. Scientific data are transmited to the Laboratorio de Astrofísica Espacial y Física Fundamental in Villafranca del Castillo (Madrid) the site of the Scientific Operations Center (SOC). The SOC allocates daily the observing time for the three instruments and processes and distributes the scientific data to their respective scientific teams.

EURD consists of two spectrographs optimized respectively for short (350-800 Å) and long (550-1100 Å) wavelengths. The detector is a stack of low noise multichannel plates, and the holographic grating yields a spatial resolution better than 0.1° and spectral resolution of 6 to 8 Å. With its large field of view of 26° × 8°, it is specially designed to detect diffuse radiation. A detailed description of the instrument can be found in Morales et al. (1996) and Bowyer et al. (1997).

2. Scientific Objectives

There are five scientific objectives for this project: the detection of the emission line spectrum from the hot phase of the interstellar medium and of the upper atmospheric airglow, the detection of a line that would be produced by neutrino decay, the spectrum of OB stars in the far ultraviolet, and the full Moon spectrum.

It is generally believed that the Solar System is inside a bubble of extremely tenuous and hot gas, called the Local Bubble. The temperature of this gas, obtained from indirect observations, is estimated to be between 10^4 and 10^6 K. Therefore it should have a very rich line spectrum in the extreme and far ultraviolet (Landini Fossi, 1990). However, due to the very low density (0.01 cm^{-3}) of the interstellar medium inside the bubble we estimated that at least 2000 hours of observing time are needed to detect these lines with EURD.

The neutrino decay line, originally proposed by Denis Sciama (1990), should appear in the wavelegth range of EURD (Sciama, 1998, Bowyer et al. 1995) but even longer integration times may be needed for its detection.

The spectrum of the full Moon that covers the whole range of the two spectrographs has been used to improve the ground calibration in wavelentgh and flux. Simultaneous observations of the full Moon have been taken with EURD and the EUVE satellite, and an absolute inflight calibration has been obtained (Edelstein et al., in preparation).

3. First Results

In order to reduce geocoronal noise, EURD was intended to always observe in the anti-sun direction and only when the satellite is in orbital eclipse. However, due to the wealth of lines which could potentially be detected from the upper atmospheric nightglow, we extended the observing time five minutes before and after the eclipse to get the very interesting airglow emission spectrum of the dawn and dusk.

MINISAT-01 flies at an altitude of 595 Km, where the atmosphere is basically composed by oxygen, atomic hydrogen and helium, hence we detect the brightest lines of these three elements that can be seen in Fig. 1, where we show a typical spectrum of one day of observation. The lower panels shows the detected images by the two spectrographs, and in the upper panels the extracted spectrum.

After more than one year of observations we have already obtained the best spectrum to date of the nightglow in the ultraviolet (López Moreno et al. 1998), detecting for the first time the complete Lyman series of hydrogen, several lines of the Lyman series of helium and two bright lines of oxygen at 834 Å and 898 Å, as well as the oxygen recombination continuum at 911 Å. We are studying the behavior of these lines with observing zenith angle and with latitude, as well as the mechanism that excites the upper atmosphere and produces the emission.

The second important result is the detection of OB stars spectra in the far ultraviolet. Very few stellar observations have been made in this wavelentgh range, with very discrepant results regarding absolute fluxes

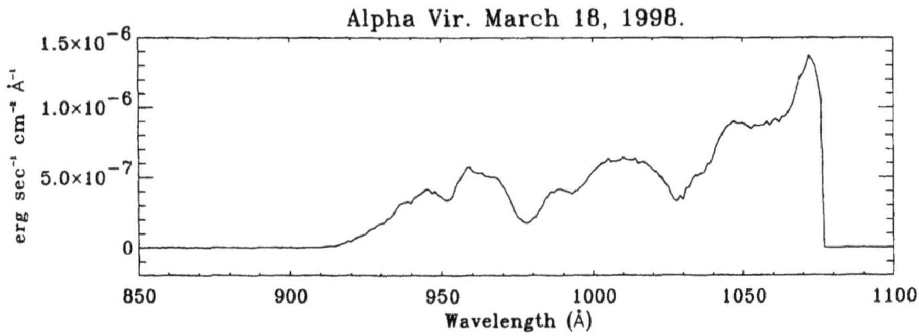

Figure 2. Flux-calibrated spectrum of Spica.

and model atmosphere fitting. In Fig. 2 we can see the image and the corresponding flux calibrated spectrum of the bright star Spica (α Vir). The diffuse emission of the upper atmosphere is seen as long vertical lines in the image, and the spectrum of the star only detectable up to the Lyman limit at 912 Å. We have found a very good agreement between our data and both Voyager data of α Vir and the corresponding Kurucz model atmosphere (Morales et al. in preparation), using the inflight calibration obtained with the full Moon spectrum. Many OB bright stars have been observed by EURD and are in the process of calibration.

References

Bowyer, S., Edelstein, J., Lampton, M., 1997, ApJ, 485, 523

Bowyer, S., Lampton, M., Peltoniemi, J.T., Roos, M., 1995, Physical Review D, 52, 3214

Edelstein et al., 1998, in preparation

Landini,M., Monsignori Fossi, B.C., 1990, A&AS, 82, 229

López-Moreno, J.J., Morales, C., Gómez, J.F., Trapero, J., Bowyer, S., Edelstein, J., Lampton, M., Korpela, E., 1998, Geophys. Res. Lett., 25, 2937

Morales, C., Bowyer, S., Gómez, J.F., Trapero, J., Ulibarrena, M., Gutiérrez, F., López-Moreno, J.J., Pérez Mercader, J., Lampton, M., Edelstein, J., 1996, in *International Conference on Small Satellites: Missions and Technology* INTA, Madrid), pp. 61

Morales et al., 1998, in preparation

Sciama, D.W., 1990, ApJ, 364, 549

Sciama, D.W., 1998, A&A, 335, 12

DYNAMICS AND CONTROL FOR A SMALL TELESCOPE

J.A. BERNÁ, M. PÉREZ AND G. BERNABEU
Dpto. de Física, Ingeniería de Sistemas y Teoría de la Señal
Universidad de Alicante
Apartado de correos 99, E-03080 Alicante (Spain)

Abstract. Our main goal is to determine the dynamic equations of a certain complex system, as is the case of the mechanical system for a small aperture telescope. Causes of this complexity are: the lack of documents about the operation of the elements belonging to the system, and the variation of dynamics with respect to the time and the position of the telescope. To check that we have obtained a valid set of dynamic equations, we will design a computer control system that will implement a self-guide system for the telescope.

1. Instrumentation

The optical system consists of a 40 cm diameter reflecting telescope and a 150 mm refracting telescope guide. The instruments are arranged on a German equatorial type mounting fitted with synchronous moving motors in R.A. and Dec. and controlled with a handbox. The images are obtained by using a SBIG CDD camera model ST4. The camera communicates with a PC through a serial port with a maximum baud rate of 57600. The image processing and system's control is performed by using a PC 486 100 MHz platform with a Windows 3.1 operating system.

2. The system's architecture

The designed self-guide system visualizes the image obtained with the CCD camera on the PC monitor. The object that we wish to track is selected on the image and the system will then obtain consecutive images, so that if there is any variation in the object's position the appropiate control signals are sent out to the telescope motors in order to rectify it.

Astrophysics and Space Science is the original source of publication of his article. It is recommended that this article is cited as: *Astrophysics and Space Science* **263**: 397–400, 1999.

2.1. THE TELESCOPE MOTORS HARDWARE CONTROL MODULE

In order for the PC to control each of the telescope's movements a card which could be connected to an ISA expansion bus has been developed. The card has a series of electromechanical relays that replace the handbox relays, so that the PC can control the telescope motors by sending a data byte to the card.

2.2. IMAGE ACQUISITION MODULE

Following the especification of the communication protocol between the CCD camera and a PC, a software module was developed to aid the functions of acquisition and image reading for the CCD camera. Because of it is a serial communication and it has been developed under Windows 3.1 operating system, the transmission time for a whole image is high, approximately 20 seconds.

2.3. IMAGE PROCESSING MODULE

This software module includes the function of processing the images in order to detect any variation in the position of an object from one frame to another. This is made by selecting the objects on an image with the same brilliance as the selected object and comparing them with the objects of the previous image, so that the module calculates the displacement between both images.

2.4. SYSTEM CONTROL MODULE

We may state the control problem like this: the telescope position will be accurately controlled through the feedback of the output signal. The output position is adjusted as the error signal requires, which is a measure of the difference between the position response, determined by the CCD camera, and the desired response of the input signal. The error signal is processed by a program implemented on the PC, so that it will satisfy a certain control criteria or specification. Figure 1 shows the system's block diagram.

2.4.1. *Experimental characterization of the telescope dynamics*

The physical model behavior of the telescope mechanical system can be very complex. An economical solution with good results is the experimental characterization of the telescope dynamics, which is made sending step signals to the synchronous moving motors of the telescope. In this paper only the positive R.A. movement will be considered. However the dynamics has been in both negative and positive R.A. and Dec. movements characterized. The

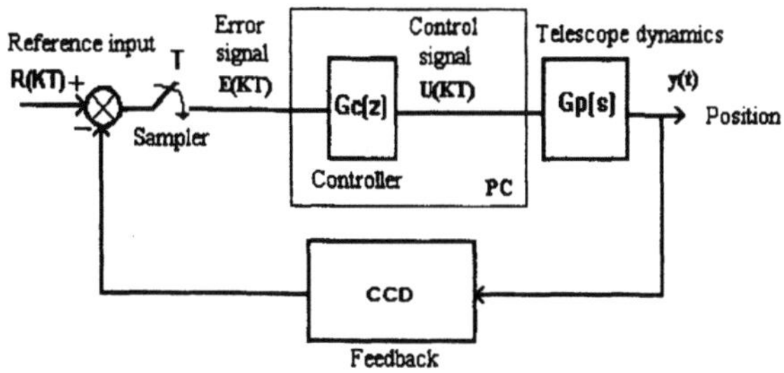

Figure 1. Closed-loop digital control system

most simple transfer function that best fit to the experimental data from positive R.A. response is,

$$G_p(s) = \frac{4.9}{s} \tag{1}$$

2.4.2. *Design of the control system*
Knowing this transfer function, the first part in the design procedure is to obtain a continuous controller $G_c(s)$ (Franklin et al, 1990; Isermman, 1992) for the R.A. movement. We have obtained the following transfer function,

$$G_c(s) = \frac{7.7681}{(s + 10.6196)} \tag{2}$$

In the second part we then convert the design to a digital control by considering $G_c(s)$ as a filter transfer function and applying the bilinear transform to obtain the corresponding Z-transform as follows,

$$G_c(z) = \frac{1.3436e^{-2}(z+1)}{(z+0.9752)} \tag{3}$$

Equation (3) is converted into a difference equation in order to implement it in a PC obtaining,

$$u(k) = -0.9752u(k-1) + 1.3437e^{-2}(e(k) + e(k-1)) \tag{4}$$

In equation (4) $u(k)$ and $e(k)$ are, respectively, the control signal and the error signal.

2.5. INTERFACE USER'S MODULE

A graphical application has been developed under Windows 3.1, allowing the user to apply the different functions which the self-guide system pro-

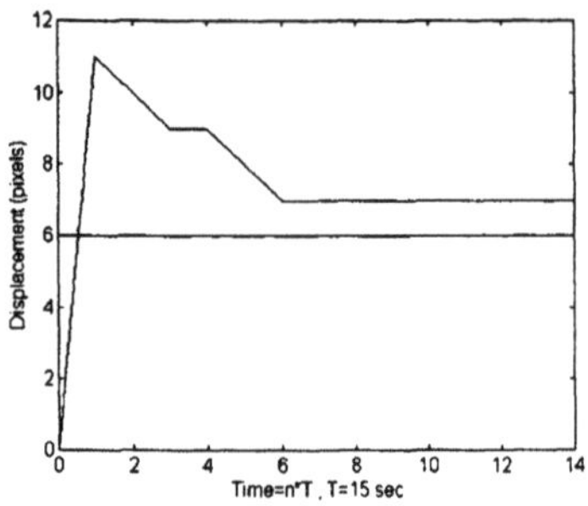

Figure 2. R.A. experimental time response

vides: to initiate the CCD camera and the synchronous motors, to keep track of and centralise objects in the CCD camera's images, to acquire and visualize images with the CCD camera, etc.

3. Conclusions

Once the system was implemented, a number of tests were carried out obtaining results that indicate a satisfactory result in the dynamics characterization, taking into account the lack of resources and the approximations carried out. Figure 2 shows the R.A. experimental time response with a six pixels amplitude step input. Due to the oscillation of the system and low resolution of the CCD camera, only a low precission is reached in the self-guide, approximately 5 arcsec. The stabilizing time is also high because the sampling time T is high. These problems can however be easily overcome by acquiring the right instrumentation which suits the system's needs.

Future outlines include determining more precise dynamics so as to reduce the error in the self-guide. This can be achived by increasing the complexity of the transfer functions and making them variable in relation to the telescope's position, as in fact occurs.

Acknowledgements

We would like to express our most sincere gratitude to the Juan Gil Albert Cultural Institute for financing this project and to the Círculo Astronómico del Mediterráneo for giving us access to its installations.

References

Franklin, G., Powell, J.D., Workman, M.L., 1990, Digital control of Dynamics Systems, Addison Wesley

Isermman, R., 1992, Digital control systems, Springer Verlag

WOMEN IN THE HISTORY OF ASTRONOMY

M. ÁLVAREZ ÁLVAREZ AND ÁNGELES I. DÍAZ
Dpto de Física Teórica, C-XI, Universidad Autónoma de Madrid, Cantoblanco, 28049-Madrid, Spain

Abstract. We think about the History of Astronomy as the History of men. As the History of a few men: Ptolemy, Copernicus, Kepler, Newton,...–men who have changed our way of looking at the sky. But the History of Astronomy is more than that, it is the History of thousand of people whose daily work has allowed the development of knowledge and scientific theories at the time they lived. This, sometimes, "tedious work" permitted the big steps. Many of these people were women, as Theano who married Pythagoras and taught mathematics and astronomy in his school, Hypatia who managed the Library of Alexandria and wrote several astronomical treatises, Hildegard of Bingen who developed a theory on the origin and structure of the Universe in the 12th century, or Sofie Brahe who, worked with her famous brother. And so many privileged women who after a long process of study, were able to develop their scientific interests in spite of been excluded from most of the educational installations and formal and informal groups of men scientist. Most of the works done by these women have been ignored, or wrongly attributed to men throughout History. It often happens that although they have been recognized as good scientists in their own times, women have been discredited by posterior historians who refused to believe that important women scientist ever existed. Here we intend to make a short summary on the lives of some of these women and their astronomical works.

1. Why women...?

...If all great astronomers throughout History, the most prominent minds, those who have achieved the most famous scientific aims, have been men? So, what is the meaning of women's work?

To understand it, we should look at that continuous flux called progress in a new way:

Astrophysics and Space Science is the original source of publication of his article. It is recommended that this article is cited as: *Astrophysics and Space Science* **263**: 401–404, 1999.

The History of Astronomy, the History of every science, is not an unconnected sequence of facts, biographies, or dates, but a gentle flow, fast or slow, but always continuous, of ideas along History. This ideas have sometimes been given up, and sometimes accepted, according to the believes of the historical time they belong to. Many times they have suffered deep changes before arriving to the form in which we know them. In this context is where names and facts fade away and hard daily work gets its real meaning. This is the work that, often, has been carried out by women.

It would not be fair to evaluate scientific works without looking at the historical time they belong to. This matter is even more crucial when talking about women. Their contributions to Astronomy not only depend on wars, hunger periods, or political affairs, as it happened with their male colleagues, but also on the social and family roles females had to play, and what was supposed they ought to do about science.

Therefore, the presence of women in Astronomy is, until the twentieth century, an on-and-off affair, a history of sudden and dramatic cuttings and historical biases, sometimes real and sometimes promoted by later historians who ignored their works. Female contributions were important to the History of Astronomy, but their most important work was that, in spite of all difficulties, they were there.

2. Antiquity

When man first developed agriculture, it was in women's hands, and to get a profitable production, some kind of time measure was needed. Therefore, ancient women must have learnt to connect astronomical facts with the growing of vegetables. With time production increased and men were freed of hunting, giving more and more time to farming, so the importance of women in production and society became to decrease. But their first contributions were not fully forgotten, and in ancient popular tales we still can find female characters as Shestat, who in the old Egypt was the stars Goddess and helped temple builders to align structures with stars, or Urania and Isis, muse of Astronomy.

In Babylonian and Egyptian civilizations, women had as many privileges as men, so it is fair to think that they played similar roles in astronomical observations.

Centuries later, at Pythagoras' School in Cotona, 28 women worked together with men (Menagius, *Historia Mulierum Philosopharum*). The most famous of them was Theano, who married Pythagoras. Some of her letters have been preserved and can still be read. We must wait until nearly the end of Greek science, in Alexandria, to find another important woman astronomer: Hipatia. She managed the Museum of Alexandria and wrote the text books on mathematics and astronomy she used in her lessons. She also wrote a treatise on the geometry of conic sections and collaborated in the Astronomical Cannon written by her father Teon. In her Sinesio's letters, she describes the plane astrolabe and how to use it to measure star positions and compute time and zodiacal sign.

3. The Middle Age

After the destruction of the Library of Alexandria (642 a.D.), we have to wait more than five hundred years to see the reborn of Astronomy in Western Europe. The few learned people who lived in this centuries belonged, mostly, to religious congregations, between them we can find the so called "Erudite Abbesses" with two main representatives:

The first one is Hildegard of Bingen (1098-1179). Her most important activity was medicine, but she also wrote several books on a wide variety of subjects. Talking about her cosmological theories, her most important works are: "Scivias", published in 1141, and "Liber divinorum operum simplicis homini", published in 1162. Between the writing of both, Hildegard should have had notice of the Greek astronomical works that were being translated from Arabic, because while in her first book she presents a quite ingenuous model of the Universe, in the second one she describes something like the Aristotle's spheres. This fact positions her at the head of the science of her time. The other known abbess is Herrad of Landsberg, who lived at the time of Hildegard in Alsacia. Her main work is called "Hortus Delicarium", in which appears a drawing of the zodiac signs and a table computed to determine holidays and Easter for a period of 532 years, from 1175 to 1706.

4. The Scientific Revolution

With the new scientific developments, new relations between women and science appeared. The Dames of Science were rich women, usually sisters or wives of the new men scientist, who could afford expensive toys as telescopes or microscopes and contributed to the divulge of science instead of its development. Besides these wealthy women, there were also women astronomer whose work represents an important part of science, between them: Sofie Brahe (1556-1643), who worked in Uraniborg with her famous brother. Marie Cunitz (1610-1664), the first woman to attempt a correction of Kepler's tables. She published "Urania Propitia" in 1650. Elisabeth Korpmann Hevelius (1647-1693), J. Hevelius' wife. She improved Cunitz's work, was her husband's assistant and after his death she published "Firmamentum Sobieskanum" and "Prodromus Astronomicae" together with the most complete star catalogue at the time, with 1888 objects. Marie Winckelmann Kirch (1670-1720) married to Gottfried Kirch, royal astronomer in Berlin, she worked with him until his death, after which Marie applied for his place, but was refused because she was a woman. The job went to her son, whom she had trained as an astronomer.

And the best known of all, Caroline Herschel, William's sister. She worked all her life with her brother, helping him with the observations an other scientific work. During her life in England she worked on comets and published with William several catalogues of stars an nebulae. She was the first woman to be awarded the Gold Medal of the Royal Society of Astronomy. There were also women who made theoretical works as Mme. Lepaute or Mme. du Piery, who besides their own contributions, collaborated with male scientist as F. Viete or J. Lalande.

All these works, as many others, are the beginning of a tradition that culminated in the group of women who worked in Harvard in the nineteenth and twentieth centuries.

5. The Women Computers in the Nineteenth Century

The first woman astronomer in America was Maria Mitchell. She was also the first woman to join the American Society for the Advance of Astronomy. Later on, when E. C. Pickering became manager of the Harvard College Observatory and began the huge program of analyzing and cataloguing as many stellar spectra as possible, he had to face the problem of getting many qualified astronomers with a short budget. He solved the problem by offering the jobs to women, who received less money than men for the same work. This is how the road of women to equality in Astronomy began.

Among the women who worked in the development of the Henry Draper Catalogue, the best known are: Williamina P. Fleming, Henrietta S. Leavitt, Annie J. Cannon and Antonia Maury. After them many others have come: as Cecilia Payne-Gaposchin, who introduced the idea that stars are mainly constituted by hydrogen and helium, or Charlotte Moore Sitterly who published the first catalogue of sunspots spectra, and all the women astronomers who nowadays work together with men in observatories and academic or scientific institutions.

References

Dronke, P., 1984, *Las Escritoras de la Edad Media*, Crítica. Cambridge University Press
Leverington, D., 1995, *A History of Astronomy*, Springer-Verlag
Alic, M., 1991, *El Legado de Hipatia* Siglo XXI S.A
Urry, C.M., et al. (Eds), 1992, *Women at Work: a meeting on the status of women in Astronomy*, Space Telescope Science Institute

CONSIDERATIONS ON THE TEACHING AND THE POPULARIZATION OF COSMOLOGY

ERIK STENGLER AND CRISTINA SILVIA HANSEN RUIZ
Departamento de Didácticas Especiales
University of La Laguna, Tenerife, Spain

Somebody who is now in the audience once said, when he was invited to give a talk at popular level, that he accepted gladly since he considered it like a debt of the scientists towards the society that sustains them. Before we concentrate on points related to the teaching of cosmology we would like to mention certain considerations on didactics in general. We think that the attitude that we have just mentioned should be present in all scientists since it is really the case that often - especially for sciences like astrophysics and particle physics - research is financed by public funds and does not bring about any economical advantage. In many environments it has been already noticed that to justify such expenses, particle physics and astrophysics must come closer to the citizen, at least by means of explaining the results of the investigations in an understandable way. NASA has already invested a lot of money and efforts in the popularisation of science and its teaching.

In associations like the Astronomical Society of the Pacific (ASP) initiatives have arisen that develop and contribute significantly to these tasks, as, for example, the magazine Mercury and the web pages included in the section 'The Universe in the Classroom' in ASP's website. It is also well known that several magazines at popular level have appeared over the last year and that in Spain and abroad projects for museums and science parks are being brought forward. In addition, the introduction in the E.S.O. of the optional course on astronomy has led many teachers to attend courses on astronomy themselves. These courses are being organised by professional and amateur astronomers alike. In fact the ideas that we will present on the didactics of cosmology are fruit of the experience gathered in the introductory courses on astronomy for teachers organised in the Canaries by the Consejería de Educación in collaboration with the IAC. In any case it can be seen that in the area of the popularisation of astronomy a lot is being done. However, this is not the case in the area of teaching, in which the educational positions are occupied by investigators that consider teaching

Astrophysics and Space Science is the original source of publication of his article. It is recommended that this article is cited as: *Astrophysics and Space Science* **263**: 405–408, 1999.

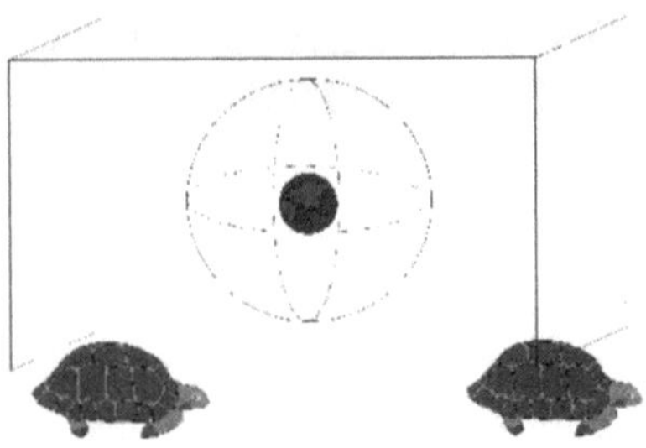

Figure 1. The KNOWN universe is a sphere, with the earth at its center

an annoying obligation to get done with as soon as possible each year. Even the educational system at university level itself fosters such a contradiction. The profile demanded for teaching positions is limited to merits related to research, such as publications and conferences, while the educational aspect is dispatched with a simple project for the subject they supposedly will teach initially and an hour of teaching practice in front of the examiners. It would be necessary to remind those holding such positions that the university pays them mainly for their teaching. Given this fault in the system itself, there are two possible ways to act. The most typical one is that of those who take advantage of their teaching positions for pursuing their research, which is what they really do with enthusiasm. In this way the feed the opinion of those who think that scientists do not care but for their own thing. In the long run, this attitude will undermine their own research, since it is the society they ignore now, whom they will need to ask for funds later. The other attitude would be that of putting an extra effort in at least being the best in teaching, given that there are no immediate financial benefits they can offer to the average citizen

Let's turn now to the didactics of cosmology. The first point refers to the little use that we make of the distinction between 'universe' and the 'known universe', and how easily we speak of the universe outside the piece that we know. In order to stress that we really should not or can not talk about what is beyond the universe we know, it can be of help to represent the universe as in figure 1 (by the way, the earth does occupy the centre of the known universe, sinc e the light speed is the same in all directions). In addition, it is inconsistent and highly confusing for the beginner in the subject to hear us stating that the 'universe' was concentrated in a tiny sphere (of the size of a golf ball, or a tennis ball, or even a single atom) and at the same time considering the possibility of the universe being infinite in size. If this were the case, it would still be infinite in size after contracting it into a 'Big Crunch', and in no way of the size of a golf or tennis ball. To avoid incurring in such an inconsistency, we should state clearly that it

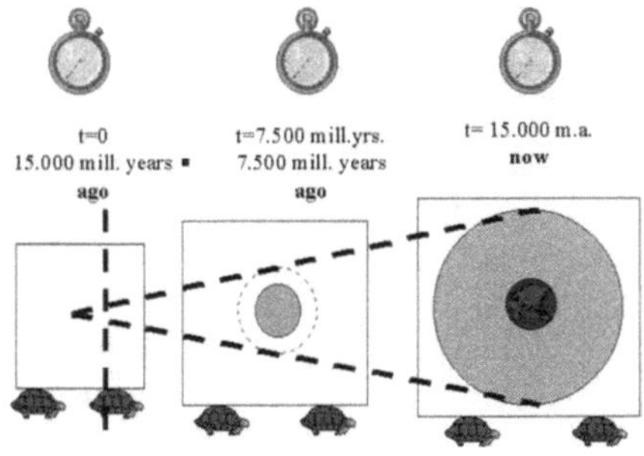

Figure 2. The space in which the universe grows is expanding at the same time

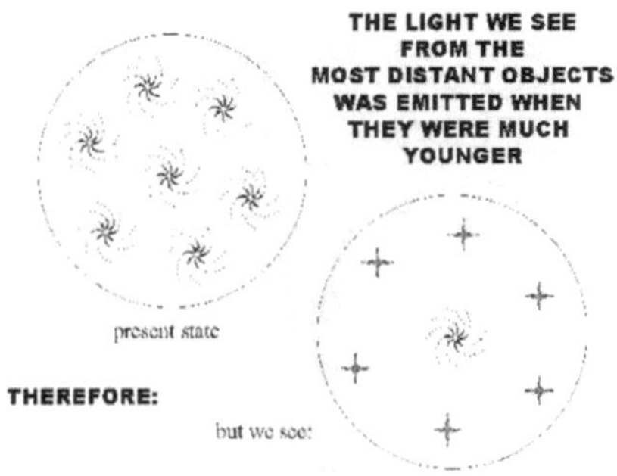

Figure 3.

is the known universe which could have been concentrated in such a tiny portion of space, as figure 2, for example, makes clear.

A second point not sufficiently mentioned and which would help a lot to understand the complex concept of the evolution of the universe is that, despite the fact that the further we look the younger we see the objects, at present all objects - all galaxies - are at the same evolutionary stage, i.e., they have the same age and look more or less the same way as ours even if we see them still as protogalaxies or as a small overdensity in the cosmic microwave background. This is of course perfectly clear for all of us who have devoted some years to research on the evolution of the universe, but we should not overestimate the intuitiveness of the audience of introductory courses or popular talks. A simple figure like figure 3 can make a big difference when it comes to understand this fact.

Getting back to the expanding universe, it is surprisingly difficult for beginne rs to understand how a uniform expansion can be compatible with - let alone imply - the recession velocities of distant galaxies being higher that that of nearby galaxies. This apparent dilemma arises sooner or later in all

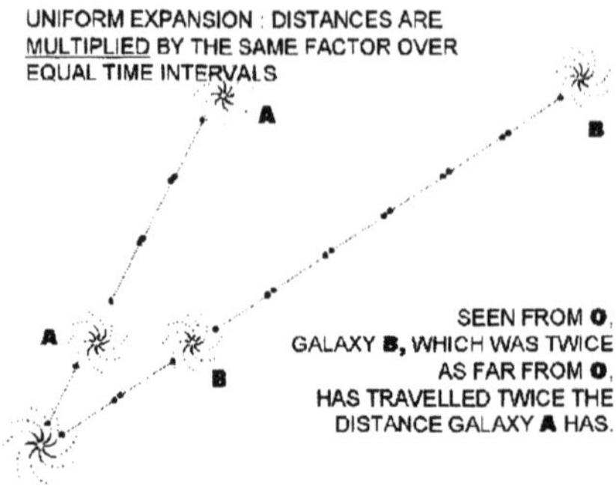

Figure 4.

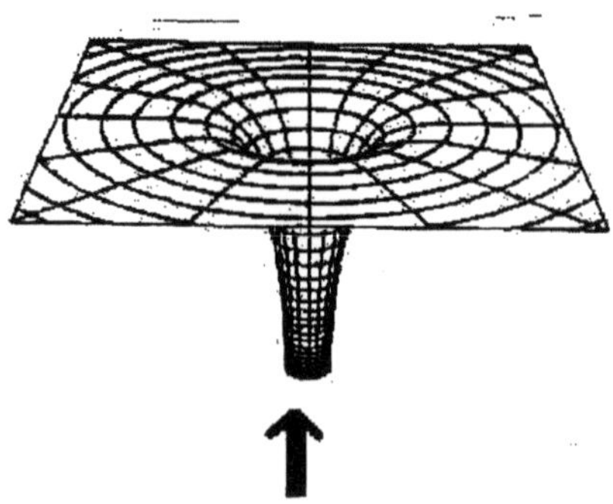

Figure 5.

kinds of audiences. And the answer is as simple as little used: the expansion is uniform in that distances are multiplied by the same factor over equal time spans. To fulfil su ch an uniformity a distant galaxy must travel a longer distance than a nearby galaxy in the same time interval. A figure like figure 4 or a simple example with three members of the audience and the tiles of the floor as distance units can help to make this clear. Finally there is the subject of black holes, so popular among students and the general public. As pointed out by J.F. Lockwood (1998), when we mention these objects we almost automatically use the example of figure 5, forgetting that it is almost impossible to expect that the listener will be able to perform a dimensional jump and to come to terms with the idea of a black hole being the three dimensional version of such a drawing. We should then not be surprised if we are asked what happens when one approaches the black hole 'from below', in the direction shown by the arrow in the figure. There are numerous examples along the same lines. But we encourage all teachers and speakers to discover them by themselves in their own talks.

AUTHOR INDEX

Acarreta, J.R. 351
Acosta-Pulido, J.A. 123
Alberdi, A. 381
Alfaro, E.J. 197
Alonso-Herrero, A. 131, 147
Alvarez, P. 355
Alvarez Alvarez, M. 107, 139, 401
Aragón-Salamanca, A. 83, 147
Aretxaga, I. 107, 139
Ascasibar, Y. 31
Axon, D. 67
Balcells, M. 361
Barcons, X. 75, 115, 381
Barrado y Navascués, D. 235, 239, 303
Beaulieu, J.P. 215
Beckman, J.E. 163
Benn, C.R. 63
Berná, J.A. 397
Bernabeu, G. 243, 259, 299, 397
Bertola, F. 163
Bonet, J.A. 343
Bonifacio, P. 79
Bouvier, J. 239, 303
Bowyer, S. 393
Brandenburg, A. 87
Bravo, E. 287, 389
Briones, F. 381
Bruzual, G. 55
Buitrago, J. 47
Carballo, R. 63, 115
Cardiel, N. 83, 159, 167
Carrera, F.J. 115
Castander, F.J. 91
Castellanos, M. 135
Castro-Tirado, A.J. 15
Ceballos, M.T. 115
Centurión, M. 79
Cepa, J. 369, 381
Cernicharo, J. 175, 205
Chen, H.W. 75
Colas, F. 351
Cornide, M. 279
Corral, L.J. 255
Corsini, E.M. 163
De Castro, E. 279
De La Padula, C. 389
De Miguel, D. 197
Díaz, A.I. 71, 107, 111, 119, 135, 139, 143, 401
Domínguez-Tenreiro, R. 35, 43
Edelstein, J. 393
Eff-Darwich, A. 347
Eiroa, C. 201, 227, 381
Elizondo, D. 31
Esteban, C. 193
Fabian, A.C. 115
Fabiani Bendicho, P. 223
Fabregat, J. 243, 259, 299
Falco, E.E. 51
Feast, M.W. 209
Fernández-Figueroa, M.J. 279
Fernández-Soto, A. 75
Ferrini, F. 71
Figer, D.F. 251
Figueras, F. 215, 315
Fonseca, V. 377
Galadí-Enríquez, D. 307
Gallego, J. 1, 95, 99, 147, 151
García Dabó, C.E. 95, 99
García-Lario, P. 283
García López, R.J. 365
García-Senz, D. 287
García-Vargas, M.L. 135
Gil de Paz, A. 147, 151
Gil-Merino, R. 47
Giovannelli, F. 389
Goicoechea, L.J. 47
Gómez, A.E. 215
Gómez, G. 295
Gómez , J.F. 393
Gómez, J.M. 351
Gómez de Castro, A.I. 263
Gómez-Flechoso, M.A. 155
Gómez-Gomar, J. 389
González-Alfonso, E. 205
González-Casado, G. 27
González Hernández, I. 327, 335, 339
González-Serrano, J.I. 63, 115
Gorgas, J. 83, 159, 167
Goupil, M.J. 215
Guirado, J.C. 311, 319
Guzmán, R. 127, 159
Hansen Ruiz, C.S. 405
Herrero, A. 171, 247, 255
Hueso, R. 351
Impey, C.D. 51
Isern, J. 389
Jiménez, A. 327, 335, 339
Jiménez-Benito, L. 111
Jiménez Reyes, S.J. 327, 335, 339
Jordi, C. 307, 315
Kneib, J.P. 55
Kochanek, C.S. 51
Koekemoer, A. 67
Korpela, E. 393
Korzennik, S.G. 347

Astrophysics and Space Science is the original source of publication of his article. It is recommended that this article is cited as: *Astrophysics and Space Science* **263**: 409–410, 1999.

Lanzetta, K.M. 75
Lecacheux, J. 351
Leeks, S.J. 205
Lefloch, B. 205
Lehár, J. 51
Lobanov, A.P. 59
López, R. 295
Luri, X. 215, 315
Mahoney, T. 373
Manchado Torres, A. 283
Manrique, A. 39
Manteiga, M. 283
Marcaide, J.M. 311, 319, 381
Marco, A. 299
Martín, E. 255, 303
Martín Mateos, I. 327, 335, 339
Mas, J.M. 381
Mcleod, B.A. 51
Mediavilla, E. 47
Miralles, J.A. 291
Molaro, P. 79
Mollá, M. 71
Montes, D. 231, 275, 279
Montesinos, B. 323
Monteverde, M.I. 171
Morales, C. 393
Morganti, R. 67
Muñoz, J.A. 51
Muñoz–Tuñón, C. 373
Najarro, F. 251
Ortiz, E. 227
Ortiz-Gil, A. 75
Oscoz, A. 47
Palacios, J. 201
Pallé, P.L. 327, 335, 339, 347
Paredes, J.M. 381
Patrón, J. 327, 335, 339
Paunzen, E. 271
Pedraz, S. 159, 167
Pelló, R. 55
Peng, C.Y. 51
Pérez, M. 397
Pérez–Fournon, I. 123
Pérez García, A.M. 103
Pérez Hernández, F. 327, 335, 339, 347
Pérez–Martínez, S. 205
Pérez–Torres, M.A. 311, 319
Pignatelli, E. 163
Pintado, O.I. 267
Pizzella, A. 163
Pons, J.A. 291
Raig, A. 27
Ramsey, L.W. 279
Reddy, S. 291
Régulo, C. 327, 335, 339
Rieke, G.H. 131
Rieke, M.J. 131
Rius, A. 381
Rix, H.W. 51
Roca–Cortés, T. 327, 335, 339, 347
Rodes, J.J. 259
Rodríguez Espinosa, J.M. 103, 355
Rojas, J.F. 351
Ros, E. 59, 311, 319
Ruiz Cobo, B. 331
Sáiz, A. 35, 43
Salvador–Solé, E. 27, 39
Sánchez, F. 355
Sánchez, S. 63
Sánchez Cuberes, M. 343
Sánchez–Lavega, A. 351
Sánchez Portal, M. 107, 119, 139
Sánchez–Salcedo, F.J. 87
Sanz–Forcada, J. 279
Sempere, M.J. 205
Serichol, N. 287
Serra, M. 47
Shukurov, A. 87
Solano, E. 219, 267, 271
Stauffer, J.R. 239, 303
Stengler, E. 405
Suárez, O. 283
Tadhunter, C. 67
Talavera, A. 263
Terlevich, E. 107, 111, 135, 139
Terlevich, R. 107, 111, 139
Thomas, J.H. 323
Tissera, P.B. 35, 43
Tomas–Calderon, L. 123
Torra, J. 215, 315
Torrejón, J.M. 243, 299
Torrelles, J.M. 381
Trapero, J. 197, 393
Trullols, E. 307
Ulla, A. 283
Varela, A.M. 373
Vázquez, M. 343
Vega Beltrán, J.C. 163
Verdugo, E. 263
Vigotti, M. 63
Vila–Vilaro, B. 123
Vilchez, J. 381
Villamariz, M.R. 247, 255
Villar–Martín, M. 67
Vladilo, G. 79
Warwick, R.S. 115
Webb, J.K. 75
Wittmann, A.D. 343
Yahata, N. 75
Yepes, G. 31
Zamorano, J. 95, 147, 151
Zeilinger, W.W. 163
Zensus, J.A. 59

MIX
Papier aus verantwortungsvollen Quellen
Paper from responsible sources
FSC® C105338

If you have any concerns about our products,
you can contact us on
ProductSafety@springernature.com

In case Publisher is established outside the EU,
the EU authorized representative is:
Springer Nature Customer Service Center GmbH
Europaplatz 3, 69115 Heidelberg, Germany

Printed by Libri Plureos GmbH
in Hamburg, Germany